# OPFOR SMARTBOOK

TheLightningPress.com

1

Second Edition
(OPFOR 1-2)

# Chinese Military

## Forces, Operations & Tactics

The Lightning Press

Norman M Wade

**The Lightning Press**

2227 Arrowhead Blvd.
Lakeland, FL 33813
**24-hour Order/Voicemail:** 1-800-997-8827
**E-mail:** SMARTbooks@TheLightningPress.com

**www.TheLightningPress.com**

# [OPFOR 1-2] OPFOR SMARTbook 1 - Chinese Military, 2nd Ed.

## Forces, Operations & Tactics



**ISBN: 978-1-935886-52-5**



**Assistant Graphic Layout and Design**: Cate Eschmann

**Photo Credits**: Shutterstock.com, where credited.

Printed in India, 2025 by Citicap Channels Limited - connect@printprinters.com

# [OPFOR 1-2] Notes to Reader

For over two thousand years, China has been surrounded by enemies, adversaries, and other competitors. Invasion, occupation, raids, and other incursions into Chinese territory were commonplace. The PLA views protecting Chinese sovereignty and security as a sacred duty. China traditionally viewed military resistance as an affair for the entire population: mass resistance, guerrilla warfare, and winning a war of attrition.

The People's Liberation Army Army (PLAA) has approximately 975,000 active-duty personnel in combat units. The PLAN is the largest navy in the world with a battle force of approximately 355 platforms, including major surface combatants, submarines, aircraft carriers, ocean-going amphibious ships, mine warfare ships, and fleet auxiliaries. The Air Force (PLAAF) and PLAN Aviation together constitute the largest aviation forces in the region and the third largest in the world, with over 2,800 total aircraft (not including trainer variants or UAVs) of which approximately 2,250 are combat aircraft (including fighters, strategic bombers, tactical bombers, multi-mission tactical, and attack aircraft). The PLA Rocket Force (PLARF) is the largest missile force in the world, operating well over 1,000 short-range, medium-range, and intercontinental ballistic missiles (ICBMs) and over 300 long-range cruise missiles.

With a force that totals approximately two million personnel in the regular forces, the People's Liberation Army (PLA) has sought to modernize its capabilities and improve its proficiencies across all warfare domains so that as a joint force it can conduct the range of land, air, and maritime operations as well as space, counterspace, electronic warfare (EW), and cyber operations. Recognizing that joint operations, information flows, and rapid decision-making are vital in modern warfare, the PRC continues to place a high priority on modernizing the PLA's capability to command complex joint operations in near and distant battlefields.

The People's Republic of China's (PRC's) military strategy is based on what it describes as "active defense," a concept that adopts the principles of strategic defense in combination with offensive action at the operational and tactical levels. Active defense is neither a purely defensive strategy nor limited to territorial defense. Active defense encompasses offensive and preemptive aspects.

** This is the second edition (OPFOR 1-2) of OPFOR SMARTbook 1, updated for 2025.*

# The OPFOR SMARTbook Series (Overview)

In today's complicated and uncertain world, it is impossible to predict the exact nature of future conflict that might involve the U.S. Army. So the Army must be ready to meet the challenges of any type of conflict, in all kinds of places, and against all kinds of threats.

In the coming years, the United States and its allies will face an increasingly complex and interconnected global security environment marked by the growing specter of great power competition and conflict, while collective, transnational threats to all nations and actors compete for our attention and finite resources.

Competition and potential conflict between nation-states remains a critical national security threat. Beijing, Moscow, Tehran, and Pyongyang have demonstrated the capability and intent to advance their interests at the expense of the United States and its allies. China increasingly is a near-peer competitor, challenging the United States in multiple arenas—especially economically, militarily, and technologically—and is pushing to change global norms and potentially threatening its neighbors. Russia is pushing back against Washington where it can—locally and globally—employing techniques up to and including the use of force. In Ukraine, we can see the results of Russia's increased willingness to use military threats and force to impose its will on neighbors. Iran will remain a regional menace with broader malign influence activities, and North Korea will expand its WMD capabilities while being a disruptive player on the regional and world stages.

## Contemporary Operating Environment

Today's operational environment presents threats to the Army and joint force that are significantly more dangerous in terms of capability and magnitude than those we faced in Iraq and Afghanistan. Major regional powers like Russia, China, Iran, and North Korea are actively seeking to gain strategic positional advantage. The interrelationship of the air, land, maritime, space, and the information environment (including cyberspace) requires a cross-domain understanding of an operational environment.

## Opposing Force (OPFOR)

An Opposing Force (OPFOR) is a training tool that should allow the U.S. Army to train against a challenging and plausible sparring partner that represents the wide range of possible opponents the Army could face in actual conflict. It enables training of all arms of the Army and prepares the Army for potential combat operations.

During the road to war leading up to events in a training scenario, the OPFOR may play the role of a "threat" (potential enemy) that is on the verge of becoming an enemy. However, the actual training event usually deals with a state of hostilities. Thus, once hostilities begin in the training event, the OPFOR acts as the "enemy" of the U.S. force in the training environment.

Now in it's second edition (OPFOR 1-2), OPFOR SMARTbook 1 has been updated for 2025. For more than two thousand years, China has been surrounded by enemies, adversaries, and other competitors. With a force that totals approximately two million personnel in the regular forces, the PLA views protecting Chinese sovereignty and security as a sacred duty. OPFOR1 topics and chapters include the strategic environment (defense & military strategy, strategic & operational environments, territorial disputes), force structure (PLA: Army, Navy, Marine, Air Force, Rocket Force, Former SSF), system warfare, information operations, reconnaissance and security, offensive and defensive actions, antiterorrism and stability actions, and capabilities (maneuver, fire support, air defense, aviation, engineer and chemical defense, network and communications, and special operations forces).

North Korea is one of the most militarized countries in the world and remains a critical security challenge for the United States, our Northeast Asian allies, and the international community. The Kim regime has seen itself as free to take destabilizing actions to advance its political goals, including attacks on South Korea, development of nuclear weapons and ballistic missiles, proliferation of weapons, and worldwide cyberattacks. OPFOR2 topics and chapters include the strategic environment, force structure (KPA: Ground Forces, Navy, Air & Air Defense Force, Strategic Force, Special Operations, Reserve and Paramilitary forces, Internal Security & Intel Services), functional tactics, recon & security, offensive and defensive actions, counterstability actions, electronic intelligence warfare, equipment and capabilities.

Whereas previous editions were an examination of historical Soviet-era forces, OPFOR SMARTbook 3: Russian Military, 3rd Ed. (OPFOR 3-3) is completely updated for 2025 focusing instead on present-day Russian Federation forces as observed in their third year of war on the battlefields of Ukraine. As such, OPFOR 3-3 represents perhaps the most comprehensive, up-to-date examination and analysis of Russian forces, operations and tactics available today. At 320 pages, OPFOR 3-3 chapters include the strategic environment (Euro-Atlantic security environment, Russian threat assessment, Russia's war on Ukraine, and NATO enlargement), Russian military forces, Russian ground forces (SV), reconnaissance and security, Russian "strike" actions, offensive actions, defensive actions, and Russian equipment & capabilities.

Spanning more than 4.6 million square miles, the Middle East has for millennia been a geographic and geopolitical crossroads and site of cooperation, competition, and conflict. The greatest single day-to-day threat to regional security and stability remains Iran, which challenges the U.S. and its allies by pursuing regional hegemony, breaching its JCPOA commitments, and posing a conventional threat to partner nations while facilitating and conducting coercive and malign activities. Iran will continue to use Syrian (and likely Iraqi) territory as a critical hub and resupply route for maintaining its campaign against Israel. Wielding the most advanced UASs and ballistic and cruise missiles Iran can design, build, and smuggle into Yemen, the Houthis have targeted Saudi Arabia's largest cities and its critical oil infrastructure.

A hybrid threat is the diverse and dynamic combination of regular forces, irregular forces, and/or criminal elements all unified to achieve mutually benefitting effects. Irregular forces are armed individuals or groups who are not members of the regular armed forces, police, or other internal security forces. Irregular forces are unregulated and as a result act with no restrictions on violence or targets for violence. OPFOR5 topics and chapters include irregular and hybrid threat (components, organizations, strategy, operations, tactics), insurgents and guerillas forces, terrorists (motivations, behaviors, organizations, operations and tactics), criminals (characteristics, organizations, activities), noncombatants (armed & unarmed), foreign security forces (FSF) threats, and functional tactics.

# (OPFOR 1-2) References

The following primary references were used to compile *OFPOR SMARTbook 1: Chinese Military (Forces, Operations & Tactics).* All references are open-source, public domain, available to the general public, and/or designated as "approved for public release; distribution is unlimited." *OPFOR SMARTbook 1: Chinese Military* does not contain classified or sensitive material restricted from public release.

** This is the second edition (OPFOR 1-2) of OPFOR SMARTbook 1, updated for 2025.*


Military and Security Developments Involving the People's Republic of China (2024), Annual Report to Congress, Department of Defense.

Annual Threat Assessment of the U.S. Intelligence Community, Office of the Director of National Intelligence, Mar 2025.

China Naval Modernization: Implications for U.S. Navy Capabilities, Congressional Research Service, Mar 2025.

U.S.-China Strategic Competition in South and East China Seas, Congressional Research Service, Mar 2025.

China-Philippines Tensions in the South China Sea, Congressional Research Service, Mar 2025.

China: Nuclear and Missile Proliferation, Congressional Research Service, Apr 2025.

China-Russia Relations, Congressional Research Service, Sep 2023.

China Primer: Uyghurs, Congressional Research Service, Sep 2023.

China's Military: The People's Liberation Army (PLA), Congressional Research Service, June 2021.

China Primer: South China Sea Disputes, Congressional Research Service, Feb 2021.

China Primer: The People's Liberation Army (PLA), Congressional Research Service, Jan 2021.

China-India Rivalry in the Indian Ocean, Congressional Research Service, Aug 2017.

Chinese Nuclear and Missile Proliferation, Congressional Research Service, Apr 2025.

U.S.-China Strategic Competition in South and East China Seas: Background and Issues for Congress, Congressional Research Service, Jan 2022.

ATP 7-100.3, Chinese Tactics, Aug 2021.


This publication presents PLA military theory largely as written and prescribed by the PLAA. In most cases this represents a best practice as determined by PLA leadership. Real-world practices of PLA units are largely opaque to outsiders, and they were generally not included as part of the analysis underpinning this document. Moreover, the PLA has not participated in an active conflict in nearly half a century, so real-world applications are minimal. Available information on Chinese military training exercises and the few recent examples of conflict seem to indicate that PLA practices—including those of the PLAA—conform closely to its military theory.

# (OPFOR 1-2) Table of Contents

## Chap 1 Strategic Environment

# Chap 2 Force Structure

Chap 3

# System Warfare

Chap 4

# Information Operations

Chap 5

# Reconnaissance & Security

# Chap 6 Offensive Actions

Chap 7

# Defensive Actions

Chap 8

# Antiterrorism & Stability Actions

# Chap 9 Equipment & Capabilities

# Chap 1 I. Understanding China's Strategy

*Ref: Department of Defense Annual Report to Congress: Military and Security Developments Involving the People's Republic of China 2024, pp. 1 to 6.*

Understanding the tenets of the People's Republic of China's (PRC's) national strategy is essential to understanding the drivers of the PRC's security and military strategy. This understanding, in turn, offers insights on the current and future course of People's Liberation Army (PLA) reform and modernization efforts in terms of its strength, technological advances, organization, and operational concepts—all of which could offer PRC leaders expanded military options to support national goals.

As PRC leader, Xi Jinping concurrently serves as the Chinese Communist Party (CCP) General Secretary, Central Military Commission (CMC) Chairman, and President of the PRC. The title used for Xi varies depending on whether he is acting in his capacity as party leader, military leader, or head of state.

## I. The PRC's National Strategy

### Key Takeaways

- The PRC's national strategy is to achieve "the great rejuvenation of the Chinese nation," "China Dream," and "Chinese modernization" by 2049. The strategy determinedly pursues political, social, economic, technological, and military development to increase the PRC's national power and revise the international order to support the PRC's system of governance and national interests.
- The PRC increasingly pushed the narrative that its preferred revisions to the international order are in other countries' interests as well.
- The PRC increasingly views the United States as deploying a whole-of-government effort to contain and suppress the PRC's rise, presenting obstacles to its national strategy.
- PRC leaders are expanding domestic efforts to develop new capabilities for advantages in competition and maintain independent supply chains and strategic stockpiles in the face of Western efforts to derisk supply chains.

The PRC characterizes its view of strategic competition in terms of a rivalry among powerful nation states as well as a clash of opposing ideological systems. PRC leaders believe that structural changes in the international system and an increasingly confrontational United States are the root causes of intensifying strategic competition between the PRC and the United States.

The PRC's strategy entails deliberate and determined efforts to amass, improve, and harness internal and external elements of national power that will place the PRC in a "leading position" in an enduring competition between systems.

The CCP characterizes its strategy to achieve political, social, and economic modernity as a grand national endeavor, sweeping in scope and far-reaching in how it will transform the PRC and, in turn, the world. The CCP defines this as "the great rejuvenation of the Chinese nation" or national rejuvenation as a state in which the PRC is "prosperous, strong, democratic, culturally advanced, harmonious, and beautiful." The PRC's strategy entails efforts to amplify internal and external elements of national power that will "raise China's international influence, appeal, and power to shape events to a new level." The PRC's strategy entails a long-term planning process to attain national rejuvenation that sets objectives, priorities, and milestones across all aspects of governance and policy areas, including economics, political affairs, legal systems, public order, national security, diplomacy, defense, education, science and technology (S&T), culture, and the environment. The objective of "national rejuvenation" or "Chinese modernization" is broad enough to justify almost any policy put forth by the CCP.

CCP officials have described achieving the unification of PRC and Taiwan as "a natural requirement" for national rejuvenation.

The PRC pursues its efforts to generate greater national power by defending and advancing its sovereignty, security, and developmental interests. Consequently, the PRC's national ambitions and statecraft rest on the foundation of the CCP-dominated political ideology of enhancing "Socialism with Chinese Characteristics," which the CCP views as the only path that will lead to the rejuvenation of the Chinese nation. The objective of this CCP-led strategy is perhaps best stated in what the CCP calls its "basic line," a single sentence in the CCP's constitution that serves as the mission of the CCP and the cornerstone for its policymaking. Last amended at the 20th Party Congress in 2022, it states:

"The basic line of the Communist Party of China in the primary stage of socialism is to lead all the people of China together in a self-reliant and pioneering effort, making economic development the central task, upholding the Four Cardinal Principles, and remaining committed to reform and opening up, so as to see China becomes a great modern socialist country that is prosperous, strong, democratic, culturally advanced, harmonious, and beautiful."

The 20th Party Congress, held in 2022, incorporated new developments since 2017 in "Xi Jinping Thought on Socialism with Chinese Characteristics for a New Era," after the term was adopted into the CCP's constitution. The inclusion of Xi Jinping's namesake ideology into the CCP constitution was hailed as a "guide to action for the entire Party and all the Chinese people to strive for the great rejuvenation of the Chinese nation." The 2022 changes to the CCP constitution included an obligation of party members to "uphold Xi Jinping's core position on the Party Central Committee and within the party" as a whole while "upholding the authority and centralized and unified leadership of the Party Central Committee."

Since the 20th Party Congress, the CCP has continued to stress the need for strengthening the PRC's capacity to secure its overseas interests, including improving its control over grain, energy and other resources, and key industrial and supply chains. The PRC relies heavily on imported oil. Food security has been elevated to a top national security priority by the CCP in recent years due to climate shocks, trade disruptions, uncertain global markets, and unsecure lines of transportation. The Party Congress report stressed the CCP's need to prevent digital penetration, sabotage, subversion, and separatism activities from external actors.

Throughout 2023, PRC leadership conceded that the COVID-19 pandemic posed challenges for the PRC's diplomatic, cultural, and economic influence abroad. PRC leadership took diplomatic measures to manage increased global concern about PRC rhetorical and diplomatic alignment with Russia before, immediately following, and during Russia's war against Ukraine as well as concern for the PRC's growing assertive and coercive economic and military actions. PRC leaders continue to believe that global trends, especially the perceived U.S. decline, are generally conducive to their long-term interests and, at the close of 2023, saw the "new period of turbulence and transformation" as "posing new strategic opportunities" in China's development.

As PRC leadership views a divided China as a weak China, they argue that "full reunification"—including the resolution of the "Taiwan question" by 2049 and solidifying the PRC's "overall jurisdiction" over Hong Kong—is one of the fundamental conditions of national rejuvenation. Beijing believes that the PRC must field a world-class military by 2049 that can "fight and win" and "resolutely safeguard" the country's sovereignty, security, and development interests. In support of this goal, the National People's Congress passed revisions to the PRC's National Defense Law in December 2020 to broaden the legal justification for PLA mobilization, including defense of the PRC's "development interests." The codification of this language in PRC law is intended to add legitimacy to the use of military force to protect overseas interests.

PRC leaders claim national rejuvenation requires the PRC to "take an active part in leading the reform of the global governance system" since many rules and norms were established, in the PRC's view, during a time of PRC weakness and without the PRC's consultation and input. The Party views aspects of the prevailing international rules-based system as constraining the PRC's strategic ambitions and incompatible with its sovereignty, security, political preferences, and development interests. To PRC leaders, revisions are necessary to accommodate the PRC's development and should reflect the CCP's preferred transformation in the distribution of power to forge an external environment more favorable to the PRC's political governance system and national interests. Advancing this goal has been more than a decade-long effort of Xi's since the 18th Party Congress, with many projects, such as the Belt and Road Initiative (BRI), undertaken partly in support of it.

# II. Key Objectives and Milestones

For decades, the PRC's leaders have framed their pursuit of modernity and power as advancing the PRC along a specific trajectory, with the PRC's centenary in 2049 serving as the target for achieving national rejuvenation and becoming a "great modern socialist country." From the CCP leadership's perspective, the PRC is a developing nation that must transition into a "fully developed and highly advanced" socialist society. This trajectory involves the CCP leadership shepherding the PRC through stages of gradual but systematic modernization and development. The CCP demarcates the stages of the PRC's strategy with milestones accompanied by objectives and priorities set by the CCP's long-term planning processes.

At the 19th Party Congress in 2017, Xi laid out the PRC's strategy in the "New Era" as a broad plan to achieve national rejuvenation with a timeline linked to two symbolically important centenary milestones reached in 2021 (the CCP's centenary) and 2049 (the PRC's centenary). To bridge the lengthy gap between the two anniversaries, Xi added interim national objectives for 2035 and laid out a broad two-stage modernization plan to reach 2049. Further demonstrating the CCP's confidence in the PRC's progress, Xi's objectives for 2035 moved up certain mid-century targets set by the CCP going back to 1987.

At his speech marking the 100th anniversary of the CCP on July 1, 2021, Xi declared that the PRC had "realized the first centenary goal of building a moderately prosperous society in all respects," by eliminating extreme poverty. Beyond 2021, the PRC will use the "moderately prosperous society" as the basis for Xi's "two-stage" plan to achieve national rejuvenation by the PRC's centenary in 2049. In the first stage from 2021 to 2035, the CCP aims for the PRC to "basically" meet its initial thresholds for becoming a "great modern socialist country." In this stage, the PRC will likely continue to prioritize economic development as "the central task" but, rather than rapid economic growth, it will seek to address its uneven economic development and inequalities that Beijing recognized as the new "principal contradiction" in PRC society in the "New Era." By 2035, the PRC seeks to increase self-reliance through enhancing its economic and technological strength, including "basically" completing its military modernization by becoming a "global leader in innovation" and enhancing self-sufficiency in key areas like food supply. The PRC intends to significantly strengthen its cultural "soft power" and improve its domestic rule of law and governance systems.

In the second stage from 2035 to 2049, the PRC aims to attain national rejuvenation and Chinese modernization, realizing an international status that Xi describes as a

# III. Origins of PRC National Rejuvenation

*Ref: Department of Defense Annual Report to Congress: Military and Security Developments Involving the People's Republic of China 2024, pp. 4 to 6.*

Understanding the origins of the PRC's national rejuvenation concept is crucial to understanding how the PRC will likely shape and pursue this strategic objective. PRC leaders have consistently framed their efforts as seeking to "restore" China to a preeminent place in the world after enduring what the Party characterizes as China's "century of humiliation" beginning in the 19th century as the Qing Dynasty began to disintegrate and lasting until the founding of the PRC in 1949. The threads of national renewal can be traced to China's reformers and nationalist revolutionary leaders in the late Qing Dynasty and emerged as a common nationalist theme in the fractured politics of China's Republican Era. Although the CCP's exact articulation of this goal as "the great rejuvenation of the Chinese nation" first emerged in the late 1980s, the CCP has championed the cause of rebuilding China since the 1920s. Xi describes the "great rejuvenation of the Chinese nation" as a "strategy" for "achieving lasting greatness for the Chinese nation."

The CCP's narrative of national rejuvenation speaks to the deep impressions from an era defined by the disintegration of China's polity, repeated violations of China's sovereignty by foreign powers, and the prolonged absence of physical and economic security for many Chinese people. For a country that claims a history stretching back thousands of years—much of it as one of the most powerful and advanced civilization in the world—nationalist appeals to restore China's greatness are deeply rooted. This resonance is crucial to understanding why the CCP portrays the PRC's rejuvenation as a nationalist project that the CCP "shoulders" for the country. The CCP's leaders frame "Socialism with Chinese Characteristics for the New Era" and the CCP as indispensable to the PRC overcoming its historical circumstances and attaining national rejuvenation.

The CCP claims that to perform its decisive role in guiding the PRC's development into a "great modern socialist country," it must ensure that the country advances in line with "the Four Cardinal Principles." First stated by Deng Xiaoping and later written into the CCP constitution, these principles mandate the CCP "to keep to the path of socialism, to uphold the people's democratic dictatorship, to uphold the leadership of the CCP, and to uphold Marxism-Leninism and Mao Zedong Thought." The Four Cardinal Principles are the basis for political and governance reforms pursued by the CCP and the outer boundaries of its efforts to "reform" and "open up" the country.

In addition to cultivating ideological discipline and fighting corruption in the CCP, Xi has sought to advance the PRC's strategy by strengthening the CCP's primacy across the PRC's governance systems and making the CCP more effective at managing the PRC's political, economic, and social problems. Xi's emphasis on building the CCP's institutional capacity and promoting internal unity—which he views as the means for the CCP to perform its strategic role—has become a prominent feature of his tenure. Xi has increased his authority in the CCP to levels not seen since the start of the reform era after the death of Mao. He took the unprecedented step of setting himself up for a third term as leader after presidential term limits were removed in 2018 and oversaw his status as the "core position of the Party Central Committee and the core position of the whole party" written into the CCP constitution at the 2022 20th Party Congress.

PRC leaders believe that structural changes in the international system and an increasingly confrontational United States are the root causes of intensifying strategic competition between the PRC and the United States. Throughout the post-Mao reform era and particularly after the end of the Cold War, the CCP's leaders recognized their socialist system was—and would remain over the long term—an underlying source of tension with the West. Given the CCP's ambitions to "restore" the PRC's place in the world and their assessment of the PRC's relative weakness in relation to rival states, CCP leaders recognized the PRC's growing strength could threaten to aggravate tensions with others without careful management. Deng Xiaoping's reputed approach to this dilemma, as attributed by other CCP leaders, was for the PRC to "hide our capacities and bide our time, be good at maintaining a low profile; and never claim leadership." The CCP under Xi's leadership, despite some internal dissent, no longer follows this approach and, instead, seeks to bolster the PRC's international profile through various tools of its national power.

Over time, the PRC has characterized strategic competition in terms of a rivalry among powerful nation states, most importantly the United States, as well as a clash of opposing ideological systems. Speaking to the CCP Central Committee in 2013, Xi remarked that the CCP needed to "appreciate" that "developed Western nations" would continue to possess "real, long-term advantages" over the PRC in the economic, technological, and military domains. Xi argued that the PRC would need to "prepare for a long period of cooperation and of conflict between these two social systems in each of these domains."

## The PRC's Views on Sovereignty, Security, and Development Interests Related to Hong Kong, Xinjiang, Tibet, and Taiwan

The PRC is concerned that perceived separatist elements in Hong Kong, Xinjiang, and Tibet are pushing for independence, buoyed by so-called "external forces." The PRC views this as an unacceptable threat to its sovereignty and stability, imperiling its national security and national development.

The PRC is implementing its interpretation of "One Country, Two Systems" in Hong Kong and Macau, wherein individuals who Beijing deems "patriots" administer those regions respectively, thereby ensuring "law-based governance." The PRC is investing in the Guangdong-Hong Kong-Macao "Greater Bay" development zone, expanding growth of high-tech industry from Guangdong into areas previously used for agriculture and fishing on the outskirts of Hong Kong and Macau. This will integrate them more tightly into the PRC's overall economic development.

The PRC has continued efforts to systematically dilute, control, and eventually assimilate local cultures in Xinjiang and Tibet into PRC society. After two German corporations began pulling out of joint ventures in Xinjiang citing discovery of forced labor and other human rights abuses at plants, Foreign Minister Wang Yi called the charges "made up lies" intended to prevent PRC's development and revitalization. In more sparsely populated Tibet, Beijing has accelerated its efforts at assimilating the local population into PRC society. In November 2023, Beijing released a white paper on Tibet calling for increasing use of the Han Chinese term "Xizang" to refer to the province, which has now been adopted by much of PRC state media.

In March 2024, responding to questions at a press conference at the National People's Congress (NPC), Foreign Minister Wang Yi dismissed Taiwan's January presidential elections as "just local elections in one part of China" that did not change the fact that Taiwan was part of the PRC. Characterizing support for Taiwan independence as a threat to peace, Wang Yi stated that the PRC would continue to strive for peaceful reunification but that the PRC would never allow Taiwan to be separated from the motherland.

"global leader in terms of comprehensive national strength and international influence." A renewed PRC will have attained—among the CCP's many goals—its objectives to field a world-class military and assume a leading position in an international order revised in line with the PRC's overall foreign policy goal to establish what it refers to as a "community of common destiny" or, the PRC's preferred official English translation, "community with a shared future for mankind."

# IV. The PRC's Core Interests

The PRC considers "core interests" as issues so central to its national rejuvenation that the official position on them is not subject to negotiation or compromise. The PRC began using the term in 2003, initially in reference to sovereignty issues regarding Taiwan and later Tibet and Xinjiang. In 2009, the senior PRC official for foreign affairs publicly defined the elements of core interests: 1) preserving the PRC's political system, 2) defending the PRC's sovereignty and territorial claims, and 3) promoting the PRC's economic development.

The list of core interests has grown over the last decade with President Xi increasingly using the terms of a national security law passed in 2015 to extend the formal status of "core interest" to territorial disputes in the South China Sea (SCS), the Senkaku Islands, and the northeastern Indian state of Arunachal Pradesh. Other sources have begun listing PRC interests in new domains—space, cyber, and electromagnetic—which, in turn, must be contested. The increasing number of issues being defined as "core interests" carries the risk of more potential sources of conflict in PRC-U.S. relations or between the PRC and neighboring states.

Chap 1

# II. Defense Policy & Military Strategy

*Ref: Department of Defense Annual Report to Congress: Military and Security Developments Involving the People's Republic of China 2024, pp. 30 to 43.*

## I. Overview

### Key Takeaways

• In 2023, the PRC's stated defense policy aims remained oriented toward advancing its sovereignty, security, and development interests while emphasizing a greater global role for itself.

• PRC leaders stress the imperative of strengthening the PLA into a "world-class" military by the end of 2049 as an essential element of its strategy to rejuvenate the PRC into a "great modern socialist country."

PRC leaders continue to demonstrate a growing willingness to use military coercion and inducements to achieve its defense policy aims. PRC leaders cast the armed forces as a practical instrument to defend the PRC's expanding global interests and to advance its foreign policy goals in the framework of "Great Power Diplomacy with Chinese characteristics."

In 2020, the PLA added a new milestone for modernization in 2027: to accelerate the integrated development of mechanization, informatization, and intelligentization of the PRC's armed forces, which, if realized, could give the PLA capabilities to be a more credible military tool for its Taiwan unification efforts.

The PLA's "core operational concept," Multi-Domain Precision Warfare (MDPW), is intended to leverage a command, control, communications, computers, intelligence, surveillance, and reconnaissance (C4ISR) network that incorporates advances in big data and AI, what the PLA calls the "network information system-of-systems," to rapidly identify key vulnerabilities in the U.S. operational system and then combine joint forces across domains to launch precision strikes against those vulnerabilities.

The PRC has stated its defense policy aims to safeguard its national sovereignty, security, and development interests. PRC leaders view these interests as foundational to their national strategy. The modernization of the armed forces is an indispensable element of the CCP's national strategy to modernize the country. The PRC's defense policy and military strategy primarily orients the PLA toward "safeguarding" its perceived "sovereignty and security" interests in the region while countering to the United States.

To adapt the PRC's armed forces to long-term trends in global military affairs and meet the country's evolving national security needs, PRC leaders stress the imperative of meeting key military transformation targets set for 2027 and 2035. These milestones seek to align the PLA's transformation with the PRC's overall national modernization so that, by the end of 2049, the PRC will field a "world-class" military. In March 2023, Xi issued a series of calls to accelerate the build-up of self-reliance in S&T, make supply chains more resilient, and bolster national reserves to be "more capable of safeguarding national security." Defense S&T has been crucial in the PRC's technological innovation. Institutional reform in the PRC's national S&T enterprise will drive national defense modernization.

A key driver of the PRC's defense policy is how PRC leaders perceive the relative threats and opportunities facing the country's comprehensive national development.

During Chairman Xi's CCP centenary speech, he called for the full implementation of the CCP's idea of strengthening the army in the new era. The last defense white paper, China's National Defense in the New Era, published in 2019, reaffirmed that the PRC's armed forces are aligned with and contribute to the strategies of the CCP, stating that ongoing military reforms "ensure absolute leadership of the CCP over the military." According to the paper, Beijing views the international environment as undergoing "profound changes unseen in a century." The CCP concludes that "international strategic competition is on the rise" and expresses deep concerns at what it sees as growing sources of instability in the near term.

Beijing offers no introspection on its role in stirring geopolitical tensions through its economic practices, military activities and modernization, excessive maritime territorial claims, assertive diplomacy, or efforts to revise aspects of global governance. Rather, the PRC describes the international system as being "...undermined by growing hegemonism, power politics, unilateralism and constant regional conflicts and wars." Similarly, the PRC contends that global military competition is intensifying and that "major countries" are adjusting their security and military strategies, reorganizing their militaries, and developing new types of combat forces to "seize the strategic commanding heights in military competition."

The PRC's stated defense policy is to "resolutely safeguard" its sovereignty, security, and development interests, according to its 2019 defense white paper, which provides continuity with past statements by PRC senior leaders and other official documents. In practice, the PRC's military power is increasingly a central feature of the CCP's regional and global ambitions. The 2019 defense white paper identifies the PRC's national defense aims that support these interests, in likely order of importance:

- Deter and resist aggression.
- Safeguard national political security, the people's security, and social stability.
- Oppose and contain "Taiwan independence."
- Crack down on proponents of separatist movements, such as "Tibet independence" and the creation of "East Turkistan."
- Safeguard national sovereignty, unity, territorial integrity, and security.
- Safeguard the PRC's maritime rights and interests.
- Safeguard the PRC's security interests in outer space, the electromagnetic spectrum, and cyberspace.
- Safeguard the PRC's overseas interests.
- Support the sustainable development of the country.

Key changes in defense policy for the "New Era" include efforts to improve coordination across the party-state to leverage all organs of national power in a unified approach to support the CCP's ambitions of a global military capability. Unlike previous defense white papers, China's National Defense in the New Era explicitly stressed the PRC's armed forces' alignment and support to the CCP's broader societal and foreign policy objectives. The white paper states that the PRC's armed forces must be ready to, "provide strong strategic support for the realization of the Chinese Dream of national rejuvenation, and to make new and greater contributions to the building of a shared future for mankind." Also notable is the explicit alignment between the PRC's defense and foreign policies, particularly in the armed forces' role in protecting the PRC's overseas interests and furthering the CCP's concept of "strategic partnerships" with other countries.

The Chairman of the CMC issues military strategic guidelines to the PLA as the foundation of the PRC's military strategy. The military strategic guidelines set the general principles and concepts for the use of force in support of the CCP's strategic objectives, offer guidance on the threats and conditions the armed forces should be

# Perceptions of the External Security Environment

*Ref: Department of Defense Annual Report to Congress: Military and Security Developments Involving the People's Republic of China 2024, pp. 6 to 7.*

In the past two years, President Xi presented his thoughts on the PRC's strategic environment on multiple occasions. For example, Xi regularly warns of "growing risks" and stresses that the PRC is on the brink of "changes unseen in a century" but that the PRC will benefit from a "profound adjustment in the international balance of power." In his speech at the 100th anniversary of the CCP in 2022, Xi asserted that, as the world experienced "once-in-a-century changes," the PRC had to adopt "a holistic approach to national security that balances development and security imperatives" and implement "the national rejuvenation." Speaking to a group of delegates attending the 14th National Committee of the Chinese People's Political Consultative Conference, Xi asserted that "western countries led by the United States have implemented all-round containment, encirclement and suppression against us." PRC leaders view U.S. security alliances and partnerships, especially those in the Indo-Pacific, as destabilizing and irreconcilable with the PRC's sovereignty, security, and development interests.

Since just prior to the dissolution of the Soviet Union, PRC leaders have consistently characterized the PRC's security environment as undergoing intense changes and viewed the international order as shifting toward a multipolar system more commensurate with the PRC's development. The CCP views a shift toward a multipolar system as consistent with its perception of global power trends. This shift is vital for the PRC to advance its strategy, perceiving U.S. power as a constraint that impedes many of the PRC's goals. The PRC's leaders have eagerly embraced narratives of the West's relative decline and the inevitability of the PRC's rise as largely consistent with their strategy and evidence of the PRC's own progress. They have adopted a narrative of a "shared future for mankind," issuing a white paper advocating a formulation echoing many aspects of the "China Dream," calling for an "open, inclusive, clean and beautiful world that enjoys lasting peace, universal security, and common prosperity." Xi echoed these themes in his 2024 News Year's call with Russian President Vladimir Putin.

Beijing has expressed concerns over growing global instability and a mounting sense of insecurity that it views as instigated by the United States. The PRC's 2019 defense white paper criticized the United States as the "principal instigator" of global instability and driver of "international strategic competition." A 2023 PRC editorial in response to U.S. derisking efforts wrote that "the United States is willfully waging war, building dollar hegemony, and creating confrontation between camps." PRC leadership views U.S. policy toward the PRC as a critical factor affecting the PRC's national objectives and increasingly views the United States as more willing to confront Beijing where U.S. and PRC interests are inimical.

Given the enduring suspicion among some in Beijing that the United States' efforts to compete with the PRC are really efforts to "suppress" or "contain" the PRC, CCP leaders hold that the accrual of the PRC's comprehensive national power (CNP) will set the conditions for the PRC's ability to confront or dissuade the United States and prevent containment. As PRC leaders seek to translate the PRC's growing economic, military, diplomatic, and technological means into influence to advance their international aspirations, they must carefully balance the PRC's expanding interests across their priorities and resources.

prepared to face, and set priorities for planning, modernization, force structure, and readiness. The CCP leadership issues new military strategic guidelines or adjusts existing guidelines whenever they perceive it necessary to shift the PLA's priorities based on the CCP's perceptions of the PRC's security environment or changes in the character of warfare.

Since 2019, trends indicate the PRC has reviewed and adjusted its military strategic guidelines. In early 2019, PRC state media indicated that Beijing held senior-level meetings to "establish the military strategy of the 'New Era.'" The PRC's 2019 defense white paper states that the PLA is implementing guidelines for the "New Era" that, "…actively adapt to the new landscape of strategic competition, the new demands of national security, and new developments in modern warfare…" Documents released following the Fifth Plenum of the 19th Central Committee in October 2020 hailed progress in the "comprehensive and in-depth" implementation of the "New Era military strategic guidelines."

These developments are notable because the CCP leadership has issued new military strategic guidelines or adjusted its guidelines only a few times since the end of the Cold War. In 1993, the CMC under Jiang Zemin directed the PLA to prepare to win "local wars" under "high-tech conditions" after observing U.S. military operations in the Gulf War. In 2004, the CMC under Hu Jintao ordered the military to focus on winning "local wars under informationized conditions." In 2014, the CMC placed greater focus on conflicts in the maritime domain and fighting "informatized local wars."

The CMC directs the PLA to be ready and able to perform specific missions and tasks to support the CCP's strategy and defend the PRC's sovereignty, security, and development interests. The PLA's missions and tasks in the "New Era" include safeguarding the PRC's territorial sovereignty and maritime rights and interests, maintaining combat readiness, conducting military training under real combat conditions, safeguarding the PRC's nuclear weapons and its interests in the space and cyberspace domains, countering terrorism and maintaining stability, protecting the PRC's overseas interests, and participating in emergency response and disaster relief.

In his speech at the 20th Party Congress, Xi detailed the PLA goals of enhancing party loyalty in the military while simultaneously strengthening the military through reform, S&T, personnel training, mechanization, informatization, and modernized military strategies.

## Modernization Goals

In a March 2021 speech, Xi detailed that the 2027 modernization goal is the first step in a broader modernization effort. The PLA's 2027 modernization goal aligns with the 100th anniversary of the PLA's founding. PLA writings note the "three-step" modernization plan connects "near-, medium-, and long-term goals" in 2027, 2035, and 2049, respectively.

The PRC's goals for modernizing its armed forces in the "New Era" are as follows:

- By 2027: "Accelerate the integrated development of mechanization, informatization, and intelligentization" while boosting the speed of modernization in military theories, organizations, personnel, and weapons and equipment.
- By 2035: "To comprehensively advance the modernization of military theory, organizational structure, military personnel, and weaponry and equipment in step with the modernization of the country and basically complete the modernization of national defense and the military …"
- By 2049: "To fully transform the people's armed forces into world-class forces."

# II. Military Strategy: Active Defense

*Ref: Department of Defense Annual Report to Congress: Military and Security Developments Involving the People's Republic of China 2024, pp. 33 to 34.*

The PRC's military strategy is based on what it describes as "active defense," a concept that combines the principles of strategic defense with offensive action at the operational and tactical levels. Active defense is neither a purely defensive strategy nor limited to territorial defense. Active defense encompasses offensive and preemptive aspects. It can apply to the PRC acting externally to defend its interests. Active defense is rooted on the principle of avoiding initiating armed conflict but responding forcefully if challenged. The PRC's 2019 defense white paper reaffirmed active defense as the basis for its military strategy.

Contemporary PRC writings describe the tenets of active defense as follows:

**Adhere to a Position of Self-Defense and Stay with Striking Back.** This describes the basic principle for the use of military force under active defense. The PRC's 2019 defense white paper describes this principle as, "We will not attack unless we are attacked, but we will surely counterattack if attacked." Active defense may entail defensive counterattacks in response to an attack or preemptively striking an adversary that the PRC judges is preparing to attack.

**Combine Strategic Defense with Operational and Tactical Offense.** This aspect offers two approaches to warfare influenced by Mao Zedong's notion of using defense and offense in turns. First, active defense may involve offensive campaigns, operations, and tactical actions in support of the strategic defense. These may occur rapidly and along "external lines." Second, it uses active defense involves the use of strategic defense to weaken the enemy and set the conditions to transition into strategic offense in order to secure victory.

**Taking the Operational Initiative.** This aspect emphasizes the effective use of offensives at the operational and tactical levels, avoiding enemy strengths, and concentrating on building asymmetric advantages against enemy weaknesses to "change what is inferior into what is superior."

**Strive for the Best Possibilities.** This calls for thorough peacetime military preparations and planning based on fighting the most challenging threat under the most complicated circumstances "in order to get the best results." This aspect stresses the importance of setting conditions in advance and suggests it is preferable to be prepared and not fight, than to fight unprepared.

**The Dialectical Unity of Restraining War and Winning War.** This tenet seeks to resolve the dilemma that using too little force may protract a war instead of stopping it while the unconstrained use of force may worsen a war and make it harder to stop. Calling for the "effective restraint of warfare," this tenet seeks to avoid war first through sufficient military preparations and powerful conventional and strategic forces that act in concert with political and diplomatic efforts to "subdue the enemy's troops without fighting." If war is unavoidable, however, this aspect calls for restraining war by taking the "opening move" and "using war to stop war."

**Soldiers and the People are the Source of Victory.** This tenet integrates the concept of active defense with the concept of "people's war." People's war comprises subordinate military strategies, "guerrilla war" and "protracted war" that Mao saw as a means to harness the capacity of China's populace as a source of political legitimacy and mobilization to generate military power. Contemporary PRC writings link the people's war to national mobilization and participation in wartime as a whole-of-nation concept of warfare.

The 5th Plenum communique holds that the 2027 goal means that the PRC military should comprehensively push forward the modernization of military theories, military organizational form, military personnel, and weapons and equipment. PRC media, citing a military source, connected the PLA's 2027 goals to developing the capabilities to counter the U.S. military in the Indo-Pacific region and compel Taiwan's leadership to the negotiation table on the PRC's terms.

Although PRC leaders view building military strength as a strategic imperative, they place important caveats on these objectives. For example, Chairman Xi's direction to the PLA to "basically complete" modernization by 2035 should occur "in step with the modernization of the country." These qualifications serve several purposes that highlight the interlocking nature of the CCP's strategic planning. First, as the PRC's interests continue to expand, the CCP expects the PLA to keep pace with the country's evolving interests and be ready and able to defend its progress. Second, linking the PLA's transformation to the country's transformation allows CCP leaders to signal the scope and scale of the internal changes they expect the PLA to implement, particularly given its historic resistance to reforms that challenge its risk-adverse organizational culture or threaten vested bureaucratic interests. These qualifications provide flexibility to the CCP's leaders to calibrate military resources and defense objectives based on the conditions of the country's overall development. This lets PRC leaders adapt to changing economic or international conditions and ensure military investments support—rather than compromise—the strategy.

The CCP has not defined what it means by its ambition to have a "world-class" military by the end of 2049. Within the context of the PRC's national strategy, however, it is likely that the PRC will seek to develop a military by mid-century that is equal to—or, in some cases, superior to—the U.S. military and that of any other great power that Beijing views as a threat to its sovereignty, security, and development interests. It is unlikely that the CCP would aim for an end state in which the PRC would remain in a position of military inferiority to the United States or any other potential rival. However, this does not mean that the PRC will aim for the PLA to mirror the U.S. military in capacity, capability, or readiness. The PRC will likely seek to develop its "world-class" military in a manner that it believes best suits the needs of its armed forces to defend and advance the country's interests and how the PLA—guided by the CCP—adapts to the changing character of warfare.

The PLA increasingly views warfare as a confrontation between opposing operational systems, rather than annihilation of opposing mechanized military forces. Following this logic, PLA writings refer to systems destruction warfare as the next way of war, transforming from mechanized warfare to an informatized and intelligentized style. Although not a new PLA approach, systems destruction warfare likely continues to be the principal theory guiding its way of war.

In November 2020, the CMC issued the "Chinese People's Liberation Army Joint Operations Outline (Trial)." The outline establishes a system for the PLA's joint operations and focuses on clarifying basic issues regarding the organization and implementation of joint operations, command rights and responsibilities, and the principles, requirements, and procedures for joint operations, combat support, national defense mobilization, and political work. According to PLA writings, the outline describes how the combat style of the PLA will integrate joint operations under the unified command of a joint operations command system. PLA writers emphasize that winning future wars will require a high degree of joint integration of various combat forces and combat elements from across the PLA services and other arms and across all domains, with jointness deepened at the operational and tactical levels. The PLA's "operational regulations" were last updated in 1999 and PLA leaders and PLA-affiliated academics have pointed to the lack of updated doctrine, which is out of step with the 2015-era structural command and organizational reforms and an obstacle to advancing the next steps in building a unified joint PLA.

# III. Relationship Between PRC Foreign Policy and Defense Policy

*Ref: Department of Defense Annual Report to Congress: Military and Security Developments Involving the People's Republic of China 2024, pp. 14 to 19.*

Managing that security environment takes a variety of forms. With some of its strategic partners, the PRC routinely engages in military cooperation (e.g., high-level military visits, combined exercises, and exchanges) to maintain or strengthen bilateral ties it perceives can help achieve its foreign policy goals. The PRC's extensive military relationships with Pakistan and Russia, for example, support partnerships it considers key to success in strategic competition. Likewise, its growing military relationship with Gabon in 2023—conducting a PLAN goodwill visit and seeking to secure military access in the Gulf of Guinea—reflects Beijing's sustained interest in winning African support for its global policy proposals. With other strategic partners, however—such as Saudi Arabia or the United Arab Emirates—the PRC gives less priority to military cooperation, probably, in part, because it finds diplomatic and economic channels as more effective in achieving its foreign policy goals.

With countries like the United States, which the PRC views as a competitor, the PRC focuses military cooperation—primarily military diplomacy—on conflict avoidance.

Examples provided include:

- **PRC-Russia Relationship.** In 2023, the PRC continued partnering with Russia to counter the United States and advance its preferred vision for the international system. Most notably, PRC support for Russia's war against Ukraine, including significant trade in dual-use goods to rebuild Russia's defense industrial base, remained robust.
- **Evolving PRC-DPRK-Russia Relationship.** Since the fall of the Soviet Union, the PRC has been the Democratic People's Republic of Korea's (DPRK's), or North Korea's, principal supporter. The PRC constitutes the vast majority of the DPRK's external trade and North Korea depends on the PRC for energy, fertilizer, and food imports. Beijing values the DPRK as a buffer between the PRC and South Korea and the U.S. forces stationed there.
- **Relationship and Engagement with the Pacific Island Region.** Since 2015, the PRC has probably viewed engagement and deliberate corruption in the Pacific Island Countries (PICs) as an opportunity to expand its regional influence, press countries to switch diplomatic recognition from Taipei to Beijing, increase influence with regional security forces, and advance the PRC's responsible great power narrative.
- **Relationship with Iranian Proxies.** The PRC balances its broader regional interests with limited engagement of Iranian proxies, such as Hezbollah in Lebanon, the Houthis in Yemen, or Iranian-backed militia groups in Iraq.
- **Military Engagements with Taliban Government in Afghanistan.** There were no known PLA military engagements with the Taliban government in 2023. However, in May 2023, the PRC foreign minister met in Islamabad with his Afghan and Pakistani counterparts to discuss stability in Afghanistan, regional peace, counterterrorism, security for PRC nationals, institutions, and projects in the region.
- **Stance on Israel-Hamas War.** Since the outbreak of the Israel-Hamas conflict, the PRC has sought to position itself as an advocate for peace in the region and aligned with the Global South in support of the Palestinian cause. In press statements and international fora, the PRC has repeatedly called for an immediate ceasefire in Gaza and negotiations leading to a two-state solution to the Palestinian crisis.

# IV. Multi-Domain Precision Warfare (MDPW)

In 2021, the PLA began discussing a new "core operational concept," MDPW. MDPW is intended to leverage a C4ISR network that incorporates advances in big data and AI, what the PLA calls the "network information system-of-systems," to rapidly identify key vulnerabilities in the U.S. operational system and then combine joint forces across domains to launch precision strikes against those vulnerabilities. MDPW is meant to sit atop an "operational conceptual system-of-systems," suggesting the PLA will develop additional subordinate operational concepts and use simulations, war games, and exercises to test, evaluate, and improve these future-oriented operational concepts. The timing of MDPW's appearance in the PRC's updated doctrine and military strategic guidelines suggests that MDPW serves as a connection between them, likely amplifying themes and guidance in both while focusing on the contours of what the PLA must do to win future wars.

PLA writings have long emphasized the importance of joint firepower strikes as a component of large-scale operations. Joint firepower strikes include multiple services combining to use their firepower capabilities to create substantial effect and have been explicitly tied to a Taiwan invasion in PLA writings. During the August 2022 CODEL visit to Taiwan, the PLARF fired multiple ballistic missiles into impact zones in waters around Taiwan, including at least four missiles that overflew Taiwan, which was unprecedented. Also during the 2022 CODEL visit, the PLAA conducted live fire events using the new PCH191 long-range rocket artillery system, which can strike Taiwan from mainland China. The military drills afforded the PLA an opportunity to train simulated joint firepower strike operations.

# V. War Control

PLA strategists' interrelated concepts of "effective control" and "war control" describe a multifaceted effort for controlling the timing, pace, scope, and scale of escalation from peacetime through crisis and war. At the 20th Party Congress as he did at the 19th Party Congress, Xi emphasized the necessity of the PRC being able to execute effective control's basic formulation—"shape our security situation, contain crises and conflicts, and win local wars"—a validation of the concept as national policy, not just PLA theory. PLA writings on war control highlight the importance of avoiding war when possible, ensuring one fights a winnable war, and, if war is unavoidable, seizing the initiative and minimizing the cost and duration. During crisis, war control calls on the PLA to prevent war but seize opportunities—which could inadvertently escalate the crisis. Alongside modernizing the PLA's capabilities and organizational reform, PRC's leaders have identified enhancing the combat readiness of the armed forces as an important element in developing the PRC's military strength. In recent years, Xi and senior military leaders have emphasized the need to evolve how the PLA trains to increase force readiness, focusing on more rigorous and realistic training as well as addressing issues in the PLA's training and education systems relating to conducting complex joint operations and adapting to other aspects of modern warfare. It probably has led to an increasing standardization of a combat readiness system across the PLA to enable the PRC to quickly transition to a wartime posture.

Along with the CCP leadership's focus on improving the PLA's combat readiness, in recent years, PLA media outlets have noted shortcomings in the military's training and education systems that reportedly left some commanders—particularly at the operational level—inadequately prepared for modern warfare. In response to perceived personnel deficiencies in the PLA, Xi approved and issued a new series of regulations in July 2022 regarding the management of PLA soldiers seeking to improve recruiting, training, promotions, benefits, and demobilization efforts for NCOs. In recent years, PLA media outlets have identified the need for the military to address the "Five Incapables" problem: that some commanders cannot (1) judge situations, (2) understand higher authorities' intentions, (3) make operational decisions, (4) deploy forces, and (5) manage unexpected situations.

# Chap 1 III(a). Strategic & Operational Environment

*Ref: ATP 7-100.3, Chinese Tactics, (Aug '21), chap 1.*

**PRC Defense Spending**

A survey of multiple models of the PRC's defense budget estimates that Beijing spends 40% to 90% more than it announces in its public defense budget, which equates to approximately $330 billion–$450 billion in total defense spending for 2024. The consensus among experts is that the PRC's publicly announced defense spending figure does not contain the entirety of PRC investment in its defense, so alternative approaches are used to assess the total value of this spending.

## I. The Strategic Environment

China's view of the strategic environment mirrors that of the United States in many ways. There are, however, key differences in both analysis of the strategic environment and the application of this analysis that underpin important differences in perspective between the two countries. Both the People's Republic of China—commonly referred to as China—and the United States assess the key elements of the strategic environment below:

**U.S. obligations to allies and partners in the Western Pacific will continue in perpetuity.** Key U.S. allies include Japan, South Korea, Australia, New Zealand, and the Philippines. The United States will likewise maintain a strong strategic interest in the Western Pacific.

**The United States will maintain strong, though unofficial, support to the Republic of China—** also known as Taiwan. The former will employ a policy of deliberate ambiguity to deter potential Chinese aggression against the island while maintaining a position respectful of Chinese policy toward the same.

**China will continue to seek improved relations with Russia and India, with Russia likely proving a more amenable partner**. China views improving these two relationships—particularly with Russia—as very important both politically and economically. Border tensions with India complicate bilateral relations and are often perceived by India as aggressive, requiring strong responses and adjustments to its defense posture.

Increasing competition for limited resources and the effects of global climate change will fuel conflict among both state and non-state actors. As populations increase, providing adequate quantities of basic necessities such as water, energy, food, and medical care becomes increasingly difficult. In addition, competition for human capital and raw materials for industry will increase as the international economy expands. Conditions such as natural disasters, extreme weather events, and their second- and third-order effects will have considerable impacts across the globe and will significantly influence political and military strategies worldwide. The global commons—the earth's unowned natural resources, such as the oceans, the atmosphere, and space—will be increasingly important to the global economy and will thus likely become contested. The PLA characterizes many tasks related to these conditions under the heading of nontraditional security measures or military operations other than war.

**As the global economy and disparate societies become increasingly interconnected, friction points continue to emerge.** State and non-state actors are already using widely proliferated but immature technologies—such as social networking— worldwide to influence populations, politics, and policy. Actors will continue to leverage emerging connections to manipulate public opinion and influence leaders; the digital world will become an increasingly important front to contest. As national economies continue to intertwine themselves, competition for jobs, natural and man-made resources, and new or emerging markets will continue to intensify.

**Technological advances will continue to enhance the lethality of capabilities across all domains.** Soldiers at all echelons on the future battlefield will face unprecedented dangers as both new and proven technologies are integrated into warfighting. For example, precision munitions will continue to proliferate and become increasingly affordable as technology costs drop. At the same time, limited defense budgets coupled with the high expense and relative rarity of modern weapons systems—particularly aircraft, surface ships, and missiles of all types—will heavily influence future major combat scenarios.

**China faces both nontraditional and hybrid threats.** Such threats include criminal organizations, terrorist organizations, and fringe or rogue states. These threats effectively exploit complex terrain, such as dense urban and cyber environments, and are not necessarily constrained by international law or protocols regulating conflict.

*Note. The chapters and appendixes of this publication address topics from the Chinese perspective. As such, the terms friendly refers to Chinese units or the units of Chinese allies. Enemy refers to units opposing China or its military. This may be a belligerent force or element within China, or an external actor. Parties are neutral regarding China. Threat uses the standard dictionary definition as opposed to that of U.S. doctrine. An opponent may be against either the U.S. or China, with context determining the correct interpretation.*

**Four strategic trends will influence future Chinese operations.** These trends are population growth, urbanization, population growth along major bodies of water, and human connectedness and interrelations.

*Note. Chinese concepts and doctrinal phrases contained in this document are used in accordance with PLA definitions as provided by the translation of Army Combined Tactics under Conditions of Informationization and Infantry Unit Tactics. Some of these concepts and phrases may mirror the names of U.S. Department of Defense (DOD) and U.S. Army terms. In such cases, the U.S. term is being referenced only if the name appears in italics.*

# II. Chinese Political Objectives

A useful analysis of the tactics of the PLA must be underpinned by an understanding of Chinese political and strategic priorities and an understanding of the complex relationship between the PLA and the Communist Party of China (CPC). China's history is deeply intertwined with its military—more so than any Western nation and possibly more so than any other country, except North Korea. The PLA was the vanguard of the Chinese Communist Revolution; its history, from the Chinese perspective, is one of glorious struggle over imperialist and capitalist oppressors. In contrast to Western militaries, the PLA is deeply politicized. It retains significant ties to Maoist and Marxist-Leninist political thought and has generations-deep connections with the CPC. While Western militaries are generally apolitical and are divorced from political parties, the PLA is officially the armed wing of the CPC.

Despite China's gradual economic liberalization and movement toward a capitalist, free-market economy, the PLA ostensibly remains generally steadfast in its support of Marxist-Leninist and Maoist philosophy. Though the PLA long ago abandoned more extreme Maoist experiments such as "rankless" force structures and eliminating rank insignia, the idea of People's War—an armed struggle of a population against a militarily superior adversary—still colors PLA thinking. This concept eventually took the form of People's War under Modern Conditions, an adaptation of Chinese Communist populist warfare that accounts for modern military capabilities. This in turn evolved to People's War in Conditions of Informationization in 2015. This evolution suggests that the PLA now sees itself as having acquired sufficient information capabilities to successfully employ them in a limited or regional military capacity. Marxist-Leninist and Maoist thinking still influence PLA operations all the way down to the tactical level, where a mix of autocratic statism and communal leadership are practiced even within small organizations.

Chinese national political objectives can be broken into two basic categories: security and development. Security objectives include the protection of the CPC as China's ruling party, the protection of Chinese sovereignty, protection of borders, internal security, and nuclear deterrence. Development objectives include the protection of Chinese economic interests at home and abroad, ensuring freedom of navigation for Chinese goods, procuring important commodities such as energy and raw materials, and establishing new export markets for Chinese goods. As China has transitioned from a closed, unstable, post-revolutionary internal political environment to a more open and stable one, strategic priorities have gradually shifted from security to development.

Today, with the primacy of the CPC virtually assured and few internal security threats, the People's Armed Police (PAP) has taken over much of the internal security mission, while the PLA focuses primarily on development-related objectives. The coordination of military modernization with economic development is a basic tenet of Chinese national strategy, resulting in significant annual defense budget increases for the past two decades. This principle is now manifest in its Military-Civil Fusion program, which seeks greater cooperation between military and civilian elements in achieving shared objectives. An important objective of Military-Civil Fusion is leveraging civilian assets and capabilities as a cost-efficient method of managing limited military resources. Military-Civil Fusion can be thought of as an extension of People's War, seeking to make best use of civilian assets and natural resources to underpin the growth and modernization of the PLA.

The ongoing Chinese movement toward development objectives requires a significant enhancement of PLA capabilities. Throughout much of its history, including its active conflicts in Korea (1950-53) and Vietnam (1979), the PLA had little to no expeditionary capability, minimal mechanization, low-technology systems, and a severe lack of military professionalism. Objectives that support development, however, generally require meaningful force projection capabilities: a blue-water navy, a modern air force, long-range targeting and strike capabilities, enhanced long-range air and sea strategic lift capabilities, and a well-trained, well-equipped, professional ground force. Due to one of the largest and longest economic growth cycles in history, China was able to fund the reform and development of the PLA, modernizing it over the last three decades from a force of poorly armed and poorly trained conscripts into a viable modern military.

# III. The New Chinese Grand Strategy

China desires status as a world power, likely using Imperial Great Britain or post-World War II United States as a model. This end state helps to achieve many smaller goals important to the Chinese: global influence, economic development, internal security, and CPC primacy. China views this final objective as incremental: it seeks to be a "prosperous society" by 2035, with the CPC still remaining the dominant political entity in China. This objective likely implies China being a regional hegemon in the Western Pacific, with a robust middle class and fully modernized military. The country desires to transition to a "leading world power" by the year 2049 (the centennial anniversary of the country's founding), complete with a "world-class military." These objectives are not nebulous, nor are they fanciful thinking— they are written into the CPC Constitution. Of note, despite the nature of these objectives, the Chinese government does not describe itself as seeking either regional or global hegemony.

Throughout Chinese history, beginning in the third century B.C. through the Second Sino-Japanese War (1937-45), incursions, invasions, and occupations along China's vast land border and coastline were constant. This fostered a deep-set national sense of sovereignty and border sanctity among the Chinese. Even in the Imperial Age (pre-1912), Chinese leaders sought to offset external threats through a combination of political savvy and military strength. China was historically the dominant economic and military power in East Asia, and Chinese leaders used this power to try and ensure domestic security and tranquility.

Much of this economic and military power was lost during the colonial and world-war periods, a time the Chinese refer to as a "century of humiliation." China's transition to a Communist political and economic system was a long and enormously destructive process that further marginalized its once-dominant power. It was not until the late 1970s that the country's resurgence really began. During this recent period of enormous economic growth and political change, China again began to seek a position as a world power. Though the global environment has changed through the years, the Chinese approach to security and economic growth has not.

China today seeks what it views as a restoration of its position as a global power through what are described by the CPC as peaceful and relatively unprovocative means. Bilateral and multilateral relationships are viewed as optimal, where all parties to a matter see benefit from an arrangement. China seeks a positive relationship with the West—especially the United States—underpinned by a massive exchange of trade and economic interdependence. China also seeks to modernize its military—not through aggressive, short-term arms buildup, but through long-term investment and development. It also seeks courteous—if not cordial—relations with its neighbors in East Asia. In short, China wishes to be a good neighbor and a good global citizen.

This approach has thus far yielded mixed results. Territorial disputes across the Western Pacific and East Asia, aggressive cyber activities, uncooperative diplomacy, questionable trade practices, and a horrendous human rights record all undermine the Chinese goal of being seen as a benevolent superpower. Activities that China views as fundamentally defensive—such as the establishment of artificial island airbases in the South China Sea—are seen as aggressive and provocative by virtually every other party in the region. Chinese cyber activities, seen by the nation as simple surveillance and deterrence activities, have caused several high-profile international incidents.

# IV. Chinese Strategic Objectives

China's strategic objectives are informed by its political objectives and generally support the broad political goals of Western Pacific dominance by 2035 and becoming a leading world power by 2049. They are also informed by certain elements of Chinese culture and history: the importance of status and honor, the desire for peace through power, and the belief in Chinese Communism, among other aspects. Chinese strategic objectives are to—

- Maintain internal security and stability.
- Secure and protect land borders and coastlines.
- Maintain regional stability.
- Maintain freedom of navigation.
- Resolve maritime territorial disputes.
- Establish positive conditions for potential hostilities.

*See facing page.*

# Chinese Strategic Objectives

*Ref: ATP 7-100.3, Chinese Tactics, (Aug '21), pp. 1-5 to 1-6.*

## Maintain Internal Security and Stability

Foremost in maintaining internal security and stability is ensuring the position of the CPC as the dominant political entity in China. PLA headquarters at provincial, county, and city levels are military components of the corresponding local government and are responsible for recruitment, demobilization, and other support to local governments. While maintaining domestic stability is the primary mission of the civilian police force and PAP, in extreme situations the PLA may be required to assist these forces in internal security operations.

## Secure and Protect Land Borders and Coastlines

China's long historic struggle with border security manifests itself today in a vast array of border security measures. Of particular note is the border with North Korea: while ostensibly friendly with the North Koreans, China is deeply concerned with the possible military and humanitarian crisis that might erupt in the event that the North Korean regime falls or war breaks out on the Korean Peninsula. In addition, China's shared border with India continues to be a source of friction and conflict.

## Maintain Regional Stability

One of the main benefits China sees to regional dominance is the maintenance of peace and order in the Western Pacific and East Asia. In addition to the complex political and diplomatic challenges, this requires the PLA to be capable of conducting shaping and deterrence operations throughout the region, able to deploy and defeat regional threats as required.

## Maintain Freedom of Navigation

China's economy is largely dependent on exports, most of which travel via the world's oceans. Maintaining safe and free passage through the global commons is a necessary component of Chinese political stability and economic development. Particular focus areas include the main shipping channels of the Western Pacific, such as the Strait of Malacca and the South Indian Ocean.

## Resolve Maritime Territorial Disputes

China views several land masses and their surrounding territorial waters in the South China Sea— and elsewhere in the Pacific—as strategically important. These include, but are not limited to, the Senkaku Islands, the Spratly Islands, and the Paracel Islands. Their importance is derived not only from their proximity to important global shipping lanes, but also due to their potential usefulness as military bases— particularly for naval and coast guard ships, fixed-wing aircraft, antisubmarine warfare capabilities, and land-based antiaircraft and antiship missile systems.

## Establish Positive Conditions for Potential Hostilities

China views conflict along a continuum, ranging from steady-state deterrence operations in peacetime through major combat operations. By meticulously—and sometimes covertly—conducting peacetime military and intelligence operations against potential opponents, China seeks to place its military in an advantageous position should active hostilities break out. Such operations include political and diplomatic efforts, offensive and defensive cyber actions, information operations, and covert intelligence operations. Central to this objective is the expansion of overseas basing for the PLA, enabling force projection outside of Chinese borders and giving Chinese leadership greater flexibility in choosing how and where to employ military force.

# V. Operational Environments (PMESII-PT)

*Ref: ATP 7-100.3, Chinese Tactics, (Aug '21), pp. 1-6 to 1-7.*

U.S. military analysis of an operational environment, including a composite environment created for training, professional education, and leader development purposes, focuses on eight interrelated operational variables: political, military, economic, social, information, infrastructure, physical environment, and time—collectively referred to as PMESII-PT. The following is a list of PMESII-PT conditions and trends found in China. It is not comprehensive, but it suggests a number of possible factors that U.S. exercise planners might use when constructing scenarios:

## P - Political

- The fractious relationship between China and Taiwan.
- Maritime disputes in the South and East China Seas.
- Complex sociopolitical interactions between China and North Korea.
- Land border disputes and frictions.
- Friction between the CPC and the growing quasi-capitalist ultra-wealthy class.
- Expansion of Chinese influence in emerging markets, particularly in Africa and Southwest Asia.
- Human rights violations, particularly against internal political opposition and minorities, and the harassment and mistreatment of journalists.
- Separatism in Western China.
- Major anticorruption efforts at every level of government.

## M - Military

- Growing use of high-technology weapons systems such as fifth-generation and low-observable aircraft, ballistic and cruise missiles, precision munitions, and networked warfare.
- Ongoing top-to-bottom reform of the PLA, including widespread professionalization.
- Expansion of standoff precision munitions and other antiaccess capabilities.
- Establishment of hardened military facilities on islands, both natural and manmade, throughout the Western Pacific.
- Gradual expansion of PLA expeditionary capabilities, particularly throughout the Western Pacific and Indian Oceans.

## E - Economic

- Continued liberalization of formerly hardline Marxist-Leninist and Maoist economic policy.
- Threats to critical shipping lanes and overland trade routes linking China with its export markets.
- Export and trade restrictions on raw materials, particularly on rare but critical metals.
- Major anticorruption efforts at every level of government.
- Complex economic relationship with North Korea.

## S - Social

- Increased internal frictions with minority groups, particularly in and around distressed border regions.
- Continued adoption of Western cultural products, especially by younger generations.

- Social resistance to heavy-handed governmental approach to internal security.
- Increased frictions between quasi-capitalist Chinese oligarchs and more-traditional CPC supporters due to failure to quell corruption.

## I - Information

- Extensive use of cyber activities—official, unofficial, and third party—to influence conditions domestically and abroad.
- Ongoing active People's Republic of China (PRC) cyber activities attempting to extract sensitive or classified information from foreign networks, particularly those in defense industries.
- Sophisticated information operations campaigns to influence both global and regional politics.

## I - Infrastructure

- Continued development of domestic infrastructure with a focus on the export economy.
- Heavy investment in overseas infrastructure, particularly in emerging markets for Chinese export goods.

## P - Physical Environment

- Ongoing effects of global climate change reducing availability of arable land and threatening low-lying coastal areas.
- Increasing frequency and severity of tropical storms, which affect military and economic development in the Western Pacific.
- Shortsighted environmental policies creating public health crises due to air and water pollution and rapid depletion of shared international resources, such as fisheries.

## T - Time

- The Chinese have historically taken a much longer view of time than the United States.
- The PLA has deliberately chosen to adopt a more Western view of time as part of its ongoing military reforms.

# VI. PRC Global Initiatives

*Ref: Department of Defense Annual Report to Congress: Military and Security Developments Involving the People's Republic of China 2024, pp. 10 to 13.*

The PRC's foreign policy framework is based around its entrenched perception that the East is rising and the West is in decline while noting that the concentration of power in international governance and institutions in the West creates an imbalance. As such, Beijing aims to promote and accelerate a transformation in the global order, including changing the distribution of power among states, revising the principles of interstate relations, and reforming global governance structures to better align with Beijing's vision. Since President Xi's first term in office, he and other PRC officials have described a desire to create a "Community of Shared Future for Mankind," contrasting this vision against the current global governance arrangement. According to then PRC State Councilor and Foreign Minister Wang Yi, global governance suffers from "weaknesses and deficiencies," highlighted by the COVID-19 pandemic, necessitating a new approach. To "seize opportunities" for reform, the PRC is actively pursuing the creation of new multinational organizations, forums, and a series of key global initiatives to uphold the authority of the CCP and the PRC's national sovereignty, security, and development interests.

The PRC has **three major initiatives** to realize this global vision—**the Global Development Initiative (GDI), Global Security Initiative (GSI), and the Global Civilization Initiative (GCI).** Since 2021, the PRC has reclassified its historical efforts under these new broad umbrellas.

## Belt and Road Initiative (BRI)

First announced in 2013, the BRI is Xi's signature foreign and economic policy that rebranded and further expanded the PRC's global outreach. Beijing uses the BRI to support its strategy of national rejuvenation by seeking to expand global transportation and trade links to support its development and deepen its economic integration with nations along its periphery and beyond. The PRC implements BRI by financing, constructing, and developing transportation infrastructure, natural gas pipelines, hydropower projects, digital connectivity, and technology and industrial parks worldwide. As of 2023, at least 149 countries had signed BRI cooperation documents—up from 147 countries in 2022, 146 in 2021, 138 in 2020, and 125 in 2019.

As the BRI has expanded, Beijing has introduced three sub-efforts: the Health Silk Road (HSR), the Green Silk Road (GSR), and the Digital Silk Road (DSR), which it describes as "sub-initiatives" to expand beyond the BRI's traditional infrastructure focus.

- The **HSR** is the PRC's World Health Organization–supported initiative for providing medical assistance through BRI transportation networks. In the future, it may help the PRC expand the international market share of PRC medical products, strengthen its bid for a role as a global public health leader, and identify—and justify—the need for new BRI projects in traditional infrastructure and healthcare investment.
- The **GSR** aims to support low-carbon infrastructure, energy, and finance projects; this initiative aligns with the PRC's own goal of achieving carbon neutrality before 2060 and presents Beijing as a responsible party in working toward the 2030 Agenda for Sustainable Development Goals established by the United Nations General Assembly.
- The **DSR** is one of the primary ways Beijing seeks to transfer PRC technology to partner countries to build strategic dependencies, which the PRC leverages to propagate its own technology standards as it seeks to set global standards for next-generation technology.

Chap 1

# III(b). Threat Assessment (2025)

*Ref: Office of the Director of National Intelligence, Annual Threat Assessment of the US Intelligence Community (Mar '25), pp. 9 to 16.*

*Several major state actors present proximate and enduring threats to the United States and its interests in the world, challenging U.S. military and economic strength, regionally and globally. China stands out as the actor most capable of threatening U.S. interests globally, though it is also more cautious than Russia, Iran, and North Korea about risking its economic and diplomatic image in the world by being too aggressive and disruptive. Growing cooperation among these actors expands the threat, increasing the risk that should hostilities with one occur, it may draw in others.*

## I. Strategic Overview

*President Xi Jinping and the People's Republic of China (PRC) want to achieve "the great rejuvenation of the Chinese nation" by 2049. The PRC will seek to increase its power and influence to shape world events to create an environment favorable to PRC interests, obtain greater U.S. deference to China's interests, and fend off challenges to its reputation, legitimacy, and capabilities at home and abroad.*

- Beijing is deeply suspicious of U.S. intentions and views Washington's measures against China as part of a concerted, whole-of-government effort, working with U.S. allies and partners, to contain China's development and rise, undermine CCP rule, and prevent the PRC from achieving its aims. PRC leaders are most concerned about strong unified opposition from the United States and its allies, and are responding, in part, by strengthening ties with partners like Russia and North Korea.
- At the same time, China's leaders will seek opportunities to reduce tension with Washington when they believe it benefits Beijing, protects its core interests, and buys time to strengthen its position.

*The PRC will likely continue posturing to be in a position of advantage in a potential conflict with the United States. The PRC will continue trying to press Taiwan on unification and will continue conducting wide-ranging cyber operations against U.S. targets for both espionage and strategic advantage. China will likely struggle to sufficiently constrain the activities of PRC companies and criminal elements that enable the supply and trafficking of fentanyl precursors and synthetic opioids to the United States, absent greater law enforcement actions.*

- China's military operations to project power over Taiwan and its efforts to assert sovereignty claims in the South and East China Seas occur routinely with confrontations that increase concern of miscalculations potentially leading to conflict.
- China has demonstrated the ability to compromise U.S. infrastructure through formidable cyber capabilities that it could employ during a conflict with the United States.

*Beijing will continue to strengthen its conventional military capabilities and strategic forces, intensify competition in space, and sustain its industrial-and technology-intensive economic strategy to compete with U.S. economic power and global leadership.*

# II. Military

*China presents the most comprehensive and robust military threat to U.S. national security.* The People's Liberation Army (PLA) is fielding a joint force that is capable of full-spectrum warfare to challenge intervention by the United States in a regional contingency, projecting power globally, and securing what Beijing claims is its sovereign territory. A major portion of China's military modernization efforts is focused on developing counter-intervention capabilities tailored against all aspects of U.S. and allied military operations in the Pacific. Beijing will focus on meeting key modernization milestones by 2027 and 2035, aimed at making the PLA a world-class military by 2049.

- Examples of PLA advances in 2024 include the PLA Navy's third carrier (CV-18 Fujian) beginning sea trials and likely being ready to enter operational service in 2025. The PLA Rocket Force probably is fielding the DF-27 ballistic missile, with a hypersonic glide vehicle payload option and an estimated range of between 5,000 and 8,000 kilometers. The PLA ground forces are also fielding its most advanced multiple rocket launcher, the PCH191, increasing its long-range, precision strike capability.
- The PLA has improved its force structure, readiness, and training. The PLA probably has made particular progress in critical areas, such as modernizing key ground forces, expanding its navy with more modern combatants, and fielding a wide variety of new missile systems; it has also improved its electronic warfare (EW) capabilities.

*The PLA has the capability to conduct long-range precision-strikes with conventional weapons against the Homeland's periphery in the Western Pacific, including Guam, Hawaii, and Alaska.* China has developed a range of ballistic and cruise missiles with conventional payloads that can be delivered from its mainland as well as by air and sea, including by nuclear-powered submarines. China may be exploring development of conventionally-armed intercontinental range missile systems, which, if developed and fielded, would allow China to threaten conventional strikes against targets in the continental United States.

The PLA will continue to pursue the establishment of overseas military installations and access agreements to project power and protect China's interests abroad. Beijing may also pursue a mixture of military logistics models, including preferred access to commercial infrastructure abroad, exclusive PLA logistics facilities with pre-positioned supplies co-located with commercial infrastructure, and bases with stationed forces, to meet its overseas military logistics needs.

*China is using complex, whole-of-government campaigns featuring coercive military, economic, and influence operations short of war to assert its positions and strength against others, reserving more destructive tools for full-scale conflict.* Beijing will likely expand these campaigns to advance unification with Taiwan, project power in East Asia, and reverse perceived U.S. hegemony.

- Beijing has pushed back against U.S. military operations, such as reconnaissance and bomber flights, freedom of navigation operations, and exercises around PRC borders and maritime claims. The PLA regularly intercepts and shadows U.S. forces and sometimes conducts unsafe maneuvers in their vicinity.

# III. Taiwan and Maritime Flashpoints

*In 2025, Beijing will likely apply stronger coercive pressure against Taiwan and perceived increases in U.S. support to the island to further its goal of eventual unification.* The PRC calls for a peaceful unification with Taiwan to resolve the Civil War that drove Taiwan's separation, even as it threatens to use force to compel unification if necessary and counter what it sees as a U.S. attempt to use Taiwan to undermine China's rise.

*Excerpted, see facing page for full discussion.*

# Taiwan and Maritime Flashpoints

*Ref: Office of the Director of National Intelligence, Annual Threat Assessment of the US Intelligence Community (Mar '25), pp. 11.*

*In 2025, Beijing will likely apply stronger coercive pressure against Taiwan and perceived increases in U.S. support to the island to further its goal of eventual unification.* The PRC calls for a peaceful unification with Taiwan to resolve the Civil War that drove Taiwan's separation, even as it threatens to use force to compel unification if necessary and counter what it sees as a U.S. attempt to use Taiwan to undermine China's rise.

A conflict between China and Taiwan would disrupt U.S. access to trade and semiconductor technology critical to the global economy. Even without U.S. involvement in such a conflict, there would likely be significant and costly consequences to U.S. and global economic and security interests.

*Beijing is working to isolate Taipei by pressuring states to downgrade diplomatic ties and support China's unification goal.* Since 2016, Taiwan's official diplomatic relationships have dropped from 22 to only 12, and several of the remaining ones are vulnerable to Chinese pressure.

China is advancing military capabilities for a cross-Strait campaign while also using its armed forces to exert steady state pressure on Taiwan. The PLA probably is making steady but uneven progress on capabilities it would use in an attempt to seize Taiwan and deter—and if necessary, defeat—U.S. military intervention, and it is intensifying the scope, size, and pace of operations around Taiwan.

*Beijing will continue to pressure Taipei with economic coercion and probably will increase it if it sees Taiwan taking steps toward formal independence.* It could suspend preferential tariff terms, selectively ban Taiwan imports to China, and arbitrarily enforce regulations.

*Beijing's aggressive efforts to assert sovereignty claims in the South and East China Seas are heightening tensions that could trigger a broader conflict.*

- In 2024, PRC tactics in the South China Sea led to the loss of the Philippines' unilateral access to some disputed areas, and forced talks between Beijing and Manila in which the Philippines agreed to concessions in exchange for access. However, Manila is unlikely to relinquish its claims, creating potential for escalation by either side.
- Tension between China and Japan over the Senkaku Islands last flared up a decade ago. Since then, Chinese ships have remained in proximity of the disputed islands, occasionally entering the territorial zone, and driving responses from Japan's Self-Defense Force to monitor the activity.

*See pp. 1-35 to 1-44 for discussion of the Taiwan Straight.*

# IV. Cyber

*The PRC remains the most active and persistent cyber threat to U.S. government, private-sector, and critical infrastructure networks. The PRC's campaign to preposition access on critical infrastructure for attacks during crisis or conflict, tracked publicly as Volt Typhoon, and its more recently identified compromise of U.S. telecommunications infrastructure, also referred to as Salt Typhoon, demonstrates the growing breadth and depth of the PRC's capabilities to compromise U.S. infrastructure.*

- If Beijing believed that a major conflict with Washington was imminent, it could consider aggressive cyber operations against U.S. critical infrastructure and military assets. Such strikes would be designed to deter U.S. military action by impeding U.S. decision-making, inducing societal panic, and interfering with the deployment of U.S. forces.

# V. Economics

*The PRC seeks to compete with the United States as the leading economic power in the world. To do so, the strategy calls for a centralized, state-directed, and nationally resourced approach to dominating global markets and strategic supply chains, limiting foreign competitors, and making other nations dependent on China.* PRC leaders are applying the same strategy to bolster China's position and become more globally dominant in critical supply chains, both in upstream inputs it can provide more cheaply than others and in downstream production at wider scale.

- China's weak domestic demand, coupled with its industrial policies, such as manufacturing subsidies, have enabled a surge in cheap Chinese exports in sectors such as steel, harming U.S. competitors and fueling a record PRC trade surplus.
- China's dominance in key supply chains enables its use of economic coercion against countries that adopt policies Beijing opposes. Beijing is developing an institutionalized framework enabling more assertive and centrally controlled trade retaliation. PRC leaders are using ostensibly unofficial or technical trade and investment barriers, administrative regulations, logistics, and symbolic sanctions in a targeted way against individuals, firms, and sectors, in parallel with messages to warn and deter.
- PRC leaders appear to be preparing for more economic friction with the United States, and probably are weighing options with the new U.S. administration while looking for leverage and other ways to prevent a major escalation and decoupling.

*China's dominance in the mining and processing of several critical materials is a particular threat, providing it with the ability to restrict quantities and affect global prices.* Beijing has shown a willingness to restrict global access to its mineral resources—sometimes in response to geopolitical disputes—as with its banning of exports to the United States of metals used in semiconductor manufacturing, such as gallium, germanium, and antimony in December 2024 in response to U.S. export controls on advanced semiconductors and chip-making equipment. Other examples include when the PRC temporarily stopped rare earth element exports to Japan in 2010, and Beijing's creation of new laws codifying its authority to restrict mineral exports. A prolonged cessation in supplies controlled by China could disrupt critical inputs needed for U.S. industry and technological advancements.

*China has similar aims in global shipping and resource access, including in the Arctic, where melting sea ice is creating opportunities for expanded maritime transport and energy exploitation, especially along the Northern Sea Route (NSR) off Russia's coast.* China seeks access to the Arctic's potentially vast natural resources, including oil, gas, and minerals, even though China is not among the eight Arctic countries that control territory in the region. Beijing seeks to normalize more direct and efficient maritime shipping routes to Russia and other Northern Hemisphere areas, as a way to fuel its economic growth and energy security and reduce its dependence on Middle East energy. China has gradually increased engagement with Greenland mainly through mining projects, infrastructure development, and scientific research projects. Despite less active engagement right now, China's long-term goal is to expand access to Greenland's natural resources, as well as to use the same access as a key strategic foothold for advancing China's broader and economic aims in the Arctic.

# VI. Technology

*China is using an aggressive, whole-of-government approach, combined with state direction of the private sector, to become a global S&T superpower, surpass the United States,* promote self-reliance, and achieve further economic, political, and military gain. Beijing has prioritized technology sectors such as advanced power and energy, AI, biotechnology, quantum information science, and semiconductors, further challenging U.S. efforts to protect critical technologies by tailoring restrictions narrowly to address national security concerns. China is accelerating its S&T progress through a range of licit and illicit means, to include investments, intellectual property acquisition and theft, cyber operations, talent recruitment, international collaborations, and sanctions evasion.

- Some forecasts indicate China's technology sectors will account for as much as 23 percent of its gross domestic product by 2026, more than doubling since 2018. In addition to private funding, the PRC government is investing hundreds of billions of dollars in priority technologies, such as AI, microelectronics, and biotechnologies, in pursuit of its self-reliance goals.

*China almost certainly has a multifaceted, national-level strategy designed to displace the United States as the world's most influential AI power by 2030.* China is experiencing a boom in generative AI with the rapid emergence of a large number of PRC-developed models, and is broadly pursuing AI for smart cities, mass surveillance, healthcare, S&T innovation, and intelligent weapons. Chinese AI firms are already world leaders in voice and image recognition, video analytics, and mass surveillance technologies. The PLA probably plans to use large language models (LLMs) to generate information deception attacks, create fake news, imitate personas, and enable attack networks. China has also announced initiatives to bolster international support for its vision of AI governance.

- China has stolen hundreds of gigabytes of intellectual property from companies in Asia, Europe, and North America in an effort to leapfrog over technological hurdles, with as much as 80 percent of U.S. economic espionage cases as of 2021 involving PRC entities.

*China also sees biotechnology as critical to becoming a dominant economic power and intends to grow its domestic bioeconomy to $3.3 trillion this year.* Beijing is investing heavily in collecting health and genetic data both at home and abroad in pursuit of these goals, and has shown it can be globally competitive in certain low-cost, high-volume commodities, such as biomanufacturing and genetic sequencing. Beijing has identified genetic data as a national strategic resource and is expanding state control over the country's gene banks and other genetic repositories, positioning it to potentially lead in precision medicine and agricultural biotechnology applications.

China has made progress in producing advanced 7-nanometer (nm) semiconductor chips for cryptocurrency mining and cellular devices using previously acquired deep ultraviolet (DUV) lithography equipment, but will face challenges achieving high-quality, high-volume production of these chips without access to extreme ultraviolet lithography tools. PRC researchers also continue to explore applying advanced patterning techniques to DUV machines to produce semiconductor chips as small as 3nm. China leads the world in legacy logic semiconductor (28nm and up) production, accounting for 39.3 percent of global capacity, and is expected to add more capacity than the rest of the world combined through 2028. These legacy chips are vital to producing automobiles, consumer electronics, home appliances, factory automation, broadband, and many military and medical systems.

# VII. Weapons of Mass Destruction (WMD)

*China remains intent on modernizing, diversifying, and expanding its nuclear posture. China's nuclear weapons and advanced delivery systems pose a direct threat to the Homeland and are capable of delivering catastrophic damage to the United States and threatening U.S. military forces here and abroad.*

*China most likely possesses capabilities relevant to chemical and biological warfare (CBW) that pose a threat to U.S., allied, and partner forces as well as civilian populations.*

# VIII. Biosecurity

*China's approach to and role in global biological, medical, and other health-related global priorities present unique challenges to the United States and the world.* The COVID-19 pandemic that ultimately led to the death of more than one million Americans—and multiples more worldwide—began in China, which Beijing still refuses to acknowledge. China's strict censorship and repression of free speech prevented doctors treating the earliest of patients in Wuhan from warning the world of a far more serious contagion than Beijing wanted told, slowing the world's preparedness and response. To this day, Beijing refuses to fully cooperate with the rest of the international community trying to definitively pinpoint the precise cause of the disease so it can head off and prepare for any new disease.

Regarding COVID-19 origins, IC agencies have continued to evaluate new information from classified and open sources, revisit previous reporting, and consult with diverse technical experts to increase our understanding of the cause of the pandemic. These efforts have led CIA to assess that a research-related hypothesis is more likely than a natural origin hypothesis.

The other hypothesis for COVID-19—natural origin—includes many scenarios in which humans could have been infected with SARS-CoV-2—the virus that causes COVID-19—or a close progenitor through exposure to wild or domestic animals. China is home to a diverse body of naturally occurring coronaviruses found in a wide geographic area, and there is precedence for these viruses to emerge within human populations far from reservoir locations. For example, the coronavirus that is the closest known relative to SARS-CoV—the virus that causes severe acute respiratory syndrome (SARS)—probably originated in Yunnan Province, according to scientific studies, even though the first SARS outbreak detected in humans in 2003 occurred in Guangdong Province, hundreds of miles away.

- The research-related incident hypothesis also considers a broad range of potential initial human-infection scenarios from events in research facilities, such as government or university laboratories, to research-related activities in the field, such as collecting samples from wild animals.

*The PRC's dominance in pharmaceutical and medical supply production combined with lower quality safety and environmental standards than those of the United States positions Beijing to potentially restrict such exports for leverage over Washington and others in trade or security disputes.* The PRC plays an increasingly important role in supplying pharmaceuticals and related medical supplies to the United States, as well as the rest of the world.

- U.S. imports of Chinese pharmaceuticals—defined as medicines, vaccines, blood, organic cultures, bandages, and organs—grew almost five-fold between 2020 and 2022 alone, from \$2.1 billion to \$10.3 billion.

- The PRC also might look to uniquely provide such supplies and medical aid to countries, more cheaply and at scales competitors cannot match, as a way to boost its global influence at the expense of the United States. The PRC's "vaccine diplomacy" during the COVID-19 pandemic—it provided vaccines to 83 countries—was driven at least in part by geopolitical considerations, such as currying favor for a new port in Burma.

# IX. Space

*China has eclipsed Russia as a space leader and is poised to compete with the United States as the world's leader in space by deploying increasingly capable interconnected multi-sensor systems and working toward ambitious scientific and strategic goals.* China has achieved global coverage in some of its intelligence, surveillance, and reconnaissance (ISR) constellations and world-class status in all but a few space technologies.

- China's Beidou constellation is a world-class position, navigation, and timing capability that competes with U.S. GPS and Europe's Galileo service. The PLA ISR architecture and satellite communications are areas the PLA continues to improve upon to close the perceived gap between itself and the U.S. military.
- China's successful lunar sample return mission in June 2024 contributes to Beijing's technological prowess and national prestige while supporting its effort to land astronauts on the Moon by 2030 and establish the first lunar base by 2035.
- China's commercial space sector is growing quickly with aspiration to be a major global competitor to U.S. and European space companies. For example, China launched the first set of satellites in its low Earth orbit (LEO) proliferated constellation last year for its own satellite Internet service to compete with Western commercial satellite Internet services.

*Counterspace operations will be integral to PLA military campaigns, and China has counterspace-weapons capabilities intended to target U.S. and allied satellites. China already has fielded ground-based counterspace capabilities, including EW systems, directed energy weapons (DEWs), and antisatellite (ASAT) missiles intended to disrupt, damage, and destroy target satellites.*

- China also has conducted orbital technology demonstrations, which, while not counterspace weapons tests, prove its ability to operate future space-based counterspace weapons. China has also conducted on-orbit satellite inspections of other satellites, which probably would be representative of the tactics required for some counterspace attacks.

# X. Malign Influence Activities

*Beijing will continue to expand its coercive and subversive malign influence activities to weaken the United States internally and globally, as well as counter what Beijing sees as a U.S.-led campaign to tarnish China's global relations and overthrow the CCP. Through these efforts, the PRC seeks to suppress critical views and critics of China within the United States and worldwide, and sow doubts in U.S. leadership and strength. Beijing is likely to feel emboldened to use malign influence more regularly in coming years, particularly as it fields AI to improve its capabilities and avoid detection.*

- PRC actors have increased their capabilities to conduct covert influence operations and disseminate disinformation. For example, pro-China online actors in 2024 used AI-generated news anchors and fake social media accounts with AI-generated profile pictures to sow divisions on issues such as drug use, immigration, and abortion.

# XI. China's Challenges

*Ref: Office of the Director of National Intelligence, Annual Threat Assessment of the US Intelligence Community (Mar '25), p. 16.*

*China faces daunting challenges that will impair CCP leaders' strategic and political achievements. China's leaders probably are most concerned about corruption, demographic imbalances, and fiscal and economic struggles because they threaten the country's economic performance and quality of life, two key factors underpinning CCP legitimacy.* Despite an acute economic slowdown, China's leaders probably will resist making needed structural reforms and instead maintain statist economic policies to steer capital toward priority sectors, reduce dependence on foreign technologies, and enable military modernization.

- China's growth probably will continue to slow because of low consumer and investor confidence. China's birth and marriage rates continue to decline, reinforcing negative population trends and a shrinking labor force.

Xi's focus on security and stability for the CCP and securing other leaders' personal loyalty to him is undermining China's ability to solve complex domestic problems and will impede Beijing's global leverage. Xi's blending of domestic and foreign security threats is undermining China's position and standing abroad, reducing Beijing's ability to shape global perceptions and compete with U.S. leadership.

# IV. Theater Commands



*Ref: Department of Defense Annual Report to Congress: Military and Security Developments Involving the People's Republic of China 2024, pp. 17 to 19.*

## I. Theater Commands

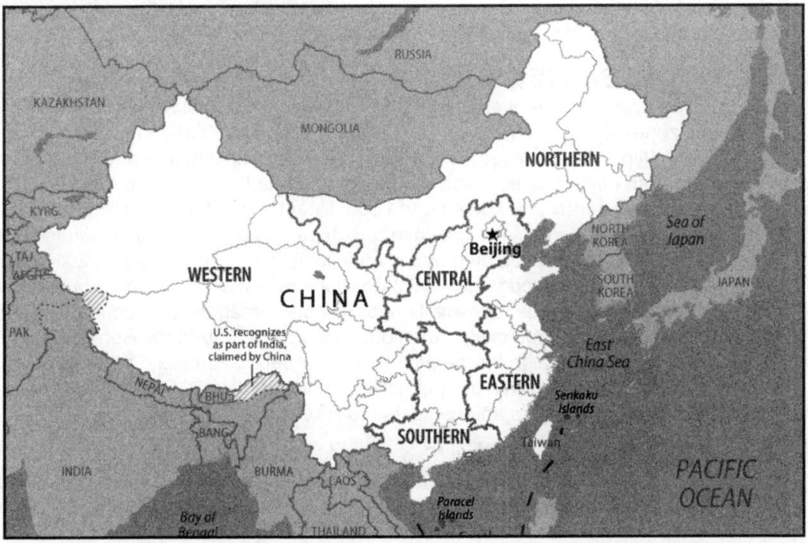


*Ref: China's Military: The People's Liberation Army (PLA), Congressional Research Service. Map of China and PLA Theater Commands. (Map by Amber Wilhelm.)*

The PRC steadily improves joint operations in the five theater commands, which were established in early 2016. Each theater command receives direction from the CMC; has operational authority over assigned PLA, PAP, and militia forces in its theater; and is responsible for all conventional combat and non-combat operations in its area of responsibility. Theater commands develop theater-specific strategies to prepare to fight and win against an adversary, develop joint operational plans and military capabilities, respond to crises, and safeguard the sovereignty and stability of claimed territories. The strategic directions of the theater commands are based on PRC perceptions of peripheral threats:

- **Eastern Theater Command (ETC)** – Taiwan, East China Sea *(See p. 1-35 to 1-43).*
- **Southern Theater Command (STC)** – South China Sea; Southeast Asia border security; territorial disputes *(See pp. 1-45 to 1-50.)*
- **Western Theater Command (WTC)** – India, Central Asia, "counterterrorism" in Xinjiang and Tibet *(See pp. 1-51 to 1-58.)*
- **Northern Theater Command (NTC)** – Korean Peninsula, Russia border security *(See pp. 1-59 to 1-64.)*
- **Central Theater Command (CTC)** – Capital defense; surge support to other theaters *(See pp. 1-65 to 1-66.)*

*See following pages (pp. 1-32 to 1-33) for discussion of related territorial disputes.*

# II. China's Territorial Disputes (Select)

*Ref: Department of Defense Annual Report to Congress: Military and Security Developments Involving the People's Republic of China 2024, pp. 17 to 19.*

The PRC's use of force in territorial disputes has varied widely since 1949. Some disputes led to war, as in border conflicts with India in 1962 and Vietnam in 1979. The PRC's contested border with the Soviet Union during the 1960s raised the possibility of nuclear war. In recent cases involving land border disputes, the PRC has sometimes been willing to compromise with and offer concessions to its neighbors. Since 1998, the PRC has settled 11 land-based territorial disputes with six of its neighbors. However, in the last decade, the PRC has employed a more coercive approach to deal with disputes over maritime features, rights to potentially rich offshore oil and gas deposits, and border areas.

## East China Sea (ECS) *(See pp. 1-35 to 1-44 for the Taiwan Straight.)*

The PRC and Japan have overlapping claims to the continental shelf and the exclusive economic zones (EEZs) in the East China Sea (ECS). The ECS contains natural gas and oil, although hydrocarbon reserves are difficult to estimate. Japan maintains that an equidistant line from each country involved should separate the EEZs while the PRC claims an extended continental shelf beyond the equidistant line to the Okinawa Trench. The PRC continues to assert sovereignty over the Japan-administered Senkaku Islands and reiterate the importance of abiding by the four-point consensus signed in 2014, which states both sides acknowledge divergent positions over the ECS dispute but prevents escalation through dialogue, consultation, and crisis management mechanisms. PRC government ships were detected in the contiguous zone around the Senkakus on 352 days in 2023—a record high for the period since Japan nationalized the islands in 2012. Japan remains concerned with the persistent deployment of CCG ships and fishing vessels in disputed ECS waters and contests the PRC's claim of sovereignty.

## South China Sea (SCS) *(See pp. 1-46 to 1-50.)*

The SCS plays an important role in security considerations across East Asia because of the region's reliance on the flow of oil and commerce through SCS shipping lanes, including more than 80% of the crude oil to the PRC, Japan, South Korea, and Taiwan. The PRC maintains sweeping sovereignty claims in the SCS, including unlawful maritime claims to waters in an ambiguous "dashed line" circumscribing the vast majority of the SCS. These claims extend over the Spratly and Paracel Islands and other land features, which are disputed in whole or part by Brunei, the Philippines, Malaysia, and Vietnam. In 2016, a unanimous Arbitral Tribunal constituted under the 1982 Law of the Sea Convention determined that the PRC's expansive SCS maritime claims are inconsistent with international law. The arbitral decision is final and legally binding. The PRC continued to employ the PLAN, China Coast Guard (CCG), and China Maritime Militia (CMM) to patrol the region and aggressively enforce the PRC's unlawful claims throughout 2023. In 2023, the PRC frequently adopted unsafe and unprofessional measures, such as firing water cannons, aggressive maneuvers and ramming, and directing military-grade lasers against SCS claimant nation ships—most prominently against the Philippines—lawfully operating in the SCS, resulting in injuries to crew members and damage to government and civilian ships.

## India *(See pp. 1-51 to 1-56.)*

Tensions with India along the Line of Actual Control (LAC)—a disputed border between the PRC's western provinces and India's northern provinces—sparked a standoff between PRC and Indian forces in mid-May 2020, which escalated on June 15, 2020, after a skirmish ensued in the Galwan Valley between Indian Army and PLA forces that ended with 20 Indian soldiers and four PRC soldiers dead. The two sides agreed to tactical pull-backs from most of the contested areas in early 2021 but standoffs continue at two locations in Ladakh on the Western Sector of the LAC. In late 2022, PRC and Indian forces engaged in an unarmed clash near Tawang along the Eastern Sector of the LAC separating Tibet and the Indian state of Arunachal Pradesh. These engagements coincide with a significant and sustained escalation in military infrastructure to support a long-term presence on the LAC. In late 2023, India hosted the 20th round of commander-level meetings with the PLA and have maintained open lines of communication.

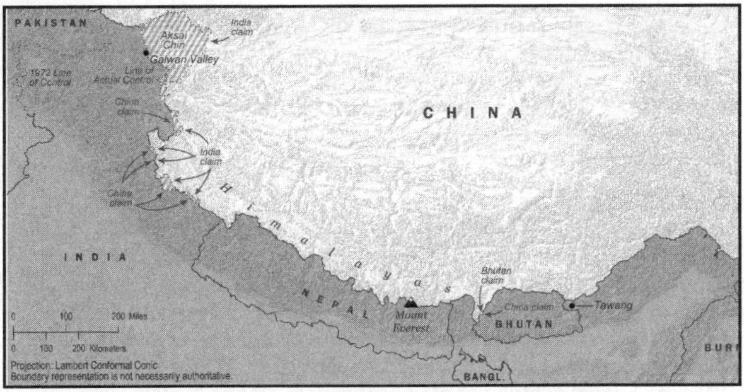


*Ref: Graphic: Disputes of China-India-Bhutan, p. 19.*

# Exclusive Economic Zone (EEZ)

The PRC has long challenged foreign military activities in its claimed EEZ in a manner inconsistent with customary international law reflected in the UNCLOS. The PRC routinely shadows and challenges U.S., allied, and partner military assets operating outside the territorial sea of any SCS claimant and occasionally conducts risky and coercive maneuvers, such as close intercepts of aircraft or dangerous at-sea maneuvering to exert Beijing's extraterritorial sovereignty claims. The PLA has used lasing, aerobatics, acoustic devices, the discharge of objects, and other activities that impinged on the ability of nearby aircraft and vessels to maneuver safely where high seas freedoms apply. PRC survey ships are extremely active in the SCS and frequently operate in the claimed EEZs of other nations in the region, such as the Philippines, Vietnam, Indonesia, and Malaysia.

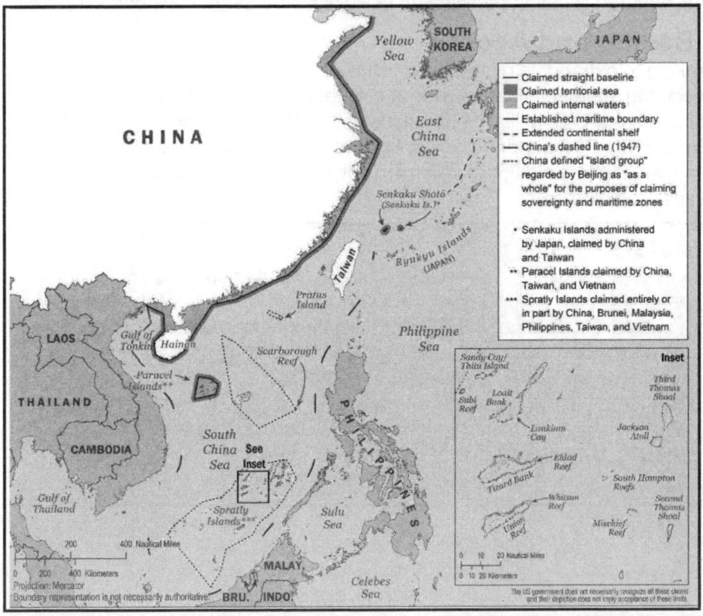


*Ref: Graphic: China's claimed maritime limits and boundaries, p. 18.*

# III. PLA's Growing Global Presence

*Ref: Department of Defense Annual Report to Congress: Military and Security Developments Involving the People's Republic of China 2024, pp. X to XI.*

PLA concepts and capabilities focus on projecting power far from China's shores. The PLAN's evolving focus from "offshore defense" to "open seas protection" and the PLAAF's interest in becoming a "strategic" air force reflect the PLA's interest in conducting operations beyond the PRC and its immediate periphery. The PLA has embraced its concept of non-war military activities (NWMA) as an effective way to secure the PRC's global interests with military force while gaining valuable operational experience.

## PLA Overseas Military Activities

Beijing implements its global counterterrorism strategy to develop the PLA's operational experience, secure greater access to overseas theaters, and prevent terrorist attacks in China and against PRC citizens and economic projects abroad. Citing its counterterrorism mission, since 2008, the PLA has dispatched more than 40 naval escort task forces (NETFs) to the Gulf of Aden, granting PLAN sailors important experience in overseas operations. In 2017, the PLA established an overseas base in Djibouti. The PRC is the largest contributor of peacekeepers among the five UN Security Council permanent members, having deployed about 50,000 personnel over the last 31 years. The PRC uses international outreach to garner the assistance of partner governments, routinely lobbies foreign partners to extradite alleged Uyghur extremists, and seeks public endorsement of its counterterrorism efforts in multilateral forums.

## Military Cooperation

The PRC uses bilateral and multilateral exercises to normalize its overseas presence and deepen defense ties. Beginning in 2014, the PRC has markedly increased participation in bilateral and multilateral exercises. Since then, Beijing has held recurring exercises, including the FALCON STRIKE air exercise with Thailand (2015), MARITIME SECURITY BELT naval exercise with Russia and Iran (2019), and BLUE SWORD with Saudi Arabia (2019). The PRC has expanded combined exercises with Russia, especially since 2018, when Beijing first participated in Russia's VOSTOK capstone exercise.

## Overseas Basing and Access

The PRC is seeking to expand its overseas logistics and basing infrastructure to allow the PLA to project and sustain military power at greater distances. A global PLA logistics network could disrupt U.S. military operations as the PRC's global military objectives evolve. Beyond the PLA support base in Djibouti, the PRC is very likely already considering and planning for additional military logistics facilities to support naval, air, and ground forces projection.

### Arms Transfers

The PRC uses foreign suppliers to overcome limitations in its domestic production capabilities, particularly for helicopters and aircraft engines. As its aerospace industry improves over the next decade, the PRC very likely will decrease its foreign acquisitions to maintain only an import relationship with foreign suppliers positioned to quickly fill niche gaps in the PRC's inventory. As of 2023, the PRC is the fourth-largest arms supplier in the world and sells nearly every category of conventional military equipment including unmanned aerial vehicles, man-portable air-defense systems (MANPADS), submarines, naval surface vessels, surface-to-air missile systems, and fighter aircraft to customers worldwide.

Chap 1

# IV(a). Eastern Theater Command (ETC)

*Ref: Department of Defense Annual Report to Congress: Military and Security Developments Involving the People's Republic of China 2024, pp. 113-115 and 124-132.*

**Key Takeaways**
- The ETC is oriented toward Taiwan and the East China Sea (ECS).
- The ETC likely would oversee a Taiwan Campaign.
- The ETC was responsible for executing exercise JOINT SWORD, aimed at pressuring Taiwan in April 2023.

The Eastern Theater Command (ETC) has responsibility for the East China Sea (ECS) and likely executes operational control over military matters related to Taiwan and Japan, including contingencies in and around the Taiwan Strait and the Senkaku Islands. *(See p. 1-32 for related discussion of East China Sea issues.)*

During 2023, the ETC engaged in a series of training events and exercises to improve joint operations and combat readiness, organizing exercises and drills consisting of long-distance training and mobilization, aerial combat, live-fire training, and the use of modified civilian ferries to augment transportation.

Major PLA units in the ETC include the PLAA's 71st, 72nd, and 73rd GAs comprised of 18 combined arms and amphibious combined arms BDEs; two PLAN naval bases, two submarine flotillas, and two destroyer flotillas; the Marine Corps headquarters and two BDEs; two air bases, 13 fighter/ground attack BDEs, one transportation BDE, one special mission aircraft division and one bomber division; and 11 PLARF missile BDEs and a PLARF combat missile base. During a contingency, the ETC likely exercises command over some then-SSF units in theater and receives strategic intelligence support elements of the former SSF to improve battlefield awareness and facilitate joint operations in the theater. The ETC likely commands all CCG and maritime militia ships conducting operations related to the ongoing dispute with Japan over the Senkaku Islands.

## I. The Taiwan Strait

**Key Takeaways**
- In 2023, the PRC continued to increase diplomatic, political, and military pressure against Taiwan.

Throughout 2023, Beijing continued to erode longstanding norms in and around Taiwan by employing a range of pressure tactics against Taiwan, including maintaining a naval presence around Taiwan, increasing crossings into Taiwan's self-declared centerline and ADIZ, and conducting major military exercises near Taiwan. Since 2022, the PLA has increased its military presence around Taiwan, primarily in the air and naval domains, and through named exercises that serve to punish Taipei for perceived provocations. This increase in normal PLA activity around Taiwan erodes warning times for future coercive military activity but there probably would be unambiguous indicators of larger hostile actions against Taiwan.

Although Beijing reaffirms that "peaceful reunification" is its preferred course of action, the PRC has never renounced the use of force and continues to signal its willingness to use military force against Taiwan. The PLA has a range of options to coerce Taipei based on its increasing capabilities in multiple domains. For example, the PRC could signal its readiness to use force or conduct punitive actions against Taiwan. The PLA could conduct a range of cyberspace, blockade, and kinetic campaigns designed to force Taiwan to capitulate to unification or compel Taiwan's leadership to the negotiation table on the PRC's terms.

## Summary of PLA Exercises Around Taiwan in 2023

The PRC continues to develop the force strength and composition necessary to secure its national interests and assert its territorial claims along its periphery. PLA training and exercises increasingly focus on military contingencies in the Taiwan Strait while simultaneously delaying or denying third-party intervention. During 2023, the PLA conducted exercises around Taiwan to improve joint operations and combat readiness with simultaneous land, air, and sea exercises. In April 2023, the PLA responded to then-President Tsai Ing-Wen's transit of the U.S. with a 3-day exercise, JOINT SWORD. As part of the exercise, the PLA encircled the main island of Taiwan with naval ships, including an aircraft carrier group east of Taiwan.

The PRC would seek to deter potential U.S. intervention in any Taiwan contingency campaign. Failing that, the PRC would attempt to delay and defeat intervention in a limited war of short duration, which could include some activities in the space and cyber domains, such as EW, network attacks, and IO. In the event of a protracted conflict, the PLA might choose to escalate cyberspace, space, or nuclear activities in an attempt to end the conflict, or it might choose to fight to a stalemate and pursue a political settlement. The PRC's perception of domestic and international receptivity to military action, the expected impact on its economy of resulting sanctions, political trends in Taiwan, and its level of confidence in the PLA's capability to conduct a successful invasion of Taiwan will determine which military option the PRC chooses during crises. The PLA practiced elements of these military options during its August 2022 large-scale military exercise aimed at pressuring Taiwan and again in April 2023 in response to Taiwan President Tsai Ing-wen's transit of the United States. The PLA could offer Xi the following military options against Taiwan, individually or in combination, with varying degrees of feasibilities associated risk.

Since 2016, the PRC has suspended formal communications with Taiwan and steadfastly maintains that Taiwan must accept the PRC's view of the "1992 Consensus" to restart such engagement. PRC leaders have directly equated the "1992 Consensus" to the PRC's "One China principle," which was reaffirmed by President Xi in a January 2019 address to "compatriots" in Taiwan.

## PLA Aircraft Entering Taiwan ADIZ

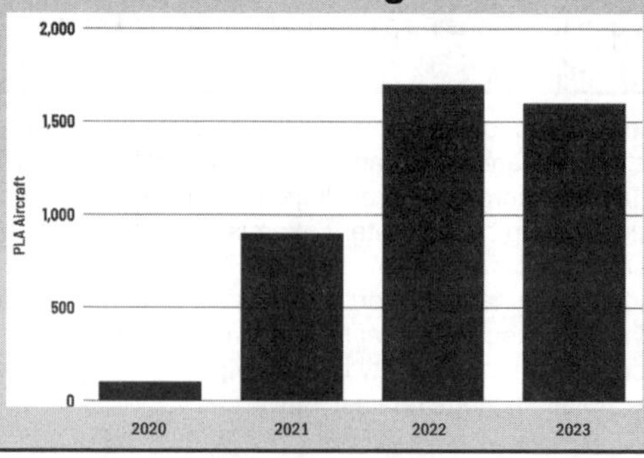


*According to Taiwan Ministry of National Defense (MND) data, the PLA entries into the Taiwan ADIZ persisted throughout the year, with a slight decrease in 2023 to 1,641 events, compared to 1,733 in 2022. Concurrently, PLA aircraft crossed the Taiwan Strait centerline at least 712 times in 2023, up from 552 in 2022.*

# The Eastern Theater

*Ref: Department of Defense Annual Report to Congress: Military and Security Developments Involving the People's Republic of China 2024, p. 117.*

## Major Military Units

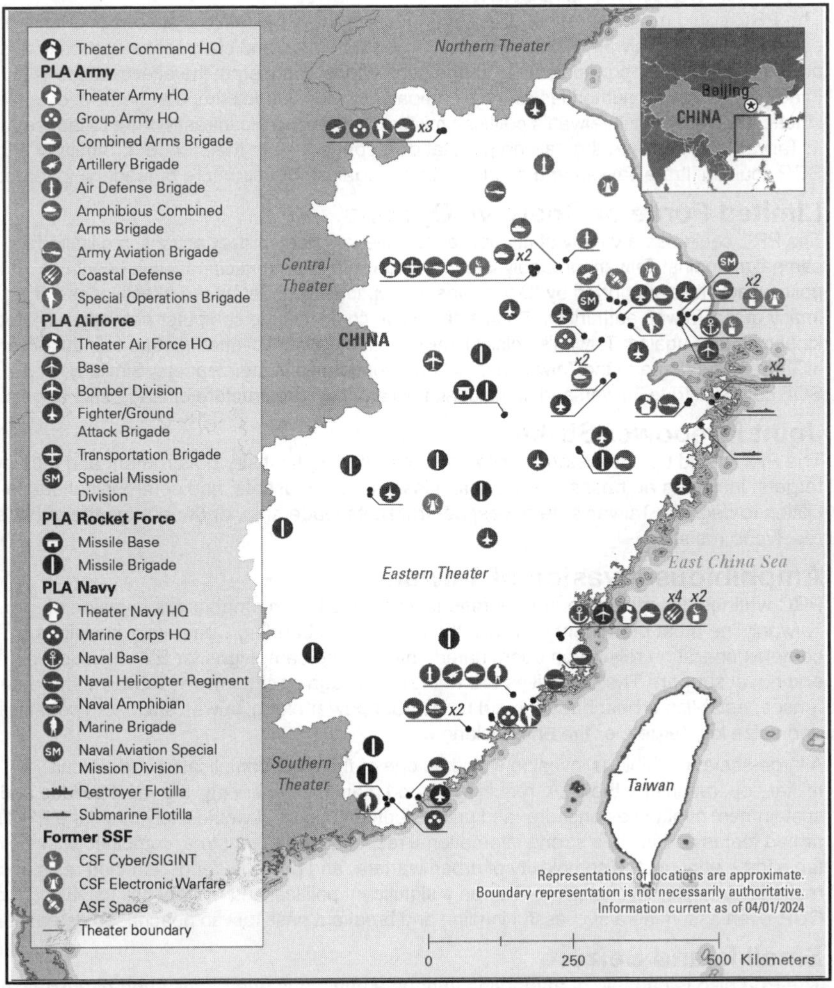


Eastern Theater Command, headquartered in Nanjing, with responsibility for central eastern China and the East China Sea, including the Strait of Taiwan. The Eastern TC likely has operational responsibility for matters involving Taiwan, Japan, and anything related to the East China Sea.

# A. PRC Courses of Action Against Taiwan

*Ref: Department of Defense Annual Report to Congress: Military and Security Developments Involving the People's Republic of China 2024, pp. 125-127.*

The PLA could offer Xi the following military options against Taiwan, individually or in combination, with varying degrees of feasibilities associated risk.

## Air and Maritime Blockade

The PRC could use a variety of disruptive, punitive, or lethal military actions in a limited campaign against Taiwan, probably in conjunction with overt and clandestine economic and political activities supported by IO to shape perceptions or undercut the effectiveness or legitimacy of the Taiwan authorities. Such a campaign could include computer network or limited kinetic attacks against Taiwan's political, military, and economic infrastructure to induce fear in Taiwan and degrade the Taiwan population's confidence in their leaders. Similarly, PLA SOF could infiltrate Taiwan and conduct attacks against infrastructure or leadership targets.

## Limited Force or Coercive Options

The PRC could use a variety of disruptive, punitive, or lethal military actions in a limited campaign against Taiwan, probably in conjunction with overt and clandestine economic and political activities supported by IO to shape perceptions or undercut the effectiveness or legitimacy of the Taiwan authorities. Such a campaign could include computer network or limited kinetic attacks against Taiwan's political, military, and economic infrastructure to induce fear in Taiwan and degrade the Taiwan population's confidence in their leaders. Similarly, PLA SOF could infiltrate Taiwan and conduct attacks against infrastructure or leadership targets.

## Joint Firepower Strike

The PRC could use precision missile and air strikes against key government and military targets, including air bases, radar sites, missiles, space assets, and communications facilities to degrade Taiwan's defenses, neutralize its leadership, or undermine the public's resolve to resist.

## Amphibious Invasion of Taiwan

PRC writings describe different operational concepts for an amphibious invasion of Taiwan. The most prominent of these, the Joint Island Landing Campaign, envisions a complex operation relying on coordinated, interlocking campaigns for EW, logistics, air, and naval support. The objectives are to break through or circumvent Taiwan's shore defenses, establish a beachhead, build up combat power along Taiwan's western coastline, and seize key targets or the entire island.

A large-scale amphibious invasion would be one of the most complicated and difficult military operations for the PLA, requiring air and maritime superiority, the rapid buildup and sustainment of supplies onshore, and uninterrupted support. It would likely strain the PRC's armed forces and invite a strong international response. These factors, combined with inevitable force attrition, the complexity of urban warfare, and potential for determined resistance, make an amphibious invasion of Taiwan a significant political and military risk for Xi and the CCP, even assuming a successful landing and breakout past Taiwan beachhead defenses.

## Small Island Seizure

The PLA also is capable of attempting various amphibious operations short of a full-scale invasion of Taiwan. With few overt military preparations beyond routine training, the PRC could launch an invasion of small Taiwan-occupied islands in the SCS, such as Pratas or Itu Aba. A PLA invasion of a medium-sized, better-defended island, such as Matsu or Kinmen, is within the PLA's capabilities. Such an invasion would demonstrate military capability and political resolve while achieving tangible territorial gain and simultaneously showing some measure of restraint. However, this kind of operation involves significant, and possibly prohibitive, political risk because it could galvanize pro-independence sentiment in Taiwan and generate powerful international opposition.

The PLA is preparing for a contingency to unify Taiwan with the PRC by force if perceived as necessary by Beijing while simultaneously deterring, delaying, or denying any third-party intervention on Taiwan's behalf.

As part of a comprehensive campaign to pressure Taiwan and the upcoming Lai administration and signal its displeasure at deepening Washington-Taipei ties, the PRC has persistently conducted military operations near Taiwan and military training for a Taiwan contingency. Throughout 2023, Beijing continued to erode longstanding norms in and around Taiwan by employing a range of pressure tactics against Taiwan, including maintaining a naval presence around Taiwan, increasing crossings into Taiwan's self-declared centerline and ADIZ, and conducting major military exercises near Taiwan.

According to Taiwan Ministry of National Defense (MND) data, the PLA entries into the Taiwan ADIZ persisted throughout the year, with a slight decrease in 2023 to 1,641 events, compared to 1,733 in 2022. Concurrently, PLA aircraft crossed the Taiwan Strait centerline at least 712 times in 2023, up from 552 in 2022, indicating a greater percentage of PLA ADIZ incursions included centerline crossings, according to the same data. *(See chart on p. 1-36.)*

In August 2023, the PLA conducted a low-intensity response to then-presidential candidate and Vice President William Lai Ching-te's transit of the U.S. The PLA sent 25 aircraft across Taiwan's claimed centerline while nine ships were on patrol around Taiwan. In April 2023, the PLA responded to then-President Tsai Ing-Wen's transit of the U.S. with a three-day exercise, JOINT SWORD. As part of the exercise, the PLA conducted an island-encirclement that included an aircraft carrier group to the east of Taiwan.

In January 2024, Taiwan held presidential and general elections, during which incumbent Vice President Lai Ching-te won the presidential election. Following the election, on January 17, the PLA conducted small-scale military maneuvers around Taiwan, consisting of joint air and naval patrols, during which 11 aircraft crossed the Taiwan Strait centerline. The PLA has continued its heightened pressure operations on Taiwan post-election.

In February 2024, a PRC fishing boat capsized near Taiwan's claimed Kinmen Islands, which resulted in the drowning of two of the PRC fishermen. Taiwan claimed in response, five CCG ships entered the restricted waters around the Kinmen Islands. According to the CCG's Fujian division, the CCG will regularly monitor the waters off the southern coast of Xiamen—which is only approximately 5 km from the Kinmen Islands.

In May 2024, the PLA ETC conducted exercise JOINT SWORD-2024A in response to Taiwan president Lai Ching-te's inauguration. The ETC announcement noted that the 2-day exercise will focus on joint sea-air-combat-readiness patrols, joint seizure of comprehensive battlefield control, and joint precision strikes on key targets. For the exercise, PLA aircraft, navy vessels, and coast guard vessels encircled the island while operating around Taiwan's outer islands.

The circumstances under which the PRC has historically indicted it would consider the use of force has evolved over time. These circumstances include the following:

- Formal declaration of Taiwan's independence
- Undefined moves toward Taiwan independence
- Internal unrest in Taiwan
- Taiwan's acquisition of nuclear weapons
- Indefinite delays in the resumption of cross-strait dialogue on unification
- Foreign military intervention in Taiwan's internal affairs.

# B. PLA's Current Posture for a Taiwan Conflict

*Ref: Department of Defense Annual Report to Congress: Military and Security Developments Involving the People's Republic of China 2024, pp. 127-130.*

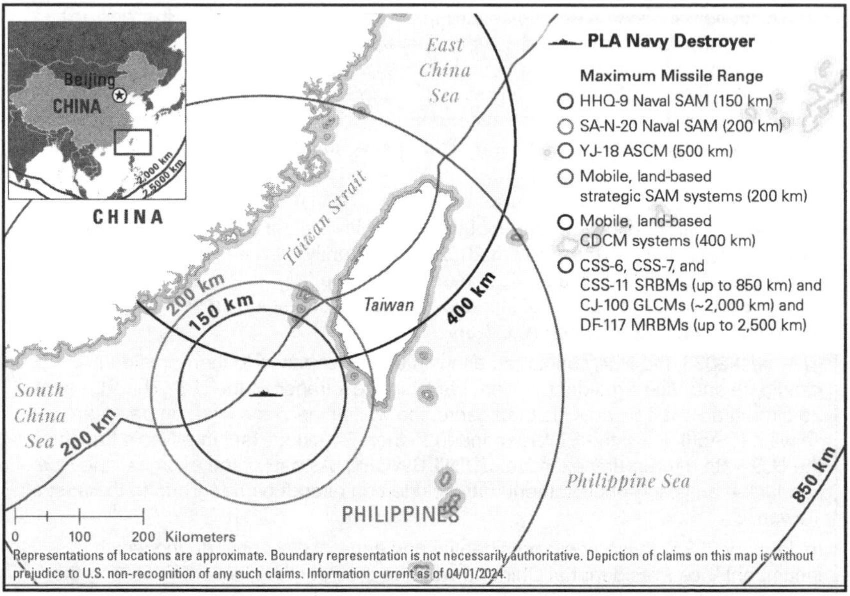


## PLA Army (PLAA)

The PLAA has increased its posture in the ETC and along the Taiwan Strait, providing the PLAA with enhanced firepower, mobility, and rapid strike capabilities. Significant reorganizations and amphibious assault training in recent years likely indicate that the Taiwan contingency is a high priority for the Army. Major PLAA contributions to a Taiwan invasion scenario likely include extensive amphibious, army aviation, SOF, and air assault operations.

The PLAA fields six amphibious combined arms BDEs—four in the ETC (nearest Taiwan) and two in the STC. PLAA units continued amphibious assault training as a single service and with joint service counterparts in 2023. Training events refined the tactics of rapid loading, long-distance transport, beach assault under complicated sea situations, and logistic support capabilities. Press reports claimed that the PLA extensively used sea, air, and ground UAS in support of the amphibious assault operation. PLAA amphibious BDEs reportedly conduct realistic, large-scale amphibious operations that are almost certainly aimed at supporting a Taiwan invasion scenario.

## PLA Navy (PLAN)

The PLAN is improving its anti-air, anti-surface, and ASW capabilities, is introducing new multi-mission platforms capable of conducting diverse missions during peace and war, and has increased its posture surrounding Taiwan since August 2022. New attack submarines and modern surface combatants with anti-air capabilities and fourth-generation naval aircraft entering the force are designed to achieve maritime superiority in the FIC and deter and counter any potential third-party intervention in a Taiwan conflict. The PLAN commissioned its third YUSHEN-class LHA in 2022 and, as of December 2023, a fourth YUSHEN-class ship was launched and began sea trials.

The PRC's amphibious fleet has, in recent years, focused on acquiring a modest number of ocean-going LPDs and LHAs ships. There is no indication the PRC is significantly expanding its number of tank landing ships and medium-sized landing craft at this time. Although the PLAN has not invested in the large number of landing ships and medium landing craft that analysts believe the PLA would need for a large-scale assault on Taiwan, it is possible the PLA assesses it has sufficient amphibious capacity and has mitigated shortfalls through investment in other operational capabilities, such as civilian lift vessels and rotary-wing assets to address this gap. The PLA may have confidence in the PRC's shipbuilding industry's massive capacity to produce the necessary ship-to-shore connectors relatively quickly.

## PLA Air Force (PLAAF)

The PLAAF has maintained a ready force posture for a variety of capabilities necessary in a Taiwan contingency. It has acquired many advanced aircraft capable of conducting operations against Taiwan without requiring refueling, providing it with a significant capability to conduct air and ground-attack operations. Long-range air defense systems provide a strong layer of defense against attacks on key military installations or population centers on China's mainland. The PRC's development of support aircraft provides the PLAAF with improved ISR capability to support PLA operations. The PLAAF has improved refueling capabilities, expanding its ability to operate further from the PRC and increasing its ability to threaten third-party intervention. In 2023, PLAN Aviation transferred many of its land-based aircraft to the PLAAF, including 300 fighter jets, so that the PLAN could focus on improving carrier-based air operations.

## PLA Rocket Force (PLARF)

The PLARF is prepared to conduct missile attacks against high-value targets, including Taiwan's C2 facilities, air bases, and radar sites, in an attempt to degrade Taiwan's defenses, neutralize Taiwan's leadership, or break the public's will to fight. As of 2023, the PLARF is increasing its presence along the Taiwan Strait with new missile BDEs, possibly indicating an increasing number of deployed missiles.

## Former Strategic Support Force (SSF)

As of April 2024, the SSF was dissolved into the now CMC-subordinate ASF, CSF, and new ISF. PLA doctrinal writings emphasize the importance of space and cyberspace domains in joint operations, suggesting that these collective units maintain the responsibility for the use of EW and cyberspace operations during a Taiwan contingency, as one of the missions of the force was to seize and maintain information dominance. The former SSF 311 Base would be responsible for political and psychological warfare, such as disseminating propaganda against Taiwan to influence public opinion and promote the PRC's interests.

The former SSF would play a strategic information and communications support role, centralizing technical intelligence collection and management and providing strategic intelligence support to theater commands involved in a Taiwan contingency. Following the April 2023 President Tsai transit, the PRC used IO in its responses, such as cyber operations, to cast doubt in Taiwan's public. However, it is currently unknown whether the now CMC-subordinate CSF will continue to maintain the psychological warfare mission under the 311 Base, due to its unclear alignment to the CSF's primary cyber and EW missions.

## Joint Logistic Support Force (JLSF)

The JLSF's primary goal is to provide joint logistics support to the PLA's strategic and campaign-level operations, such as a Taiwan contingency, by conducting C2 of joint logistics, delivering materiel, and overseeing various support mechanisms. The JLSF participates in joint, theater-level exercises, becoming most relevant when units operate far from their home garrisons and beyond their organic logistics capabilities.

# C. Taiwan's Defensive Capabilities

*Ref: Department of Defense Annual Report to Congress: Military and Security Developments Involving the People's Republic of China 2024, pp. 130-132.*

## Key Takeaways

- Geopolitical events in 2023 continued to accelerate Taiwan's development of asymmetric concepts and capabilities to counter the PRC's improving capabilities.
- However, Taiwan seeks to balance asymmetric capabilities with conventional ones, the latter of which it relies on to defend against PRC gray-zone operations.

In 2022, Russia's invasion of Ukraine and the PRC's forceful response to then-U.S. Speaker of the House's CODEL to Taiwan in August increased the urgency with which Taiwan is pursuing defense reforms, which continued throughout 2023. In 2024, then President-elect "William" Lai Ching-te indicated that Taipei would seek continuity in President Tsai's defense reform policies while making efforts to enhance defense capabilities. Taiwan's elected president, who is the commander in chief of Taiwan's armed forces, oversees the National Security Council and the MND, which altogether formulate and implement national security and defense policy.

Taiwan is taking steps to address the military threat posed by the PLA, including building its war reserve stocks, growing its defense-industrial base, improving joint operations and crisis response National Defense Report stresses enhanced asymmetric and joint capabilities to deter an invasion in illustrating the military's strategy for defending the island, though Taiwan seeks to balance these asymmetric capabilities with conventional ones, such as fighter jets and surface combatants, which it uses to defend its waters and airspace from PRC gray-zone pressure operations.

In 2023, Taiwan prioritized a whole-of-society approach as a means to improving defensive resilience. The All-Out Defense and Mobilization Agency leads this mission by coordinating efforts between military and civilian agencies to accomplish holistic readiness against PRC gray-zone activities and a potential PLA attack. As part of its whole-of-society defense approach, Taiwan finalized conscription policy changes, which extended conscription from 4 months to 1 year in the beginning of 2024, to enhance military training duration and quality. Separately, Taiwan government and non-government organizations took steps to address PRC disinformation operations through public awareness campaigns.

The Taiwan military comprises three main services—the army, navy, and air force—with approximately 180,000 active-duty personnel across all services. As of 2021, the MND had accomplished its goal of filling 90% of the active-duty billets (169,000) with volunteers. Taiwan's transition to an all-volunteer force increases overall manpower-related costs needed to attract and retain personnel, though these costs are partially offset by cost savings from manpower reductions, which provided some margin to improve individual pay and benefits, housing, and incentive pay.

Taiwan continues to increase its defense budget to support defense acquisitions and strengthen its forces against PRC military pressure. In 2023, the Tsai administration announced defense spending was at a historic high. In August 2023, Tsai's administration proposed total defense spending of about $19 billion for 2023, a 3.5% increase from 2022, which represents about 2.5% of Taiwan's GDP. A Taiwan defense-related think tank poll reported that more than 60% of people in Taiwan support increasing the defense budget to 3% of GDP. During her tenure in office, President Tsai Ing-wen sought to double defense spending to deter PRC aggression and improve Taiwan's ability to defend itself. President Lai Ching-te seeks to continue expanding the defense budget and named it one of his top priorities. Starting in 2021, President Tsai's administration allocated two multi-year supplemental budgets totaling $16 billion for conventional and asymmetric systems. The supplemental budgets cover the purchase of 66 F-16 V fighter jets, missile ships, and anti-ship weapons, such as the Hsiung Feng missile system. The PRC's official defense budget continues to grow to around $220 billion in 2023, with much of that focused on developing the capability to unify Taiwan with the PRC by force.

Starting in 2020, during President Tsai's second term, Taiwan prioritized domestic defense industrial growth and defense equipment production to expand its defense procurements options outside of the United States. Taiwan relies heavily on its domestic defense production capabilities because most countries, save for the United States, are hesitant to sell military equipment to Taiwan, due the threat of economic retaliation from the PRC.

Reservists, conscripts, and civil defense volunteers augment the volunteer active-duty military. Following their conscription period, former conscripts join the pool of approximately 1.7 million reserve personnel. To better coordinate personnel from these disparate sources, Taiwan established the All-Out Defense Mobilization Agency in 2022 to reform its emergency mobilization capability and prepare for and coordinate whole-of-society defense with civilian agencies.

Taiwan has stated that it is working to develop new cost-effective concepts and capabilities for asymmetric warfare. Specific areas of emphasis in Taiwan's strategy include offensive and defensive information and EW, high-speed stealth vessels, shore-based mobile missiles, rapid mining and minesweeping, unmanned aerial systems, and critical infrastructure protection. Taiwan has dedicated significant defense spending toward its Harpoon Coastal Defense Systems, upgrading its existing F-16 fighters and its domestically produced Hai Kun submarine, the first of which was unveiled in 2024.

## Consistent U.S. Defense Engagement with Taiwan

In response to U.S. defense engagement with Taiwan, Beijing routinely accuses the United States of not abiding by its One China Principle, which the PRC touts as the basis for all diplomacy with Beijing. U.S. defense engagements with Taiwan, as one element of the unofficial U.S.-Taiwan relationship, remain consistent with our one China policy—as guided by the Taiwan Relations Act (TRA), the Three U.S.-China Joint Communiques, and the Six Assurances. U.S. defense engagement with Taiwan has evolved over time in response to the PRC's capacity and willingness to use military coercion against Taiwan. This evolution is consistent and required by U.S. policy.

The 1979 TRA states that the United States "will make available to Taiwan such defense articles and defense services in such quantity as may be necessary to enable Taiwan to maintain a sufficient self-defense capability." In 1982, President Reagan clarified in an internal memo—which the United States made public in 2019—that the quantity and quality of U.S. defense assistance provided to Taiwan be "conditioned entirely on the threat posed by the PRC." President Reagan further emphasized that this linkage is intended as a "permanent imperative" of U.S. foreign policy.

The United States opposes unilateral changes to the cross-strait status quo by either side, does not support Taiwan independence, and expects cross-strait differences to be resolved by peaceful means. United States defense engagement with Taiwan will continue to bolster these positions and be conditioned entirely on the evolving threat posed by the PRC, as enumerated in U.S. policy.

# D. Taiwan Straight Military Balance

*Ref: Department of Defense Annual Report to Congress: Military and Security Developments Involving the People's Republic of China 2024, pp. 164-165.*

**TAIWAN STRAIT MILITARY BALANCE, GROUND FORCES**

| | PRC | | Taiwan |
|---|---|---|---|
| | Total | Taiwan Strait Area | Total |
| Total Ground Force Personnel | 1,040,000 | 427,000 | 104,000 |
| Group Armies/Army Corps | 13 | 5 | 3 |
| Combined Arms Brigades | 80 | 30 | 7 |
| Artillery Brigades | 15 | 5 | 3 |
| Army Aviation Brigades | 13 | 4 | 2 |
| Air Assault Brigades | 3 | 1 | 0 |
| Airborne Brigades | 6 | 6 | 0 |
| Marine Brigades | 11 | 5 | 2 |
| Tanks | 3,800 | 1,000 | 800 |
| Artillery Pieces | 7,600 | 2,300 | 1,100 |

**TAIWAN STRAIT MILITARY BALANCE, NAVAL FORCES**

| | PRC | | Taiwan |
|---|---|---|---|
| | Total | Eastern and Southern Theater Command Navies | Total |
| Aircraft Carriers | 3 | 1 | 0 |
| Amphibious Assault Ships | 3 | 3 | 1 |
| Cruisers | 8 | 4 | 0 |
| Destroyers | 42 | 30 | 4 |
| Frigates | 49 | 36 | 22 |
| Corvettes | 50 | 40 | 0 |
| Medium Landing Ships/ Tank Landing Ships/Amphibious Transport Dock | 58 | 51 | 51 |
| Attack Submarines | 47 | 31 | 4 |
| Nuclear-Powered Attack Submarines | 6 | 2 | 0 |
| Nuclear-Powered Ballistic Missile Submarines | 6 | 6 | 0 |
| Coastal Patrol (Missile) | 60 | 60 | 43 |
| Coast Guard Ships | 141 | N/A | 170 |

**TAIWAN STRAIT MILITARY BALANCE, AIR FORCES**

| | PRC | | Taiwan |
|---|---|---|---|
| | Total | Eastern and Southern Theater | Total |
| Fighters | 1,900 (3,100*) | 800 (950*) | 350 (400*) |
| Bombers/Attack | 500 | 300 | 0 |
| Transport | 500 | 40 | 50 |
| Special Mission Aircraft | 250 | 150 | 20 |

# IV(b). Southern Theater Command (STC)

*Ref: Department of Defense Annual Report to Congress: Military and Security Developments Involving the People's Republic of China 2024, pp. 115 to 118 and 132 to 135.*

The **Southern Theater Command** is oriented toward the South China Sea, Southeast Asia border security, and territorial and maritime disputes.

The Southern Theater Command (STC) is oriented toward the South China Seas (SCS), Southeast Asia border security, and territorial and maritime disputes. The STC responds to U.S. freedom of navigation operations in the SCS and can assume command as needed over all CCG and CMM ships conducting operations enforcing the PRC's claimed sovereignty and supporting PLA operations. In 2023, STC units performed multiple live-fire drills and amphibious training events near PRC-occupied features in the SCS.

The STC covers mainland and maritime Southeast Asia, including the SCS. This geographic area implies that the STC is responsible for securing the SCS, supporting the ETC in any operation against Taiwan, and assuring the security of sea lines of communication (SLOCs) vital to the PRC's global ambitions in the SCS. The STC is responsible for responding to U.S. freedom of navigation operations in the SCS by regularly tracking and reacting to U.S. ships operating in the PRC's dashed-line claim.

Major PLA units in the STC include the PLAA's 74th and 75th GAs comprised of 12 combined arms and amphibious combined arms BDEs and two infantry BDEs; two PLAN naval bases, a nuclear submarine base, two submarine flotillas, two destroyer flotillas, and an aircraft carrier task group; two Marine Corps BDEs and one SOF BDE; two air bases, 14 fighter/ground attack BDEs (includes Navy aviation), one transportation BDE, one special mission aircraft division, and one bomber division; and 10 PLARF missile BDEs and two PLARF combat missile bases.

The STC plays a significant role in the PLA's bilateral and multilateral exercises with countries in Southeast Asia, such as participating in bilateral military exercises with Lao People's Democratic Republic, Singapore, and Cambodia in 2023. The STC participated in the multilateral military exercise, Peace and Friendship-2023, hosted in Guangdong and attended by Cambodia, Lao PDR, Malaysia, Thailand, and Vietnam.

The PLA Hong Kong and Macao garrisons are subordinate to the STC. In August 2023, the PLA Hong Kong and Macao garrisons conducted an annual rotation of forces. Since 1997, the PLA has rotated forces by land, air, and sea from Shenzhen at night, nominally as part of the usual annual rotation. Since the 2019 pro-democracy protests, the PRC has maintained a rotational deployment of PAP forces in Hong Kong. The PAP and PLA units publicly highlight their anti-riot, counterterrorism, and disaster prevention training.

In November 2023, the STC led live-fire drills, including the deployment of howitzers and counter-battery radars at the PRC-Burma border focused on mobility, border control, and firepower capabilities. The exercise occurred shortly after the Three Brotherhood Alliance (a pro-democracy collation composed of ethnic armed organizations [EAOs] in Burma) launched Operation 1027 against the junta's State Administrative Council in the north of the country, near the PRC-Burma border. The PLA commander of the exercise noted that the exercise showed the PLA's determination and capability to defend national sovereignty and border stability. In late April, the PRC conducted additional live-fire drills in response to fighting at the border. The increased exercise activity is in response to the expansion of internal unrest in Burma following an October attack by EAOs against the military junta near the border with the PRC.

In 2023, the PRC's first domestically produced aircraft carrier, the Shandong, conducted three major far seas training events in April, September, and November. The trainings included operating near Guam, participating in Joint Sword exercises near Taiwan, and breaking the PLA's previous record for aircraft sorties in a deployment. With the PRC's other aircraft carrier, Liaoning in maintenance for 2023, the increase in training with Shandong demonstrates the PLA's capabilities further from the PRC's borders and into the Philippine Sea.

# I. South China Seas (SCS)

**Key Takeaways**

- In 2020, the PRC did not resume land reclamation or major military infrastructure construction at its seven Spratly Islands outposts.
- The PRC's Spratly outposts are capable of supporting military operations, include advanced weapon systems, and have supported non-combat aircraft; however, no large-scale presence of combat aircraft has been yet observed there.
- In 2020, the PRC continued to deploy PLAN, CCG, and civilian ships in response to Vietnamese and Malaysian drilling operations within the PRC's claimed "nine-dash-line" and Philippines' construction at Thitu Island.

## A. Developments in the Security Situation

In July 2016, an international arbitral tribunal convened at the Philippines' behest pursuant to provisions in the 1982 UNCLOS, ruled that the PRC's claims to "historic rights" to the SCS waters in its ambiguous "dashed line" were contrary to UNCLOS and unlawful. Since December 2019, Indonesia, Malaysia, the Philippines, and Vietnam have explicitly referenced the arbitral ruling in notes verbales to the UN affirming that the PRC's dashed-line claim are unlawful. Beijing, however, categorically rejects the tribunal decision, and the PRC continues to use coercive tactics, including the employment of CCG and CMM ships, with PLAN in overwatch positions, to enforce its illegal claims. The PRC does so in ways calculated to remain below the threshold of provoking conflict. In 2023, the PRC continued to selectively enforce a yearly, unilateral 3-month fishing ban that includes the waters inside the EEZ of the Philippines and Vietnam to incrementally enforce its unlawful maritime claims.

Throughout 2023, tensions between the PRC and the Philippines were notably higher. PRC maritime forces took unsafe and unprofessional measures, including ramming and forcibly boarding Philippine vessels en route to resupplying a Philippine outpost on Second Thomas Shoal.

The PRC states that international military presence in the SCS is a challenge to its sovereignty. Throughout 2023, the PRC deployed PLAN, CCG, CMM, and civilian ships to advance its illegal maritime claims around features, such as Second Thomas Shoal, Scarborough Reef, and Sandy Cay/Thitu Island, as well as in response to oil and gas exploration operations by rival claimants in the PRC's unlawful dashed-line claim.

In 2023, CCG enforced its illegal maritime claims in the SCS using water cannons, aggressive maneuvers, ramming, and floating barriers. The PRC's collisions with Philippine supply ships generated media attention that highlighted the PRC's aggressive behaviors. Since August 2023, the PRC has harassed and, in some cases, impeded Philippine ships' access to Second Thomas Shoal during at least eight resupply missions. During some of these missions, PRC and Philippine ships

# The Southern Theater

*Ref: Department of Defense Annual Report to Congress: Military and Security Developments Involving the People's Republic of China 2024, p. 117.*

## Major Military Units

Southern Theater Command, headquartered in Guangzhou, has responsibility for south-central China, including the border with Vietnam and the South China Sea. The Southern TC's primary missions are maintaining security in the South China Sea and supporting the Eastern TC in any major amphibious operation.

# B. China-Philippines Tensions in the South China Sea

*Ref: China-Philippines Tensions in the South China Sea, Congressional Research Service (Mar '25).*

In 2023 and 2024, the People's Republic of China (PRC, or China) increased pressure on the Philippines to abandon one of its nine outposts in the Spratly Islands in the South China Sea (SCS) and attempted to deny Philippine vessels access to parts of the Philippines' exclusive economic zone (EEZ) that China claims as its own territory. The escalation of long-standing Sino-Philippine tensions raises the possibility of a crisis or conflict involving China, the Philippines, and potentially the United States. The United States and the Philippines signed a Mutual Defense Treaty in 1951 and a Visiting Forces Agreement in 1998.

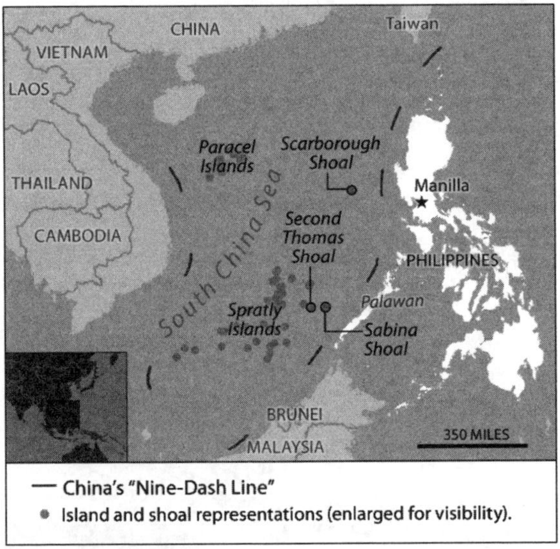


*Ref: China-Philippines Tensions in the South China Sea, Congressional Research Service (Mar '25), Fig 1. The South China Sea and Disputed Areas.*

## Rising Tensions in the South China Sea

Second Thomas Shoal, an atoll in the Spratly Islands, is a potential flashpoint in the SCS (Figure 1). A low-tide elevation and fully submerged at high tide, it is roughly 105 nautical miles (nm) from the Philippine island of Palawan, 620 nm from the PRC, and 22 nm from Mischief Reef, a low-tide elevation in the Spratlys where the PRC maintains an outpost. The Philippines posts a small cadre of marines on a now-derelict Philippine Navy ship, the BRP Sierra Madre, which it grounded on the shoal in 1999 to protect its maritime claims. Since 2013, China has increased its presence near the shoal as part of its stated attempt to end the Philippines' presence there.

In 2023, the China Coast Guard (CCG) and maritime militia vessels—ostensibly private boats that support PRC maritime claims—interfered with Philippine boats resupplying the BRP Sierra Madre. PRC officials have called occupation of disputed features by the Philippines "a red line" for the PRC. Notable incidents included the PRC's targeting of a Philippine Coast Guard boat with a military-grade laser, and using water cannons against, surrounding, and colliding with Philippine vessels. In December 2023, China engaged in what the Philippine government called a "serious escalation" in the SCS. PRC vessels reportedly swarmed the area near Second Thomas Shoal and "harassed, blocked, and executed dangerous maneuvers," resulting in a collision between a PRC coast guard ship and a Philippine boat on a resupply mission. Each side accused the other of ramming one of its vessels. A CCG spokesperson stated its December 2023 actions were "professional, standardized, legitimate, and legal."

The Philippines successfully resupplied the BRP Sierra Madre on several occasions in 2023 and 2024, despite PRC obstruction. The U.S. Department of State issued statements critical of PRC actions and supportive of the Philippines' right to operate in waters around Second Thomas Shoal.

In June 2024, CCG personnel boarded a Philippine rigid hull inflatable boat, assaulted AFP marines, and destroyed equipment. In a separate incident, a Philippine marine lost his thumb when a PRC vessel rammed his boat as it sought to resupply the BRP Sierra Madre. Following these incidents, in July 2024, the two countries agreed to a provisional agreement regulating future resupply missions. The Philippines has successfully resupplied the derelict ship in the months since, and the two countries agreed to maintain the accord in January 2025 despite ongoing PRC harassment elsewhere in the South China Sea.

In April 2024, the Philippines deployed its largest coast guard vessel, the BRP Teresa Magbanua, to Sabina Shoal, an unoccupied feature in the Spratlys, responding to reports of crushed coral. China responded by sending CCG vessels, which later rammed Philippine Coast Guard boats and prevented the resupply of the BRP Teresa Magbanua. The Philippine vessel withdrew in September, citing bad weather, lack of supplies, and sick crew.

Since 2012, the PRC has harassed and blocked Philippine fishing boats operating near Scarborough Shoal, a traditional fishing area for both countries. In December 2023, the CCG deployed a water cannon on Philippine vessels attempting to deliver provisions to Filipino fishermen near the shoal. In 2024, incidents included dangerous aerial maneuvers by PRC fighter jets around Philippine military and civilian aircraft and the deployment of the PRC's largest coast guard ship to the area.

## SCS Disputes and International Law

An arbitral tribunal convened under the United Nations Convention on the Law of the Sea (UNCLOS) ruled in 2016 that the "Nine-Dash Line," which China uses to mark its territorial claims in the SCS, has "no legal basis," and that several PRC actions in the SCS violated the Philippines' sovereign rights. The tribunal found that Second Thomas Shoal, Scarborough Shoal, and the PRC-occupied Mischief Reef fall within the Philippines' EEZ, and that China had unlawfully interfered with Philippine fishing at Scarborough Shoal and created a risk of collision. China declared the ruling "null and void." *(For more on the disputes, see CRS In Focus IF10607, China Primer: South China Sea Disputes.)* Successive U.S. Administrations have urged both the PRC and the Philippines to abide by the 2016 ruling. The United States is not a party to UNCLOS.

## U.S.-Philippines Security Cooperation

The Philippines has been one of the largest recipients of U.S. military assistance in the Indo-Pacific region, including $40 million in Foreign Military Financing (FMF) in FY2023 and assistance under the Department of Defense's Indo-Pacific Maritime Security Initiative. In 2024, the Biden Administration pledged to work with Congress to provide the Philippines with $500 million in FMF from emergency supplemental appropriations for FY2024 (P.L. 118-50). In February 2025, the Trump Administration reportedly exempted $336 million in assistance for Philippines military modernization from the Administration's freeze on foreign aid.

Under EDCA, the two countries agreed to deepen military cooperation, upgrade certain Philippine military facilities, and allow U.S. forces rotational access to these facilities. In 2023, the number of these bases expanded from five to nine. The United States and the Philippines engage in approximately 20 military exercises and events annually, including combined patrols in the SCS. A 2024 U.S.-Japan-Philippines trilateral meeting included agreements on maritime security and economic cooperation.

have collided, which caused minor damage to PRC and Philippine ships and, in two instances, injured Philippine personnel.

In December 2023, the CCG of employed water cannons, erected a floating barrier, and employed a long-range acoustic device to impede Philippine access to Scarborough Reef. The PRC stated the Philippines' subsequent efforts to remove the barrier were illegal and provoked unnecessary confrontation.

## C. Outposts in the Spratly Islands and Select Disputed Features

Since early 2018, the PRC-occupied Spratly Islands outposts have been equipped with advanced anti-ship and anti-aircraft missile systems and military jamming equipment, representing the most capable land-based weapons systems deployed by any claimant in the disputed SCS areas to date. In mid-2021, the PLA deployed an intelligence-gathering ship and a surveillance aircraft to the Spratly Islands during U.S.-Australia bilateral operations in the region. Throughout 2023, the PRC regularly used its Spratly Islands outposts to support naval and coast guard operations in the SCS. The PRC has added more than 3,200 acres of land to the seven features it occupies in the Spratlys. The PRC has added military infrastructure, including 72 aircraft hangars, docks, satellite communication equipment, antenna array, radars, and hardened shelters for missile platforms.

The PRC has stated these projects are mainly to improve marine research, safety of navigation, and the living and working conditions of personnel stationed on the outposts. However, the outposts provide airfields, berthing areas, and resupply facilities that enable the PRC to maintain a more flexible and persistent military and paramilitary presence in the area. This improves the PRC's ability to detect and challenge activities by rival claimants or third parties and widens the range of response options available to Beijing.

## D. China Maritime Militia (CMM) and Land Reclamation in the South China Seas

Since at least 2014, CMM vessels have engaged in covert small-scale reclamation activity and likely caused physical changes observed at multiple unoccupied features in the Spratly Islands, including Lankiam Cay, Eldad Reef, Sandy Cay, and Whitsun Reef. Beijing likely is attempting to covertly alter these features so that it can portray them as naturally formed high-tide elevations capable of supporting PRC maritime claims out to the farthest extent of the PRC's dashed-line claim. In contrast to the PRC large-scale reclamation program, which was overt with the original status of occupied features well documented, the less well-known historical record about many of the unoccupied features makes them more susceptible to PRC efforts to shape international opinion regarding the status of the features.

# IV(c). Western Theater Command (WTC)

*Ref: Department of Defense Annual Report to Congress: Military and Security Developments Involving the People's Republic of China 2024, pp. 118-120.*

The **Western Theater Command** is oriented toward India and counterterrorism missions along China's Central Asia borders.

The Western Theater Command (WTC) is oriented toward India and counterterrorism missions along the PRC's Central Asia borders. The WTC focuses on Xinjiang and Tibet Autonomous Regions, where the CCP perceives a high threat of separatism, terrorism, and extremism. The WTC is geographically the largest theater command in the PRC and is responsible for responding to conflict with India, border interactions with Central Asian states, and what the PRC refers to as the "three evil forces" of terrorism, separatism, and extremism in Tibet and Xinjiang.

Major PLA units in the WTC include the 76th and 77th GAs, Xinjiang Military District (84th GA), and Xizang Military District (85th GA) comprised of 16 combined arms BDEs and three infantry BDEs; three air bases, eight fighter/ground attack BDEs, one transportation division, one transportation BDE, and one UAV unit; and eight PLARF missile BDEs and two PLARF combat missile bases.

The WTC's primary focus is on securing the PRC's border with India. In recent years, differing perceptions between India and the PRC regarding border demarcations have facilitated multiple clashes, force buildups, and military infrastructure construction. In June 2020, PLA and Indian patrols clashed in Galwan Valley—the most violent clash between the two countries in 45 years. The PLA has not drawn down its positions or troop numbers since the 2020 clash and has built infrastructure and support facilities to maintain multiple BDE deployments along the LAC. WTC and Indian commanders have held 21 corps commander-level talks since the 2020 clash.

Domestically, the WTC focuses on the Xinjiang and Tibet Autonomous Regions, where the CCP perceives a high threat of separatism and terrorist activity, particularly among Uyghur populations in Xinjiang. It serves as a backstop should civilian authorities and paramilitary forces need assistance. In January 2024, the PRC State Council Information Office issued a white paper, "China's Legal Framework and Measures for Counter-terrorism," which was the PRC's first comprehensive white paper on the subject; previous publications were regionally focused. Beijing likely issued a broad counterterrorism policy to all theater commands and military districts to alleviate perceived targeting of minority populations in the WTC. In August 2023, Xi Jinping visited Urumqi in Xinjiang to reinforce his counterterrorism policies in Xinjiang and urged to "more deeply promote the Sinicization of Islam and effectively control illegal religious activities."

## I. China-India Border

Tensions with India along the Line of Actual Control (LAC)—a disputed border between the PRC's western provinces and India's northern provinces—sparked a standoff between PRC and Indian forces in mid-May 2020, which escalated on June 15, 2020, after a skirmish ensued in the Galwan Valley between Indian Army and PLA forces that ended with 20 Indian soldiers and four PRC soldiers dead. The two sides agreed to tactical pull-backs from most of the contested areas in early 2021 but standoffs continue at two locations in Ladakh on the Western Sector of the LAC. In late 2022, PRC and Indian forces engaged in an unarmed clash near

Tawang along the Eastern Sector of the LAC separating Tibet and the Indian state of Arunachal Pradesh. This was the first such clash since the 2020 skirmish although local commanders quickly defused the clash and the overall standoff did not substantively escalate. These engagements coincide with a significant and sustained escalation in military infrastructure to support a long-term presence on the LAC. In late 2023, India hosted the 20th round of commander-level meetings with the PLA. Both sides agreed to maintain communication and dialogue through military and diplomatic channels.

*Editor's Note: the remainder of discussion in this section on China-India tensions and the Uyghurs is provided for historical perspective, as referenced previously from the Office of the Secretary of Defense Annual Report to Congress: Military and Security Developments Involving the People's Republic of China 2021.*

## What Happened

Beginning in early May 2020, PRC and Indian forces faced off in unarmed clashes at multiple locations along the LAC. The resulting standoff triggered the buildup of forces on both sides of the disputed border. Each country demanded the withdrawal of the other's forces and a return to pre-standoff conditions, though neither the PRC nor India agreed on those conditions. The PRC blamed the standoff on Indian infrastructure construction, which it perceived to encroach on PRC territory, while India accused the PRC of launching aggressive incursions into India's territory.

This incident was the deadliest clash between the two nations in the past 45 years. On June 15, 2020, patrols violently clashed in Galwan Valley resulting in approximately 20 Indian soldiers and, according to PRC officials, the death of 4 PLA soldiers. In addition, on September 8, 2020, a PLA patrol fired warning shots at an Indian patrol near Pangong Lake—the first shots fired along the LAC in decades.

Throughout the standoff, PRC officials sought to downplay the severity of the crisis, emphasizing Beijing's intent to preserve border stability and prevent the standoff from harming other areas of its bilateral relationship with India. The PRC seeks to prevent border tensions from causing India to partner more closely with the United States. PRC officials have warned U.S. officials to not interfere with the PRC's relationship with India.

As of June 2021, the PRC and India continue to maintain large-scale deployments along the LAC and make preparations to sustain these forces while disengagement negotiations have made limited progress. Beginning in May 2020, the PLA launched incursions into customarily Indian-controlled territory across the border and has concentrated troops at several standoff locations along the LAC. In addition, a substantial reserve force from the Tibet and Xinjiang Military Districts were deployed to the interior of Western China to provide a rapid response. The June 2020 skirmish in Galwan Valley, which led to the deaths of 20 Indian soldiers, marked the first loss of life on the LAC since 1975. In February 2021, the Central Military Commission (CMC) announced posthumous awards for four PLA soldiers, though the total number of PRC casualties remains unknown.

As of April 2021, PLA and Indian Army (IA) representatives have held eleven rounds of Corps-level negotiations since the beginning of the standoff in May 2020, consisting of meetings between the commanders of the PLA's South Xinjiang (Nanjiang) Military District and the Indian Army's 14th Corps. The negotiations have yielded limited disengagement at specific areas along the LAC. In addition to military negotiations, on 10 September, the PRC's Minister of Foreign Affairs, Wang Yi, met with India's Minister of External Affairs, Subrahmanyam Jaishankar, on the sidelines of the Shanghai Cooperation Organization Foreign Ministers' meeting in Moscow. The two ministers released a plan, which has yet to be carried out, for resolving the dispute and expressed a desire to resolving the standoff peacefully while maintaining dialogue.

# The Western Theater

*Ref: Department of Defense Annual Report to Congress: Military and Security Developments Involving the People's Republic of China 2024, p. 119.*

## Major Military Units

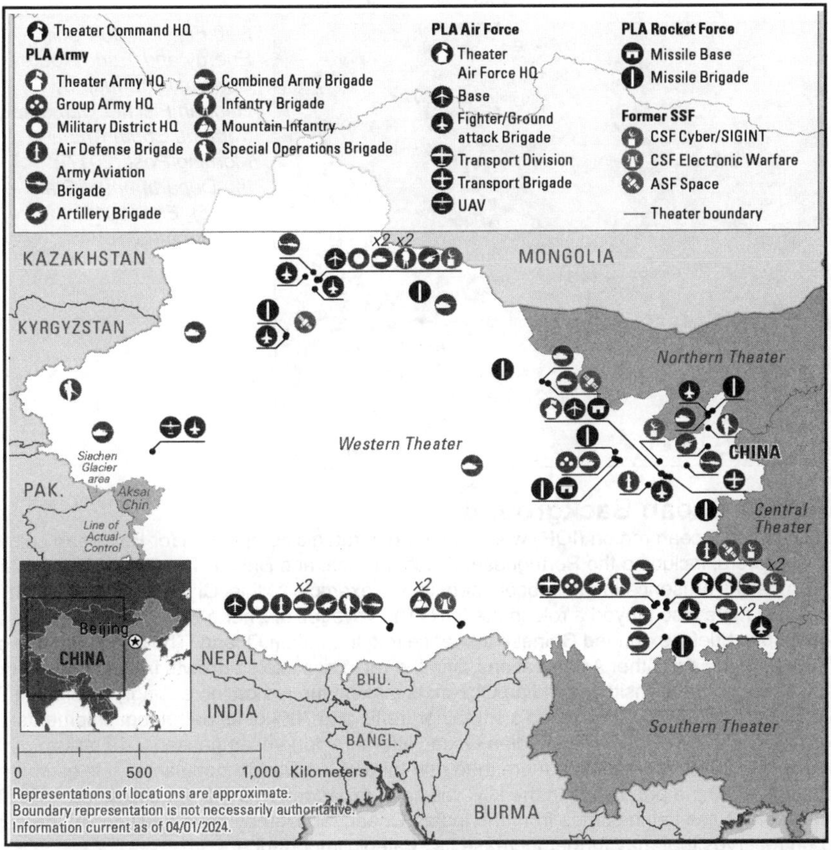


Western Theater Command, headquartered in Chengdu, has responsibility for virtually the entire western half of China, including borders with India and Russia and the regions of Tibet and Xinjiang. The Western TC's primary missions are contending with perceived separatist and terrorist threats in Tibet and Xinjiang and addressing border issues with China's two powerful western neighbors.

# II. China-India Rivalry in the Indian Ocean

*Ref: China-India Rivalry in the Indian Ocean, Congressional Research Service (Aug '17).*

Rivalry between China and India, the world's two most populous nations, appears to be increasing. While tensions over the two nations' disputed border and other diplomatic frictions have attracted attention in recent months, the China-India rivalry is also expanding into the Indian Ocean. Congress may consider the implications of this evolving strategic rivalry for U.S. interests and strategic balances of Asia.

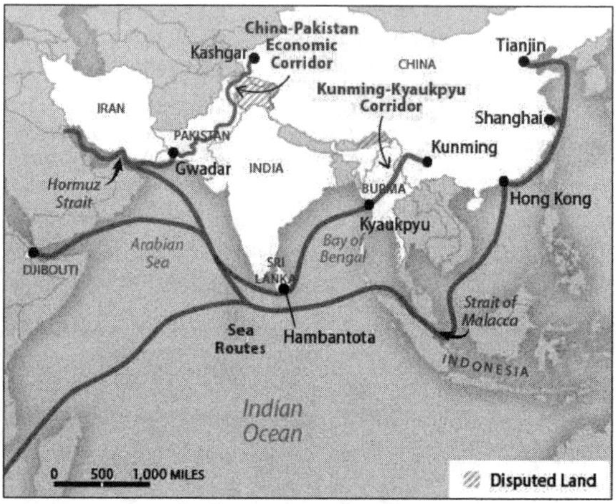


*Ref: Fig. 1. Indo-Pacific Energy and Trade Routes. Source: CRS map by Hannah Fischer with data from the South China Morning Post (2017); the Department of State (2015); Esri (2016); DeLorme (2016).*

## Indian Ocean Background

The Indian Ocean region (IOR) was a center of strategic competition for European colonial powers, including the Portuguese, Dutch, French, and British. The British East India Company's presence in the subcontinent is one example of this. China and other Asian powers have also played a role in the IOR. The voyages of Zheng He (1405-1433), for example, briefly extended Chinese influence into the Indian Ocean. The economic rise of China, India, and other Asian nations, and the rapid associated growth in seaborne trade and energy that transits the Indian Ocean, is making the region increasingly significant. Approximately 50% of the world's container traffic and 70% of its petroleum shipments transit the Indian Ocean. The Indian Ocean littoral region encompassed 2.49 billion people in 2010, representing more than one-third of the world's population. It is estimated that by 2030 the population in the IOR will have expanded by 27%. An estimated 15% of the world's fish catch comes from the Indian Ocean, and it may have additional hydrocarbon reserves that may further increase its strategic importance.

## China's Expanding Presence in the IOR

China's developing strategic presence in the IOR is manifesting itself largely through port and infrastructure projects, and through its first regional military base. China's development and investment projects have gained strategic coherence through Beijing's Belt and Road Initiative (BRI). This initiative is intended to develop trade and energy routes through the Indian Ocean and its littoral. The BRI's most high-profile project is the $46 billion China-Pakistan Economic Corridor designed to link China's west through Pakistan to Pakistan's Gwadar Port on the Arabian Sea. China has also proposed a Bangladesh-China-India-Myanmar (BCIM) Economic Corridor. The Kunming-Kyaukpyu, Burma segment of the BCIM has moved forward. China is rebuilding its ties with Burma and may augment its oil and gas linkages with that country by developing a Special Economic Zone at Kyaukpyu on the Bay of Bengal. China has agreed to sell submarines to Bangladesh and Pakistan and has also invested heavily in port facilities and other infrastructure in Hambantota, Colombo, and elsewhere in Sri Lanka. China is developing investments in port, rail, and pipeline projects in Malaysia, including at Melaka, Selankor, and Perak. Malaysia recently agreed to purchase four Chinese littoral mission ships. China also reportedly leased an island in the Maldives to develop Maritime Silkroad infrastructure. In addition, in July 2017, China sent troops to Djibouti to set up its first-ever overseas military base. The base will reportedly support Chinese naval ships operating in the region. Djibouti's location on the Horn of Africa gives it access to the strategic Bab el-Mandeb Strait, which connects the Mediterranean Sea with the Indian Ocean via the Suez Canal.

China-India dynamics in this region have further implications for U.S. interests and policy. The IOR is a crucial transit route for energy and other cargoes shipped between the Middle East and East Asia's booming economies. This strategic importance is one factor behind rising naval procurement by India, China, and their neighbors, and has led to greater Chinese naval activity in the IOR. China's economic, diplomatic, and military expansion into the region is widely viewed as a challenge for India, long the region's dominant power, and may be leading to a more outward-looking foreign policy in New Delhi, and to warming U.S.-India relations in recent years. The extensive U.S. naval presence in the Indian Ocean and adjacent waters—along with U.S. diplomatic and economic relations with the nations that rely on these transit routes—also makes these dynamics directly relevant to U.S. interests. One issue for Congress is whether to hold hearings on this issue to help the administration develop a comprehensive and integrated strategy toward the Indian Ocean or the Indo-Pacific more broadly. Another issue for Congress is whether to take into account this rivalry when considering naval procurement, or reviewing defense planning and multilateral cooperation in the region.

## Strategic Implications

Much of the Indian Ocean is within the U.S. Navy's Seventh Fleet's area of responsibility. The United States military has air, naval, and logistics hubs on the island of Diego Garcia and in Djibouti. The fight against Al Qaeda and the Islamic State in Afghanistan and Iraq has increased the strategic importance of the IOR for the United States as large numbers of U.S. military personnel have deployed through the region in recent years. The region's energy and trade routes are of strategic importance to the United States and many other nations. Any disruption of the strategic sea lanes that link the energy-rich Persian Gulf with the energy-dependent economies of Asia would be detrimental to the world's economy. Increasing U.S. energy independence makes it relatively less vulnerable to interruption of these routes than other nations. China's dependence on imported energy, and the strategic vulnerability that this presents, has been labeled its "Malacca dilemma" after the key strategic strait through which an estimated 80% of Chinese oil imports flow. Much of China's strategic maneuvering in the IOR can be viewed as an attempt to minimize this vulnerability.

The extension of Chinese influence into South Asia and the IOR, when viewed in the context of China's assertive behavior in the East China Sea and the South China Sea, raises concerns in many Asian capitals. China's contemporary naval presence in the Indian Ocean began with counterpiracy operations in the Arabian Sea. While some in India increasingly feel encircled by China's strategic moves in the region, the Chinese may also feel threatened by its limited ability to secure sea lanes, upon which China's economy depends for both trade and energy. The Chinese also emphasize the potential economic benefits of the Belt and Road Initiative to regional states. Understanding and effectively managing these evolving security and economic issues may be key to preserving regional stability and U.S. interests in the region. Some IOR states are hedging against China's rising power or developing their own defense capabilities; others are seeking more accommodative strategies or combining these approaches. Hedging strategies by Asian states include both developing intra-Asian strategic ties as well as enhanced ties between regional states and the United States. From this perspective there may be an opportunity for enhanced security collaboration between and among democratic nations such as the United States, India, Australia, and Japan.

## Competing Perceptions of the Crisis

Despite the ongoing diplomatic and military dialogues to reduce border tensions, the PRC has continued taking incremental and tactical actions to press its claims at the LAC. Sometime in 2020, the PRC built a large 100-home civilian village inside disputed territory between the PRC's Tibet Autonomous Region and India's Arunachal Pradesh state in the eastern sector of the LAC. These and other infrastructure development efforts along the India-China have been a source of consternation in the Indian government and media. In contrast, PRC has attempted to blame India for provoking the standoff through India's increased infrastructure development near the LAC. Asserting that its deployments to the LAC were in response to Indian provocation, Beijing has refused to withdraw any forces until India's forces have withdrawn behind the PRC's version of the LAC and ceased infrastructure improvements in the area.

## The PRC's Current Objectives in the Standoff

As the standoff continues, the PRC has expressed its aim to prevent the standoff from worsening into a wider military conflict. Additionally, Beijing has voiced its intent to return bilateral relations with New Delhi to a state of economic and diplomatic cooperation it had perceived to be improving since the 2017 Doklam standoff. PRC state-controlled media forcefully asserted China's intent to refuse any territorial concessions demanded by India. PRC officials, through official statements and state media, had also sought unsuccessfully to prevent India from deepening its relationship with the United States during and subsequent to the standoff, while accusing India of being a mere "instrument" of U.S. policy in the region.

# III. The Uyghurs

*Ref: China Primer: Uyghurs, Congressional Research Service (Sep. '23).*

Uyghurs (also spelled "Uighurs") are a Muslim ethnic minority group living primarily in the Xinjiang Uyghur Autonomous Region (XUAR) in the far northwest of the People's Republic of China (PRC or China). They have garnered the attention of U.S. policymakers, particularly since 2018, due to reports of the mass internment of Uyghurs in "reeducation" centers. The facilities were part of an ongoing government effort to systematically transform the thought and behavior of Uyghurs and forcefully assimilate them into PRC society, an effort some observers say is destroying Uyghur culture and identity. The U.S. government has responded by implementing targeted restrictions on trade with Xinjiang and imposing visa and economic sanctions on some PRC officials.

## Forced Assimilation

Since 2017, in tandem with a national policy referred to as "Sinicization," XUAR authorities have instituted measures to assimilate Uyghurs into Han Chinese society and reduce the influences of Uyghur, Islamic, and Arabic cultures and languages. Some experts contend the PRC government has used counterterrorism as a pretext for carrying out assimilation policies and mass detentions in Xinjiang. The XUAR government enacted a law in 2017 that prohibits "expressions of extremification," and placed restrictions upon dress and grooming, traditional Uyghur customs, and adherence to Islamic dietary laws (halal). Thousands of mosques in Xinjiang have been closed, demolished, or "Sinicized," whereby Islamic motifs and Arabic writings have been removed. The Xinjiang government has placed nearly half a million Uyghur and other minority children in state-run boarding schools, and has banned the use of Uyghur language in instruction in all schools in the XUAR.

## Mass Internment

Between 2017 and 2019, XUAR authorities, at the behest of Chinese Communist Party General Secretary Xi Jinping, arbitrarily detained over 1 million ethnic Uyghur

# China Primer: Uyghurs

*Ref: China Primer: Uyghurs, Congressional Research Service (Sep. '23).*

*CRS, using U.S. Department of State Boundaries; Esri; Global Administrative Areas; DeLorme; NGA (China Primer: Uyghurs, Congressional Research Service (Jan '22).*

Uyghurs speak a Turkic language and practice a moderate form of Sunni Islam. The XUAR, often referred to simply as Xinjiang (pronounced "SHIN-jyahng"), is a provincial-level administrative region that comprises about one-sixth of China's total land area and borders eight countries. The region is rich in minerals, produces over 80% of China's cotton, and has China's largest coal and natural gas reserves and a fifth of its oil reserves. PRC officials refer to Xinjiang as a "core hub" for China's Belt and Road Initiative, which involves Chinese-backed infrastructure projects and energy development in neighboring Central and South Asia.

All or parts of the area comprising Xinjiang have been under the political control or influence of Chinese, Mongols, and Russians for long spans of the region's documented history, along with periods of Turkic or Uyghur rule. Uyghurs played a role in the establishment of two short-lived, semi-autonomous East Turkestan Republics in the 1930s and 1940s. The PRC asserted control over Xinjiang in 1949 and established the XUAR in 1955. Uyghurs once were the predominant ethnic group in Xinjiang; they now make up less than half of the region's population of 26 million, according to official sources. The government long has provided economic incentives for Han Chinese, the majority ethnic group in China, to migrate to the region; Hans now constitute 42% of the XUAR population and the majority in Urumqi, the regional capital. Since 2017, the Xinjiang government has carried out a campaign to forcefully reduce birth rates or "illegal births" among Uyghurs and other minority groups.

Since an outbreak of Uyghur demonstrations and interethnic unrest in 2009, and sporadic clashes involving Uyghurs and Xinjiang security personnel that spiked between 2013 and 2015, PRC leaders have carried out large scale criminal arrests and intensive security measures in the XUAR, aimed at combatting "terrorism, separatism, and religious extremism." Three violent incidents in China in 2014 purportedly carried out by Uyghurs against Han civilians were described by some outside observers as acts of terrorism.

and other Muslims in reeducation centers. Detainees generally were not accused of crimes, but rather were held on the basis of past religious, cultural, scholarly, social, and online activities, as well as foreign travel, that the government later deemed "extremist," "pre-criminal," or potentially terrorist. Detainees were compelled to renounce many of their Islamic beliefs and customs as a condition for their release. Treatment in the centers reportedly included food deprivation, psychological pressure, sexual abuse, medical neglect, torture, and forced labor. Leaked Xinjiang police files, which included thousands of detainee records and images, important party directives, and police protocols, revealed the prison-like nature of the reeducation centers.

Since 2019, the XUAR government appears to have released some detainees, prosecuted many as criminals, and sent others to work as factory labor. Many forms of political indoctrination continue in homes and factories. Some reeducation centers appear to have been converted to high-security prisons and new prisons have been built. In 2022, the XUAR government reported it had prosecuted 540,826 people since 2017; human rights groups presume many were Uyghurs charged with political crimes. More than 300 Uyghur intellectuals and cultural figures are believed to be detained or serving prison sentences. Among Uyghur intellectuals handed long sentences for state security crimes are Ilham Tohti (convicted in 2014), an economics professor who had maintained a website related to Uyghur issues, and Rahile Dawut, a scholar of Uyghur folklore who was convicted in September 2023. Both were found guilty of "separatism" and sentenced to life in prison.

In August 2022, the United Nations Office of the High Commissioner for Human Rights determined that China's counterterrorism and counter- "extremism" strategies have led to "interlocking patterns of severe and undue restrictions on a wide range of human rights" and may constitute crimes against humanity. It called on China to "release all individuals arbitrarily deprived of their liberty" and "urgently repeal" all discriminatory laws and policies against Uyghurs and other Muslim minorities in the XUAR.

## Forced Labor

The PRC government has pressured many Uyghurs, including former detainees, into accepting employment in textile, apparel, agricultural, consumer electronics, and other labor-intensive industries, in Xinjiang and other provinces. Some factories utilizing Uyghur labor are tied to global supply chains. Uyghurs who refuse to accept such employment may face detention or other penalties.

## East Turkestan Islamic Movement

The PRC government has attributed some past deadly incidents in the XUAR since the 1990s to the East Turkestan Islamic Movement (ETIM), which it portrays as a Uyghur separatist and terrorist group with ties to global terrorist organizations. The U.S. government designated ETIM as a terrorist organization under Executive Order 13224 in 2002 (to block terrorist financing) and in 2004 placed ETIM on the Terrorist Exclusion List (TEL). Since 2017, the U.S. government has noted a lack of credible evidence supporting ETIM's continued presence, and stated in 2022 that it "has not seen clear and convincing evidence of ETIM's existence." In 2020, the Trump Administration removed ETIM from the TEL. United Nations sanctions monitors differ in their assessment, reporting that in 2022, ETIM had a presence in Afghanistan and Syria.

# IV(d). Northern Theater Command (NTC)

*Ref: Department of Defense Annual Report to Congress: Military and Security Developments Involving the People's Republic of China 2024, pp. 120 to 122 and 14 to 15.*

The **Northern Theater Command** is oriented toward the Korean Peninsula and Russian border security

The Northern Theater Command (NTC) is oriented toward the Korean Peninsula and Russian border security. The NTC includes the PRC's provinces bordering Mongolia, Russia, North Korea, and the Yellow Sea. It is responsible for operations along the PRC's northern periphery and border security associated with North Korean, Russia, and Mongolia.

Major PLA units in the NTC include the PLAA's 78th, 79th, and 80th Group Armies comprised of 17 combined arms BDEs; two PLAN naval bases, a nuclear submarine base, two submarine flotillas, two destroyer flotillas, and an aircraft carrier task group; three Marine Corps BDEs and an PLANMC aviation BDE; two air bases, 14 fighter/ground attack BDEs, one transportation BDE and one special mission aircraft division; and eight PLARF missile BDEs and one PLARF combat missile base.

During a contingency, the NTC likely exercises command over some former SSF units in theater and receives strategic intelligence support from the elements of the former SSF to improve battlefield awareness and facilitate joint operations within the theater. The North Theater Navy would be responsible primarily for protecting the sea approaches to northern China but could provide mission-critical assets to support other naval forces during contingencies beyond the Northern Theater. In 2023, NTC forces conducted various joint- and single-service training activities, including destroyer maritime training, submarine training, and Northern/Interaction-2023—a bilateral air and naval "strategic maneuver" with Russia in the Sea of Japan.

## I. PRC-Russia Relationship

In 2023, the PRC continued partnering with Russia to counter the United States and advance its preferred vision for the international system. Most notably, PRC support for Russia's war against Ukraine, including significant trade in dual-use goods to rebuild Russia's defense industrial base, remained robust. During 2023, Beijing promoted Russian narratives blaming the United States and NATO for the war, buoyed Russia's economy against international sanctions by purchasing energy, increasing trade to record levels, and selling Russia necessary dual-use inputs that Russia's military industries rely on, such as microelectronics, precision machine tools, spare parts, titanium and magnesium alloys, and chemical inputs for munitions production. In Europe, Beijing and Moscow have jointly called for a shift away from the continent's Western-led security apparatus toward a Eurasian-led security architecture. However, Beijing probably will struggle to pursue this aim in Europe due to its tacit support for Russia's invasion of Ukraine.

Beijing was largely unresponsive to international criticism over its support to Russia, barring a few exceptions. This year, the PRC special envoy for Eurasian Affairs, Li Hui attended the Ukraine peace conference in Jeddah, Saudi Arabia, which Russia was not invited to, and claimed to have "further consolidated international consensus." The PRC restricted and then loosened export controls on unmanned aerial vehicle (UAV) exports after international scrutiny of the PRC's supply of commercial UAVs to the Russian war effort.

Despite the strain of Russia's war against Ukraine, the PRC and Russia maintained their bilateral military exercise schedule at 2022 levels. Exercises during 2023 included two combined bomber patrols, a new naval exercise called NORTH CO-OPERATION and combined naval patrols into the Bering Sea. In September 2024, China and Russia conducted the second iteration of NORTHERN COOPERATION, likely followed by a combined naval patrol to the Sea of Japan, which likely included anti-submarine defense and sea rescue training. Although these activities probably modestly improved their capabilities and interoperability, their main effect signaled the strength of the PRC-Russian partnership. The PRC and Russia participated together in trilateral naval exercises with South Africa and separately with Iran, both aimed at improving participants' military cooperation and maritime security.

The PRC almost certainly is applying lessons from Russia's war against Ukraine toward countering what it perceives as a U.S.-led containment strategy. Diplomatically Russia's war against Ukraine probably reaffirmed to Beijing the importance of persuading countries Beijing considers the Global South in Africa, Latin America, the Middle East, and the Indo-Pacific region to accept PRC narratives about global dynamics. The PLA probably seeks to incorporate lessons from how Russia and Ukraine are employing influence operations in the conflict into its own doctrine. The effects of Western sanctions against Russia almost certainly have amplified the PRC's push for defense and technological self-sufficiency as well as economic and energy resilience.

## Key Features of the Relationship

*Ref: China-Russia Relations, Congressional Research Service (Sep '23).*

Building on the foundation of the 1991 Sino-Soviet Border Agreement, the 2001 Treaty of Good-Neighborliness and Friendly Cooperation, among other things, noted Beijing's and Moscow's satisfaction on border issues and set broad areas of cooperation ranging from economics and trade to counterterrorism. The renewal of the treaty in 2021 reflects the overall positive trajectory of relations.

The direction of the bilateral relationship appears, in part, to reflect close personal ties between Presidents Xi Jinping and Vladimir Putin. Since 2013, Xi and Putin have met numerous times and established regular dialogue mechanisms at lower levels. In 2019, PRC and Russian leaders announced their intention to develop a "Comprehensive Strategic Partnership of Coordination for a New Era," professing a "high degree of political trust" and "all-around cooperation."

Military cooperation between the PRC and Russia is significant, encompassing exchanges and joint exercises, as well as intelligence sharing and joint development of weapons systems. In November 2021, the two sides signed a Road Map for Military Co-operation for 2021-2025 to guide collaboration in this sphere. The PRC and Russia are founding members of the Eurasia-based Shanghai Cooperation Organization (SCO), an intergovernmental group mainly focused on security affairs.

The PRC and Russia also enjoy strong commercial and financial ties and are partners in their attempts to "de-dollarize" the global economy, which they see as beholden to the United States. (See CRS In Focus IF11885, De-Dollarization Efforts in China and Russia.) Both governments express opposition to the use of unilateral sanctions as tools of policy.

The PRC and Russia often cooperate and coordinate in multilateral settings, including the United Nations; the SCO; the BRICS grouping; and the Group of 20 (G20). In 2022, the PRC joined Russia in vetoing a U.S.-led draft U.N. resolution that would have tightened sanctions against North Korea over its missile launches. Where frictions may arise, the PRC and Russia have tried to harmonize the interests of overlapping ventures, such as the Belt and Road Initiative (BRI) and the Eurasian Economic Union (EEU).

# The Northern Theater

*Ref: Department of Defense Annual Report to Congress: Military and Security Developments Involving the People's Republic of China 2024, p. 112.*

## Major Military Units

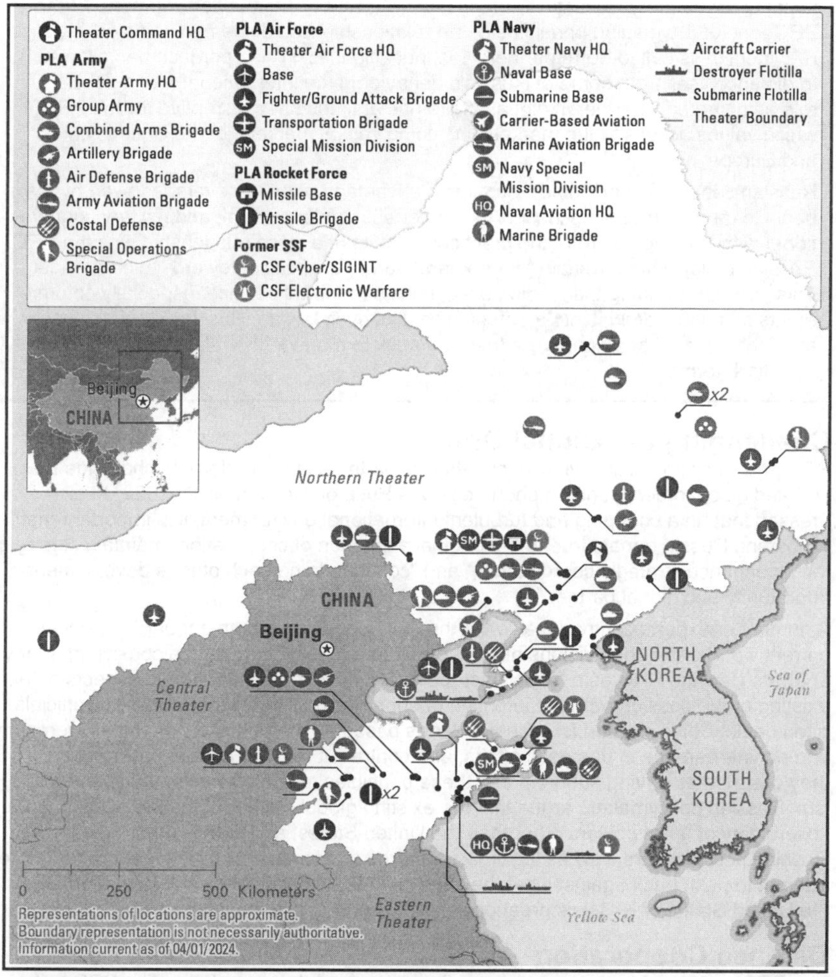


Northern Theater Command, headquartered in Shenyang, with responsibility for northeastern China along the Mongolian, Russian, and North Korean borders. The Northern TC's most prominent responsibilities are the various contingencies stemming from China's shared border with North Korea and maintaining border security with Mongolia and Russia.

# China-Russia Relations (Select Issues)

*Ref: China-Russia Relations, Congressional Research Service (Sep '23).*

## PRC and Russian Perspectives

PRC officials frequently describe the current moment as a "historic high" in the China-Russia relationship, with the two sides mutually supporting each other's positions on national sovereignty, security, territorial integrity, and economic development. For the PRC, one of the guiding principles of the relationship would seem to be flexibility, as is evident in its call to "form partnerships, not alliances." This approach has allowed it to distance itself from some of Russia's behavior in the international arena. Russia's role as a strategic partner in global affairs has at times been disruptive for the PRC, which values access to the markets, including capital markets, of the United States and Europe.

Russian concerns over China's economic advantage often are overshadowed by the benefits (and, perhaps, necessity) of greater economic, political, and military cooperation. Despite being aware of the practical benefits of a close relationship with the PRC, Russian policymakers remain guarded and harbor skepticism toward China's leaders. This includes avoiding the binding obligations of a formal alliance, especially commitments that could draw it into a potential conflict over Taiwan. Russian officials appear to reject the label of a "junior" partner and seek to manage the relationship on mutually beneficial terms.

## Challenging the Global Order

PRC and Russian officials view each other as partners in their efforts to challenge the U.S.-led global order. During a phone call with Putin on December 30, 2022, Xi expressed that "in a changing and turbulent international environment, it is important that China and Russia remain true to the original aspiration of cooperation, maintain strategic focus, enhance strategic coordination," and "continue to be each other's development opportunity and global partner."

A number of experts contend that while the PRC and Russia both appear to reject the current world order, their visions of what ought to replace it may not be consistent. Many analysts view the PRC as a "revisionist" power working to change certain aspects of the existing order, possibly with the intention of superseding it in the long run. PRC officials often depict China's vision for global order as based on the principles of state sovereignty and noninterference in domestic affairs, and contrast it with the current order, which they describe as having subordinated these principles to human rights and interventionism. Russian policymakers argue that the existing global order ignores the position and sovereignty of great powers other than the United States, and denies other powers what Russian policymakers view as legitimate spheres of influence. Russian leaders therefore attempt to push back against what they see as U.S. unipolar overreach, and to remind the United States of Russia's great power status and role in the international system.

## Defense Cooperation

The PRC and Russia are close partners in military and defense affairs. The PRC historically has been a major buyer of Russian arms to modernize its military. Defense trade has become more complex as the PRC's defense industry has grown more sophisticated, reportedly increasing Russian concerns about technology transfer and intellectual property violations. Russia's defense industry relies on importing key components and investment from China for the development of advanced weapon systems.

The two countries conduct combined military exercises, developing interoperability, and increasing defense sector cooperation and joint arms development. Russia has sought to increase its participation in military exercises with the PRC, including at the strategic and tactical level, as well as naval exercises and air patrols in the Asia Pacific. Russia arguably conducts these exercises to develop interoperability with PRC forces and to signal that Russia remains a player in the region. In August 2023, a joint PRC and Russian flotilla conducted a maritime patrol in areas including the vicinity of Alaska.

## Energy and Economics

Economic relations between China and Russia are asymmetrical; China's economy is considerably larger and more dynamic. China is Russia's second largest export market after the EU; Russia is China's ninth largest. Russia's role as a key provider of PRC energy imports is a defining feature of the bilateral relationship. According to PRC customs data, Russia tied Saudi Arabia as China's largest source of crude oil in 2022, providing approximately 17% of China's total imports. China's planned transition away from coal potentially could lead to a greater role for Russian hydrocarbons (including "green coal" and natural gas). The Power of Siberia Pipeline, which is expected to reach full capacity by 2027, could make China one of the largest individual importers of Russian natural gas. *(For more on economic relations between China and Russia, refer to CRS In Focus IF12120, China's Economic and Trade Ties with Russia.)*

### Impact of Russia's 2022 Invasion of Ukraine

Russia's invasion of Ukraine has drawn increased scrutiny to the Beijing-Moscow relationship, including questions regarding the extent to which China is willing and able to relieve sanctions pressure on Russia. China has avoided public condemnation of Russia's actions, even though some observers believe Russia's invasion strained relations. According to these observers, China was unaware of Russia's plans and remains unwilling to be drawn into the conflict. Russia, meanwhile, appears to be turning to China to circumvent Western sanctions, including selling oil and purchasing critical components for its defense industry. A July 2023 report by the Office of the Director of National Intelligence found that the PRC has become "an even more critical economic partner for Russia" since the latter's 2022 invasion of Ukraine.

Some trade data suggests some PRC firms may be providing dual-use goods to Russia that may be contributing to its war effort. However, it is unclear whether and to what extent these activities are state-directed. Media reports suggest the PRC government is selective in its engagement, allowing transactions that maximize advantages to China—such as buying oil at a discount—but is unwilling to supply lethal weaponry.

The Biden Administration has repeatedly warned China against assisting Russia's invasion of Ukraine. In January 2023, State Department Spokesperson Ned Price stated: "We've been very clear with the PRC, including in private [ ... ] about any costs that would befall the PRC should they decide to assist Russia in a systematic effort to evade U.S. sanctions or in the provision of security assistance that would then be used against the Ukrainian people in Ukraine." Sixteen firms registered in China have been sanctioned pursuant Executive Order 14024 in connection to Russia's invasion.

# II. Evolving PRC-DPRK-Russia Relationship

Since the fall of the Soviet Union, the PRC has been the Democratic People's Republic of Korea's (DPRK's), or North Korea's, principal supporter. The PRC constitutes the vast majority of the DPRK's external trade and North Korea depends on the PRC for energy, fertilizer, and food imports. Beijing values the DPRK as a buffer between the PRC and South Korea and the U.S. forces stationed there. Beijing is reluctant to economically punish North Korea because it fears punitive measures could destabilize the Kim regime.

Though it is a partner to the DPRK and Russia, Beijing has publicly distanced itself from Moscow and Pyongyang's growing defense cooperation, referring to it as a strictly bilateral matter. Beijing probably views close association with North Korea-Russia cooperation as a risk to its desired reputation as a responsible great power, especially as Russia and North Korea collaborate to violate UN sanctions that the PRC supports. The possibility that Russia might transfer weapons technology or dual-use items, which could support Pyongyang's nuclear weapons program, runs counter to Beijing's stated opposition to nuclear weapons on the Peninsula.

While distancing itself from Russia-North Korea cooperation, the PRC has deepened engagement with Pyongyang, in December 2023 holding the first high-level PRC-North Korea meeting since 2019 and announcing that 2024 would be a "Year of Friendship" between the PRC and North Korea. These efforts are probably an effort to counterbalance Russian engagement with the DPRK.

# IV(e). Central Theater Command (CTC)

*Ref: Department of Defense Annual Report to Congress: Military and Security Developments Involving the People's Republic of China 2024, pp. 122 to 124.*

The **Central Theater Command** is oriented toward capital defense and providing surge support to other theaters

CTC's mission is the defense of Beijing while providing support to other theater commands. CTC's primary responsibility is to defend the capital and CCP leadership while providing a strategic reserve to the other theater commands. It is centrally located, connecting the four remaining theater commands.

Major PLA units in the CTC include the PLAA's 81st, 82nd, and 83rd GAs comprised of 18 combined arms BDEs; two air bases, 10 fighter/ground attack BDEs, one bomber division, one bomber BDE, and one transportation BDE; and nine PLARF missile BDEs and one PLARF combat missile base. As of late 2023, the CTC may include a new Marine Corps BDE and SOF BDE reorganized from PLAA units.

In 2023, CTC-subordinate units conducted a range of training and support-related activities. In July, a combined armed BDE from the PLAA's 82nd GA exercised using reconnaissance drones and radar systems to counter rivals equipped with portable anti-aircraft and anti-tank missiles. In August, an 81st GA BDE supported disaster relief efforts in response to flooding in Beijing, transporting emergency supplies and transferring victims. In September, multiple personnel were selected from the 81st GA to join the recently established 10th PRC Peacekeeping Infantry Battalion for the UN mission in South Sudan.

*See following page (p. 1-66).*

# The Central Theater

*Ref: Department of Defense Annual Report to Congress: Military and Security Developments Involving the People's Republic of China 2024, p. 123.*

## Major Military Units

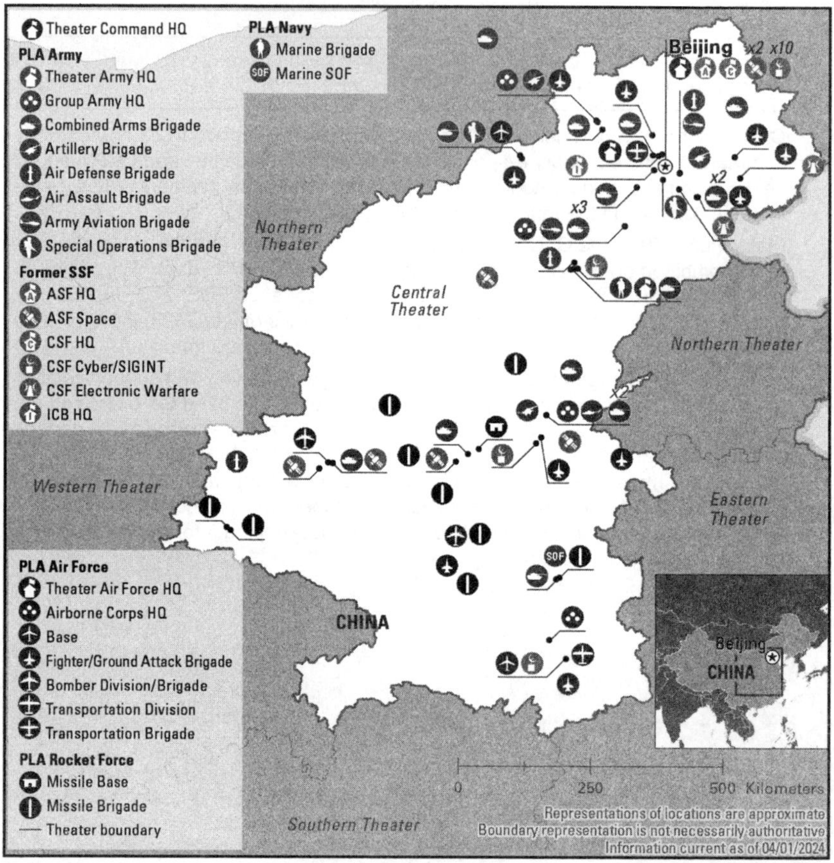


Central Theater Command, headquartered in Beijing, has responsibility for north-central China and the capital region. The Central TC's primary mission is the defense of Beijing, and it serves as the national strategic military reserve.

# I. China's Armed Forces (The PLA)



*Ref: Department of Defense Annual Report to Congress: Military and Security Developments Involving the People's Republic of China 2024, pp. 44 to 45 and ATP 7-100.3, Chinese Tactics, (Aug '21), chap 1.*

The PLA is the world's largest active-duty military force and comprised of approximately 2.035 million active, 510,000 reserve, and 500,000 paramilitary personnel for a total force of 3.045 million. In efforts to create a leaner, more mobile force, the PLAA has steadily reduced active-duty personnel in the last three decades but still outnumbers other services with 965,000 soldiers in 2023. In 2023, the PLAA transferred three brigades (BDEs) to the PLAN Marine Corps (PLANMC). The addition of these three BDEs, supplemented by previous additions of aviation and special forces BDEs, brings the PLANMC up to a total of 11 BDEs. These BDEs provide improved expeditionary force capabilities to the PLANMC.

* *These figures, which are smaller than those in the 2023 report, reflect a reversion to a previous methodology, not an assessed change in the size of the PLA.*

The PRC maintains its goal to achieve a fully modernized national defense and military force by 2035 and for the PLA to become a world-class military by 2049. The force is progressing toward its 2027 benchmark of military modernization that aligns with the 100th anniversary of the PLA's founding on August 1, 1927. The 2027 benchmark, introduced in late 2020 as leaders discussed the 14th FYP (2021–2025), represents the start of the new three-step development strategy. The original three-step modernization strategy sought to achieve mechanization by 2020; modernization of military theory, organization, personnel, and equipment by 2035; and become a world-class force by mid-century. With basic mechanization considered achieved in 2020, the 2027 goal is a short-term marker and represents a modification, not a compression in timeline, for the PRC's ambition to achieve complete military modernization of the PLA by 2035. The PLA centenary goal set by the CCP accelerates the integrated development of mechanization, informatization, and intelligentization and fielding of a combat-ready force with improved strategic capabilities to defend national sovereignty, security, and development interests by 2027.

## Key Takeaways

- The PLA has sought to modernize its capabilities and improve its proficiencies across all warfare domains to become a joint force able to conduct the full range of land, air, and maritime as well as nuclear, space, counterspace, electronic warfare (EW), and cyberspace operations.
- The PLA's evolving capabilities and concepts continue to strengthen the PRC's ability to "fight and win wars" against a "strong enemy" (referring to the United States), counter an intervention by a third party in a conflict along the PRC's periphery, and project power globally.
- In 2023, the PLA continued to adjust its military structures, field modern indigenous systems, build readiness, and strengthen its competency in joint operations.

# I. China's Security Apparatus

*Ref: ATP 7-100.3, Chinese Tactics, (Aug '21), pp. 2-1 to 2-2.*

The PLA represents the armed forces branch of China's security organizations. Other organizations under this umbrella include the Ministry of Public Security (MPS), the Ministry of State Security (MSS), the People's Armed Police (PAP), and the China Militia. These organizations are collectively referred to as the Chinese Security Apparatus. The Chinese Security Apparatus is not fully analogous to the U.S. Department of Defense (DOD): it is more of a descriptive term than a discrete organization. All Security Apparatus organizations fall under the purview of the CPC, and their missions often overlap with one another. One of the defining elements of Chinese internal politics is the complex interplay between these organizations and their leadership. In most cases, the PLA can be thought of as the most important and most influential member of the Chinese Security Apparatus.

Forces making up each organization in the Chinese Security Apparatus can be described as either civilian, paramilitary, or military. The composition of each organization varies significantly. Police, security, and some intelligence forces are described as civilian; the MPS and MSS are comprised mostly of these personnel. The PAP and the China Militia are largely paramilitary: they conduct military-like training and employ some military equipment and tactics, but they are not considered part of the PLA.

## Ministry of Public Security (MPS)

The MPS is China's national police force and the world's largest police organization. Though a national-level force, it is organized and commanded locally, typically by provincial governments. It is responsible for day-to-day law enforcement throughout China, and it is directly reportable to China's State Council.

## Ministry of State Security (MSS)

The MSS is China's primary national intelligence apparatus, having responsibility for both foreign and domestic intelligence work and counterintelligence. The MSS's counterintelligence mission requires a fairly significant "secret police" element that does not work with the MPS. The PLA generally does not integrate activities with either the MPS or the MSS, except when intelligence or security information is exchanged. The roles and responsibility of the MSS are in transition, as a new high-level command—the People's Liberation Army Strategic Support Force (PLASSF)—was recently established and appears to overlap numerous responsibilities with the MSS.

# Chinese Armed Forces

The Chinese Armed Forces consist of the PLA, the PAP, and the China Militia. The PLA's primary role within the Chinese Armed Forces is to defend China from external or international threats, while the PAP focuses on internal threats.

## People's Armed Police (PAP)

The PAP's mission is internal security and, in conjunction with the MPS, limited law enforcement. The PAP also has the official mission to support or reinforce the PLA during times of war. PAP forces are organized on a provincial basis, with two national-level antiterrorism units stationed in the capital region. The PAP features military-like organizations and extensive use of military equipment, including small arms, helicopters, and light armored vehicles. The U.S. does not have an equivalent organization.

## China Militia

The China Militia is a massive quasi-formal militia element of the Chinese Armed Forces, consisting mostly of poorly trained and equipped part-time regional military units. The militia's primary mission is to provide logistics and security support to the PLA, though disaster relief and internal security missions are also part of its mission set. The China Militia is a different entity from the People's Liberation Army Reserve. The militia can be can be considered an aspect of People's War philosophy, to be integrated with the military and other entities in times of national emergency or crisis. Like the rest of the Chinese Armed Forces, it is being downsized and modernized, with less emphasis put on manpower and more emphasis on quality training and equipment. Like the PAP, there is not a true U.S. equivalent to the China Militia.



## People's Liberation Army (PLA) - The Armed Forces of China

The PLA is the armed forces of China. **Though called an army, the PLA is analogous to the U.S. DOD in that it consists of all of the service branches: an army (People's Liberation Army Army—PLAA), navy (People's Liberation Army Navy—PLAN), air force (People's Liberation Army Air Force—PLAAF), rocket force (People's Liberation Army Rocket Force—PLARF), and strategic support force (PLASSF).** In addition, the People's Liberation Army Joint Logistics Support Force (PLAJLSF), though not a service, is a national-level PLA organization. The PLA is under the command of the Central Military Commission (CMC), as supervised by the Politburo Standing Committee and the CPC. The PLA is officially the armed wing of the CPC; all levels of the PLA are supervised by a system of political officers. The expressed primary missions of the PLA are to protect the ruling status of the CPC, ensure China's sovereignty and territorial integrity, safeguard Chinese interests at home and abroad, and help maintain global stability. See figure 2-1 for a simplified diagram of the PLA's command structure.

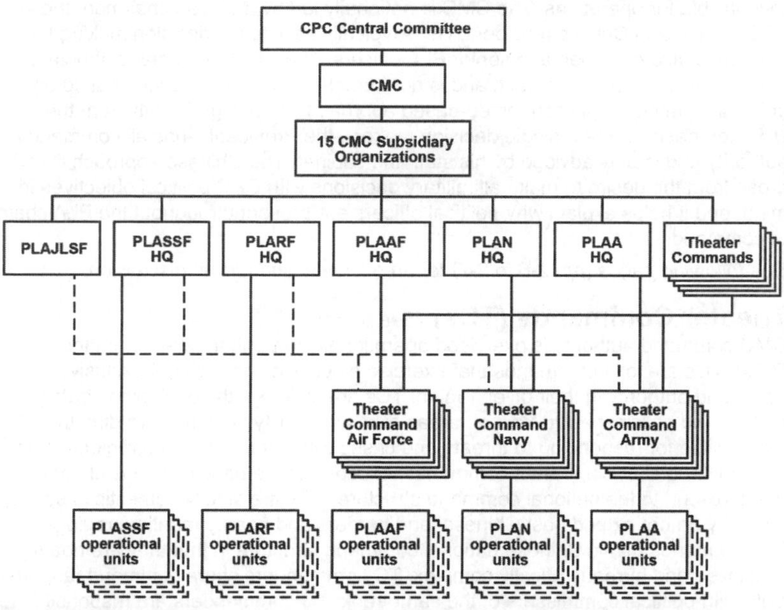


***Note.*** Theater command branches report to two entities: their respective theater commands and their headquarters. For example, the Western Theater Command Army reports to the Western Theater Command and PLAA HQ.

| | |
|---|---|
| CMC | Central Military Commission |
| CPC | Chinese Communist Party |
| HQ | headquarters |
| PLAA | People's Liberation Army Army |
| PLAAF | People's Liberation Army Air Force |
| PLAJLSF | People's Liberation Army Joint Logistics Support |
| PLAN | People's Liberation Army Navy |
| PLARF | People's Liberation Army Rocket Force |
| PLASSF | People's Liberation Army Strategic Support Force |
| ——— | command relationship |
| - - - - | support relationship |

*Ref: ATP 7-100.3 (Aug '21), fig. 2-1. Simplified PLA command structure.*

# II. Chinese National and Strategic-Level Organizations

The command structure of the PLA is complex and opaque to outsiders. Both its command structure and apparatus are deliberately complex, designed first and foremost to ensure the security of the CPC. In stark contrast to Western military philosophy, Chinese command structures do not generally appoint a single decision maker or imbue a single individual with absolute command authority: decisions are to be made via consensus, rather than by fiat. Of note, however, the current Chinese President, Xi Jinping, appears to wield much greater direct authority over the PLA than his predecessors. There are no fewer than ten different national-level command organizations in China, organized across at least three different levels of a complex hierarchy. This book will not discuss in detail the intricate web of organizations and internal politics at the top of China's command structure, but it will provide a simplified overview to enable better understanding of how these command relationships influence tactical operations.

## Central Military Commission (CMC)

The CMC is the most powerful and significant national-level command body in China. It is also the most relevant to tactical operations. Recent restructuring created no fewer than 15 functional departments, including a national-level army department and a strategic rocket department. In general, national headquarters are responsible for capabilities development, education, and training, while theater commands are responsible for operations. The CMC is notionally led by a civilian chairman: the CPC's head and China's president. He is typically assisted in decision making by numerous vice chairmen and senior PLA officers. The CMC's command structure relies on a committee approach and is not formal—it routinely changes based on political dynamics. This committee-based approach differs significantly from the U.S. approach, where a single decision maker—the President—has all command authority and is only advised by his staff and cabinet. The Chinese approach likely arose from the desire to make all military decisions with CPC political objectives in mind, and it helps explain why political officers are present throughout the PLA chain of command.

*See following pages (pp. 2-6 to 2-7) for an overview and further discussion.*

## Theater Commands (TCs) *(See p. 2-9.)*

CMC command authority is exercised operationally through theater commands (TCs). TCs are joint commands that exercise extensive—though not exclusive—command authority in their given region. TCs are tasked with developing strategy, tactics, and policy specific for their areas of responsibility, and they are directly responsible for responding to threats and crises within their assigned regions. TCs are similar to U.S. combatant commands, except that the parallel chains of command extend to the national command structure. TCs are strictly domestic in scope, and they do not extend past Chinese land borders and nearby maritime regions.

TC commands mimic Chinese national command structure in that they are heavily politicized and bureaucratically complex. TC commanders share command responsibility with political commissars of the same rank; TC commanders are responsible for operations, while political commissars are responsible for ideological functions. TC staffs participate in committee-based decision making. Each TC has a single-service subordinate headquarters responsible for that service's operations in the theater, plus an organic electronic countermeasures brigade, an information operations support brigade, and a reconnaissance and intelligence support brigade. PLAA, PLAAF and PLAN units have dual chains of command during peacetime, reporting both to their national service-level headquarters and to their assigned TC. In wartime they will likely be under complete control of their respective TCs.

*See pp. 2-8 to 2-9 for an overview and further discussion.*

# PLA's Modernization & Reform

*Ref: Department of Defense Annual Report to Congress: Military and Security Developments Involving the People's Republic of China 2024, pp. VII to IX.*

The PLA has sought to modernize its capabilities and improve its proficiencies across all warfare domains to become a joint force capable of the full range of land, air, and maritime as well as nuclear, space, counterspace, electronic warfare, and cyberspace operations. Despite its progress, the force still has significant deficiencies including in commander proficiency, long-distance logistics, and urban warfare. In 2023, the PLA continued to adjust its military structures, field modern indigenous systems, build readiness, and strengthen its competency to conduct joint operations.

## People's Liberation Army Army (PLAA) *(See pp. 2-25 to 2-38.)*

The PLAA continues its decades-long effort to modernize equipment and focus on combined arms and joint training to become a world-class military by 2049. The PLAA continued to demonstrate long-range joint fires capabilities during the April 2023 JOINT SWORD exercise.

## People's Liberation Army Navy (PLAN) *(See pp. 2-39 to 2-44.)*

Numerically, the PRC has the largest navy in the world, with a battle force of over 370 ships and submarines, including more than 140 major surface combatants. The PLAN is largely composed of modern multi-mission ships and submarines. In 2023, the PLAN continued to grow its ability to perform missions beyond the First Island Chain (FIC) by conducting the first extended area deployment of its new YUSHEN-class amphibious assault ship and three deployments with CV-17 Shandong to the Philippine Sea, a record number for any PLAN carrier in a calendar year.

## People's Liberation Army Air Force (PLAAF) *(See pp. 2-47 to 2-52.)*

The PLAAF is modernizing and indigenizing its aircraft and unmanned aerial systems rapidly, matching U.S. standards. In 2023, the PLA transferred significant portions of PLAN shore-based, fixed-wing combat aviation units, facilities, air defense, and radar units to the PLAAF. Given time, this shift will probably enable better command and control over the PRC's integrated air defense systems as well as the network of ground-based air domain awareness radars supporting the PRC's national integrated air defense system network.

## People's Liberation Army Rocket Force (PLARF) *(See pp. 2-53.)*

The People's Liberation Army Rocket Force (PLARF) organizes, mans, trains, and equips the PRC's strategic land-based nuclear and conventional missile forces as well as associated support forces and missile bases. In 2020, the PLARF advanced its long-term modernization plans to enhance its "strategic deterrence" capabilities.

## (Former) Strategic Support Force (SSF) *(See pp. 2-57 to 2-58.)*

On 19 April 2024, Beijing announced the dissolution of the SSF, and the alignment of its subordinate forces—Aerospace Force (ASF) and Cyberspace Force (CSF)—directly under the CMC. Beijing established a new Information Support Force (ISF), under the CMC, to coordinate the management of military's networks and communications systems. The updated PLA organizational structure features four theater-grade services—the PLAA, PLAN, PLAAF, and PLARF—and four deputy-theater-grade forces or service arms: the ASF, CSF, ISF, and the Joint Logistic Support Force (JLSF).

## The Joint Logistic Support Force *(See p. 2-24.)*

The JLSF is concentrating its efforts on improving joint strategic and campaign-level logistic efficiencies through training and integrating civilian products and services. The JLSF supports multimodal transportation methods to facilitate the movement of PLA forces and equipment for training. During wartime or mobilization, the JLSF probably will rely on its military representative offices and dispatch centers to route military traffic.

# China's Military Leadership

*Ref: Department of Defense Annual Report to Congress: Military and Security Developments Involving the People's Republic of China 2024, pp. 39 to 41.*

## Central Military Commission (CMC)

The CMC, the PRC's top military decision-making body, consists of a civilian chair, two vice chairs, the chiefs of the Joint Staff and Political Work Departments, and the head of the Discipline Inspection Commission; CCP General Secretary and PRC President Xi Jinping is currently serving his third term as CMC Chairman.

### China's Military Leadership

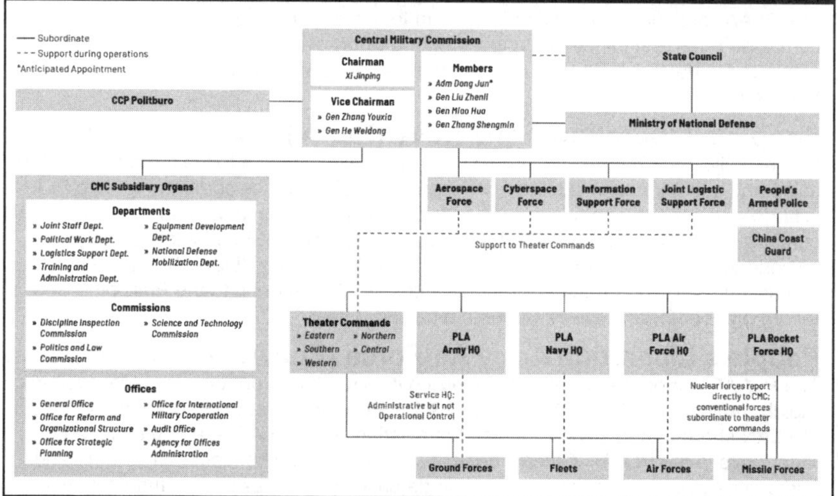


## Members of the CCP Central Military Commission

As the military's highest decision-making body, the CMC is technically a department of the CCP Central Committee. The CMC Chairman is a civilian, currently, though not always concurrently serving as the General Secretary of the CCP and President of the PRC. CMC members are appointed at Party Congresses every 5 years. In fall 2022 at the 20th Party Congress, General Zhang Youxia ascended to the first Vice Chairman position, joined by General He Weidong as the second Vice Chairman. Other CMC members include General Liu Zhenli and returning members Admiral Miao Hua and General Zhang Shengmin. General Li Shangfu was appointed to the CMC and, in March 2023, to the position of Minister of National Defense but, in October 2023, Beijing removed Li from his role without explanation, probably because of corruption relating to the PLA's weapons and acquisitions program during his previous role as head of the CMC EDD. At the end of 2023, the CMC consisted of two vice chairs, the chiefs of the Joint Staff and Political Work Departments, and the head of the Discipline Inspection Commission.

**Chairman Xi Jinping concurrently serves as the CCP General Secretary, CMC Chairman, and President of the PRC.** Xi was first appointed as Party General Secretary and CMC Chairman in 2012 and as President in the spring of 2013. Xi was reappointed to all his positions for an unprecedented third term at 2022's 20th Party Congress and the 2023 NPC. In 2016, Xi was announced as the commander in chief of the CMC's Joint Operations Command Center (JOCC) and was named "core" leader of the CCP Central Committee.

**Vice Chairman General Zhang Youxia is the PRC's top uniformed official and former junior vice chairman**. Zhang was first appointed to the CMC in 2012 as the head of the General Armaments Department—now the EDD—where he oversaw the PLA's manned space program as well as MCF and military modernization efforts. Zhang gained rare experience as a combat commander during the PRC's brief war with Vietnam in 1979. Zhang formerly commanded the Shenyang military region, which shares a border with North Korea and Russia. Zhang is one of the PLA's "princelings." His father, a well-known military figure in the PRC, served with Xi's father at the close of Chinese Civil War in 1949. Zhang, age 73 in 2023, was expected to retire following his second CMC term due to previously followed age norms.

**Vice Chairman General He Weidong is the PRC's second-most senior uniformed officer and a former commander of the PLA's Eastern Theater.** His ascent to a vice chairman position absent prior CMC membership is unusual and probably a testament to his extensive operational experience focused on Taiwan. Before his selection as vice chairman, He served briefly in the CMC JOCC, where he played a key role in planning live-fire drills in the Taiwan Strait as part of the PLA response to the then-U.S. House Speaker's August 2022 visit to Taipei.

**For a majority of 2023, Li Shangfu served as the PRC's Minster of National Defense before being removed in October 2023 for corruption-related charges (refer to the Special Topics section for more information on PLA corruption).** General Li Shangfu was appointed to the CMC at the 20th Party Congress in October 2022 and as the Minister of National Defense at the NPC in March 2023.

**In December 2023, President Xi Jinping selected Adm. Dong Jun as the PRC's new Minister of National Defense.** Adm. Dong Jun is the fifth defense minister to serve under Xi since 2012. Dong is the PLA's third-most senior officer and manages its relationship with state bureaucracies and foreign militaries. Unlike the U.S. Secretary of Defense, he is not part of the chain of command, but instead leads the PLA's international engagement. Dong is the first career naval officer to rise to the role of defense minister and previously had been commander of the PLAN since 2021.

**Joint Staff Department Chief General Liu Zhenli oversees PLA joint operations, a narrowing of the wider responsibilities held by the former General Staff Department prior to reforms initiated in 2015.** Liu is one of few remaining active-duty PLA officers with combat experience and is recognized as a combat hero for his service in the PRC's border war with Vietnam. Like his predecessor Li Zuocheng, Liu rose through the ranks of the PLAA, assuming command of the service in 2021. Beginning in 2015 as Army chief of staff, Liu guided the service through a major period of reform with ground force downgraded to an equal standing with the other branches. Liu, at age 59 in 2023, is the youngest CMC member.

**Political Work Department Director Admiral Miao Hua oversees the PLA's political work, including propaganda, organization, and education.** Miao is a former Army officer who switched services to the Navy in December 2014 when he became political commissar of the PLAN. Miao may have ties to Xi from his time serving in the 31st Group Army (GA) in Fujian Province, when his career overlapped with Xi's. Miao participated as the PLAN political commissar during the Navy's BRI cruise in mid-2017. Miao Hua, at age 67 in 2023, retained his position following the 20th Party Congress.

**Secretary of the Discipline Inspection Commission General Zhang Shengmin oversees the highest-level organization responsible for investigating military violations of CCP discipline, including corrupt practices.** Zhang is a deputy secretary and third-ranking member on the standing committee of the CCP's Discipline Inspection Commission. Zhang's reappointment reflects the CCP's continued commitment to the anticorruption campaign in the military. Zhang, at age 66 in 2023, remained on the CMC in his current position following the 20th Party Congress.

# CMC Command Authority

*Ref: ATP 7-100.3, Chinese Tactics, (Aug '21), pp. 2-5 to 2-6.*

*See pp. 1-27 to 1-58 for related discussion of the theater commands (territorial disputes).*

CMC command authority is exercised operationally through theater commands (TCs). TCs are joint commands that exercise extensive—though not exclusive—command authority in their given region. *See p. 2-4 for further discussion of the TCs.*

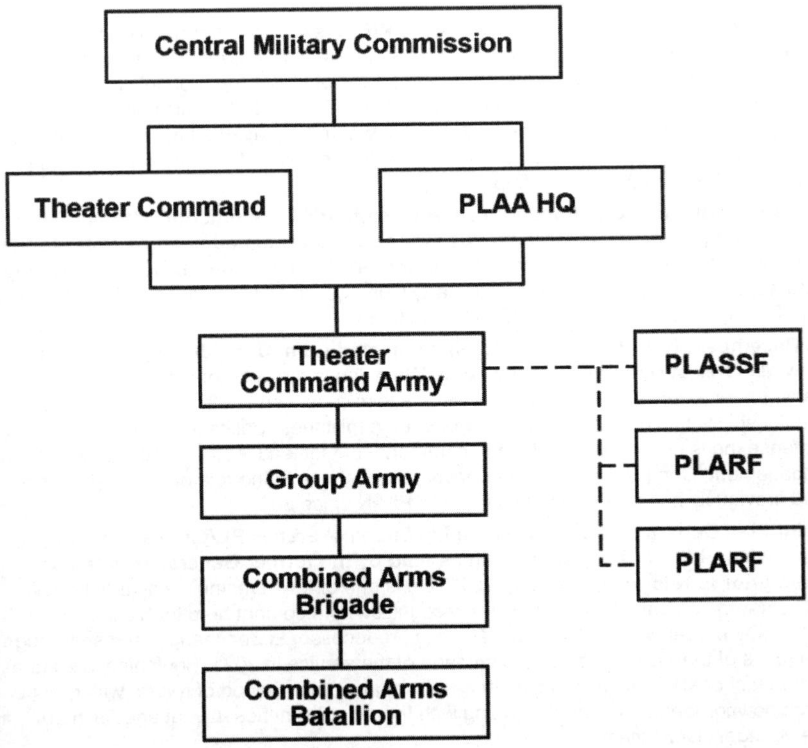


***Note.*** The theater command army has a duel reporting structure. In addition, support from other service-level organizations integrates at the theater command army level.

| | |
|---|---|
| HQ | headquarters |
| PLAA | People's Liberation Army Army |
| PLAJLSF | People's Liberation Army Joint Logistics Support Force |
| PLARF | People's Liberation Army Rocket Force |
| PLASSF | People's Liberation Army Strategic Support Force |
| ———— | command relationship |
| - - - - - | support relationship |

*Ref: ATP 7-100.3 (Aug '21), fig. 2-2. PLAA chain of command, isolated.*

# Theater Commands (TCs) *(See pp. 1-31 to 1-66.)*

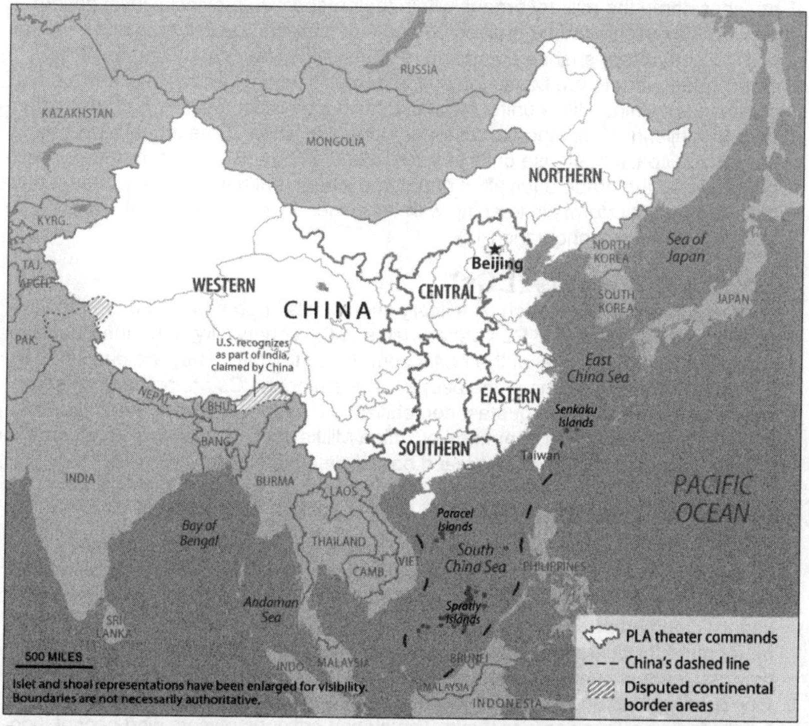


*Ref: China's Military: The People's Liberation Army (PLA), Congressional Research Service. Map of China and PLA Theater Commands. (Map by Amber Wilhelm.)*

The PRC continues to implement reforms associated with the establishment of five theater commands in early 2016. The Eastern, Southern, Western, Northern and Central Theater Commands replaced seven army-based military regions and are now the highest-ranking **"joint operations command" organizations** within their respective geographical areas. Each theater command receives direction from the CMC and has operational authority over the PLA forces within its theater. The Theater Command is also responsible for all conventional combat and non-combat operations within its area of responsibility. Theater commands are responsible for developing theater-specific command strategies aimed at preparing to fight and win against an adversary, developing joint operational plans and military capabilities, responding to crises, and safeguarding the sovereignty and stability of territories. The strategic directions of the theater commands are based on PRC perceptions of peripheral threats:

- **Eastern Theater Command (ETC)** – Taiwan, East China Sea *(See p. 1-35 to 1-43).*
- **Southern Theater Command (STC)** – South China Sea; Southeast Asia border security; territorial disputes *(See pp. 1-45 to 1-50.)*
- **Western Theater Command (WTC)** – India, Central Asia, "counterterrorism" in Xinjiang and Tibet *(See pp. 1-51 to 1-58.)*
- **Northern Theater Command (NTC)** – Korean Peninsula, Russia border security *(See pp. 1-59 to 1-64.)*
- **Central Theater Command (CTC)** – Capital defense; surge support to other theaters *(See pp. 1-65 to 1-66.)*

## Military Districts (MDs)

Military districts (MDs) correspond to Chinese provincial-level governments and are considered the military organs within their respective provincial-level governments. They use the same dual-command structure as TCs, with a commander and political commissar sharing command responsibility. The primary responsibilities of MD headquarters are conscription and demobilization, Military-Civil Fusion, and command of China Militia units. MDs also have jurisdiction over PLA border defense units in Xinjiang, Tibet, and part of Inner Mongolia, while TC headquarters command border and coastal defense units in the rest of the country. MD commanders previously carried the grade equivalent of a corps leader, but this has likely been reduced following recent reforms in keeping with the general theme of professionalization and modernization throughout the PLA.

## Local Commands (LCs)

Local commands (LCs), such as military subdistricts or garrisons at the prefecture, county, and city level, and PLA departments at the county, city, and municipal level represent the lowest level of the PLA's command structure. They are described as the military service organs of the people's government. LCs are primarily charged with meeting conscription quotas, demobilization, national defense education, local Military-Civil Fusion, and command of China Militia units. In Xinjiang, Tibet, and part of Inner Mongolia they also command border defense units.

# III. PLA Weaknesses and Deficiencies

The PLA regularly identifies shortcomings in the force's warfighting and command capabilities in official media. These critiques guide its modernization efforts. Despite its rapid progress, the force has not yet demonstrated the type and scale of sophisticated urban warfare or long-distance logistics capabilities that would likely be required for operations against Taiwan or major contingencies overseas. The way that leaders critique officers suggest challenges with decentralized command. One prominent example is the "Five Incapables" slogan the PLA uses to highlight pitfalls for the officer corps to avoid. The slogan critiques officers who cannot judge situations, understand higher authorities' intentions, make operational decisions, deploy troops, or deal with unexpected situations. The PLA is targeting modernization efforts to address some of these challenges. In 2021, the PLA announced a new training system with the goal of improving exercise realism and developing officers capable of operating in a joint environment.

# II. PLA Fundamentals



*Ref: ATP 7-100.3, Chinese Tactics, (Aug '21), chap 1.*

## I. The People's Liberation Army (Overview)

The People's Liberation Army (PLA) continues more than two millennia of Chinese military tradition. China lays cultural claim to many of the world's most famous works of military strategy and philosophy—most prominently, The Art of War by Sun Tzu—and these works are still highly influential throughout the PLA. The Chinese take military philosophy, politics, and theory very seriously: the PLA is considered the vanguard of the Chinese Communist Revolution, and its philosophical underpinnings are important to both its culture and its approach to warfighting at all levels. The PLA also has a deep commitment to Communist and Maoist philosophy.

*Note. This is an explanation of naming and acronym conventions. The proper name for China's military is the People's Liberation Army, or PLA. This organizational structure is comparable to the U.S. Department of Defense (DOD); it does not refer exclusively to ground forces, as does the U.S. Army.*

*The PLA's land-based service is the People's Liberation Army Army (PLAA), the naval branch is the People's Liberation Army Navy (PLAN), and the aerial branch is called the People's Liberation Army Air Force (PLAAF). The PLA's strategic conventional and nuclear ballistic and cruise missile service is the People's Liberation Army Rocket Force (PLARF). The People's Liberation Army Strategic Support Force (PLASSF) is a new service-level organization that incorporates national-level cyber warfare, electronic warfare (EW), space, and other operational support elements under a single headquarters, and the People's Liberation Army Joint Logistics Support Force (PLAJLSF) is a new service-level sustainment organization.*

The PLA's basic warfighting philosophy is that of active defense: a fundamentally defensive political and strategic stance, enabled—when required—by operational and tactical offense.

For over two thousand years, China has been surrounded by enemies, adversaries, and other competitors. Invasion, occupation, raids, and other incursions into Chinese territory were commonplace. The PLA views protecting Chinese sovereignty and security as a sacred duty. China traditionally viewed military resistance as an affair for the entire population: mass resistance, guerrilla warfare, and winning a war of attrition. This understanding has evolved in the modern age to where a PLA enabled by technology, well-trained personnel, and a whole-of- government, defense-in-depth approach deters conflict before it ever happens and protects China and the CPC from foreign aggression and internal tumult. The basic concept of active defense informs every level of PLA operations and acquisitions. Figure 1-1 (following page) provides a graphic depiction of the policy and theoretical underpinnings of the active defense philosophy.

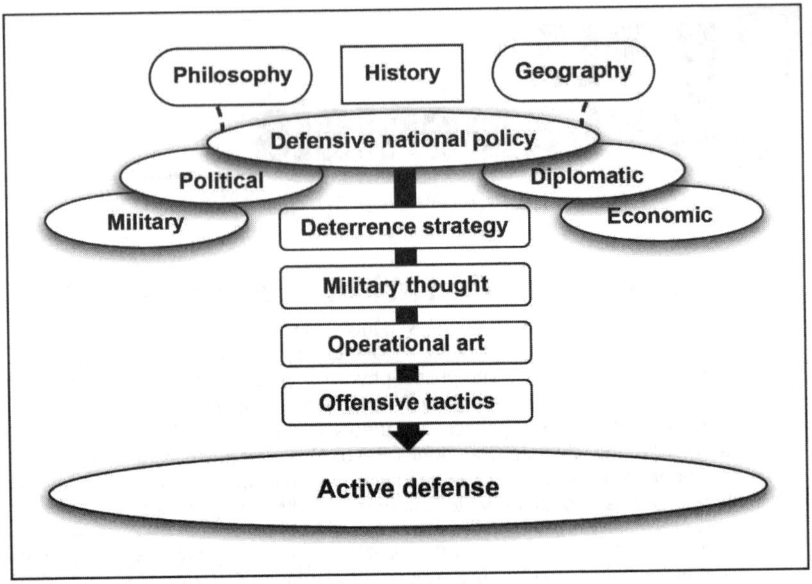


*Ref: ATP 7-100.3 (Aug '21), fig. 1-1. Active defense.*

Presently, the PLA is in the midst of a period of comprehensive reform. Central to this is the evolution of the "big army"—the dominance of the PLAA at the expense of other services—not only throughout the PLA, but also throughout Chinese politics and society. Reducing manpower and equipment levels and employing a quality-over-quantity approach is central to this effort, as is the expansion of joint integration and an expanded emphasis on domains other than just the ground. In addition, PLAA units are now nationally deployable—able to operate anywhere within Chinese borders—rather than being strongly tied to a specific region. This is intended to reduce manpower requirements and the influence of the PLA in local politics, but it requires a top-to-bottom reorganization of PLA training, doctrine, and acquisition.

Two historical figures feature prominently in the philosophy and strategy of the PLA: Sun Tzu and Mao Zedong. The PLA views these two strategists as equally important, with People's War being viewed as a modern addendum to The Art of War. Sun Tzu and Mao together created the framework from which the PLA's modern-day strategy and tactics are derived.

Influenced largely by The Art of War, the PLA—and China as a whole—look to create advantages along a vast competition continuum, ranging from diplomacy and covert operations in peacetime to major combat operations. The PLA views all available government means as a kind of defense in depth. Defense of China begins with skilled diplomacy and prudent political measures at home and abroad. State actors and agencies set conditions for military success if and when military action becomes necessary. The PLA decisively defeats opponents militarily if peaceful measures fail. The ideal outcome in every scenario is to "win without fighting"—as a philosophy taken directly from Sun Tzu. Military-Civil Fusion is an important method of incorporating civilian support to achieve military objectives in both peacetime and war. 1-32. PLA doctrine describes three basic ways to use military power: warfighting, military deterrence, and military operations other than war. Developing warfighting capabilities is the PLA's core task; prevention of conflict is the PLA's most important mission. The PLA's deterrence mission creates numerous subordinate objectives designed to demonstrate Chinese capabilities and will, all viewed as critical to preventing conflict both domestically and abroad. Though China states unequivocally that it will not initiate conflict, practices only self-defense, and will respond militarily only if attacked, its basic concept of deterrence has been broadened significantly in recent years. Active defense now includes far more politically and strategically offensive activities. Examples include, but are not limited to—

- Expanded participation in global peacekeeping operations.
- Expanded air and naval presence in the Western Pacific in an offshore defensive strategy.
- More rapid and thorough responses to non-war crises.
- Expanded operations in the cyber and space domains.
- Defense of the primacy and stability of the CPC.

PLA views on modern war allow for limited preemptive actions. People's War initially mandated few to no pre-emptive military activities, but PLA views on modern war concede that an unmitigated modern military operation may win a war before a proper response can be organized. As such, certain allowances for offensive activity under the umbrella of active defense have been made. According to PLA doctrine, the "first shot" in the area of politics and strategy must be differentiated from the first shot on the plane of tactics. If any country or organization violates the other country's sovereignty and territorial integrity, the other side will have the right to "fire the first shot" on the plane of tactics.

People's War encompasses all three levels of warfare. At the strategic level, People's Armies use their superior manpower, lesser logistic requirements, and greater political will to defeat opponents in defensive wars of attrition. At the operational level, a combination of mobility, deception, and willpower allow formations to rapidly maneuver and close with the enemy in order to engage it in close combat, offsetting limitations in firepower, technology, and training. At the tactical level, units aggressively maneuver in order to create massive advantages in combat power at key times and locations, while simultaneously preventing enemy response through effective deception operations. Infiltration techniques allow small units to close with and defeat the enemy in detail. At all levels, People's Army units rely on superior willpower, cohesion, deception, and manpower to overcome technologically advanced opponents.

People's War was initially developed as Mao's interpretation of Marxism in military conflict: a class struggle between an oppressed people and a bourgeois or imperialist professional military. The Second Sino-Japanese War and the Chinese Civil War (1945-49) strengthened this understanding: both conflicts involved a People's Army conducting mobile or guerrilla operations, on its own soil, against better-equipped and trained opponents. Both conflicts resulted in Chinese victories, but they were enormously costly, both economically and in lives lost.

People's War concept and doctrine were significantly altered beginning in the 1980s. At the end of the Mao era in the late 1970s, PLA strategists began to reassess the nature of People's War. Mao had taken Marxist-Leninist ideals to extremes in the military, adopting a military without hierarchy or rank. This approach proved ineffective, and reforms were needed to make the PLA into a modern force capable of deterring conflict and defending Chinese strategic and political interests. This evolved into People's War under Modern Conditions, a modification that accounted for the increasing lethality and complexity of the modern battlefield. Joint operations—along with mechanization and artillery—were emphasized. The concept of deterrence expanded to include not only fighting defensive wars within Chinese territory, but also the idea that a strong regional and maritime security environment might prevent fighting in Chinese territory in the first place.

The modern iteration of People's War is People's War in Conditions of Informationization. This theory incorporates thinking and strategy informed by numerous geopolitical and technological developments over the past 25 years. First and foremost is the emerging information environment: the PLA recognizes the importance of the

# II. PLA Principles—People's War

*Ref: ATP 7-100.3, Chinese Tactics, (Aug '21), pp. 1-11 to 1-12.*

PLA principles were originally written by Mao Zedong during the Long March of 1934—35 and revised during the Japanese occupation of China, beginning in 1937. These principles still serve as the basis for People's War theories, though they have been modernized periodically along with the rest of Chinese military thought. There are numerous different interpretations and translations of these principles, varying widely based on when and where they were written and translated. However, the versions are all similar, generally reflecting Communist political sensibilities, a focus on mobility and deception, and a strong understanding of basic military theory. The key themes of People's War are—

- Eliminate isolated pockets of the enemy before concentrating to fight larger forces.
- Capture small villages and towns before capturing large urban areas.
- Eliminate the enemy's fighting capacity; do not focus on territory.
- Fight no battle unprepared; develop strategy based on the worst conditions.
- Concentrate forces to achieve an overwhelming advantage in numbers, then defeat the enemy in detail.
- Choose the first battle carefully.
- Unify the command and coordinate.
- Combine mobile war, positional war, and guerrilla war.
- Employ forces and tactics flexibly.
- Fight in one's own way, and let the enemy fight in its.

## Eliminate Isolated Pockets of the Enemy Before Concentrating to Fight Larger Forces

This principle is similar to Western maneuver warfare theories developed during the World Wars. It calls for rapid maneuver and decisive attacks against weak points before finally seeking a decisive engagement against an unbalanced enemy.

## Capture Small Villages and Towns Before Capturing Large Urban Areas

This somewhat unique Communist Chinese principle was formed from the success of early PLA operations against Japanese and Republican Chinese forces. Operating in rural areas made it difficult for early PLA opponents to detect and fix PLA formations. In expeditionary warfare, this principle likely implies targeting less-challenging defensive positions first, then moving on to more well-defended and critical positions once initial gains are consolidated.

## Eliminate the Enemy's Fighting Capacity; Do Not Focus on Territory

This principle is similar to the principle in paragraph 1-45 about towns and cities. It recognizes that, for most military campaigns, the center of gravity is the enemy force, not territory. Chinese forces may temporarily—or even deliberately—cede territory in order to enable decisive action against the enemy's forces.

## Fight No Battle Unprepared; Develop Strategy Based on the Worst Conditions

This principle is similar to the general planning process of many armed forces. Commanders ensure that troops are as prepared as possible for any battle, and they conduct planning based on the enemy's most dangerous course of action, rather than its most probable.

## Concentrate Forces to Achieve an Overwhelming Advantage in Numbers, Then Defeat the Enemy in Detail

This is another principle similar to Western military theories, though it specifies the importance of numerical advantage. This principle is most often seen in PLAA tactics: tactical-level units seek to use a combination of maneuver and deception to achieve their desired numerical superiority and allow engagement and defeat of the enemy in detail. Traditionally, the PLA sought to fight using close combat techniques, taking advantage of night operations to enable infiltration whenever possible. Technological advances changed this approach, incorporating longer-range weapons and nonlethal effects to strike the enemy at greater distances, decreasing the effect of the technological gaps the PLA faces with regard to night vision and electro-optical capabilities.

## Choose the First Battle Carefully

This principle acknowledges the importance of gaining and maintaining the initiative; in this case, through a victorious first engagement. This principle extends beyond simple military battles; it also includes choosing the first decisive political or economic engagement carefully in order to set conditions for victory in a future decisive engagement. Along with "fight no battle unprepared," this principle suggests an element of caution and prudence in determining when and if to go to war.

## Unify the Command and Coordinate

As a basic and timeless military principle, PLA organization and doctrine regarding unified command differs somewhat from Western militaries. While the latter imbue command authority with a single individual—the commander—the PLA considers unity of command to be as much a political issue as an operational principle. Units feature both military and political leadership, and creating political unity of purpose is considered a critical component of People's War.

## Combine Mobile War, Positional War, and Guerrilla War

This principle outlines the three types of warfare proposed by Mao, and it resembles the more modern theory of hybrid warfare. Mobile units use a combination of maneuver and deception to achieve surprise in decisive actions; static, defensive units fix opponents and defend or secure key areas from attack and guerrilla forces conduct irregular campaigns either behind enemy lines or within enemy formations. All forms are carefully coordinated to ensure unity of purpose and situational understanding.

## Employ Forces and Tactics Flexibly

Similar to U.S. Army principles and doctrine, this principle acknowledges the importance of flexibility in any military situation. Historically, the PLA was notoriously inflexible, with subordinates often failing to employ their own judgment due to fear of reprisal or lack of professional competence. The PLA's new focus on building a professional noncommissioned officer corps and a decentralized leadership philosophy reflect its understanding of shortfalls in this area. Many PLA exercises currently stress tactical innovation on the part of commanders at all levels.

## Fight in One's Own Way, and Let the Enemy Fight in Its

Perhaps the most famous of Mao's principles, this principle reinforces China's views about the uniqueness of its approach to military operations. It stresses the importance of knowing both one's own capabilities and those of the opponent.

digital battlefield, and it has emerged as a world leader in integrating cyber warfare into its operational construct. Figure 1-2 displays the timeline of the evolution of People's War concepts.

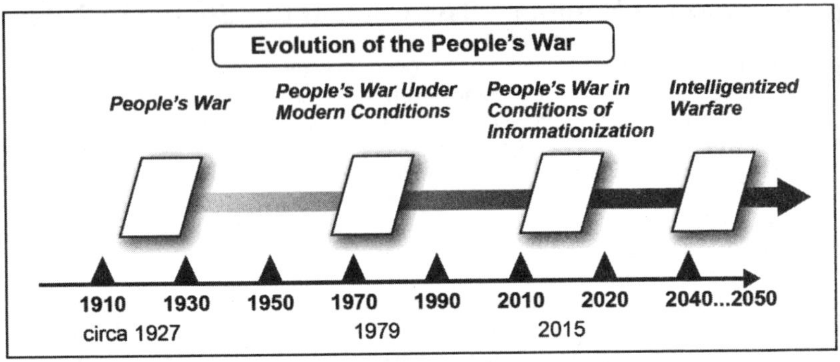


*Ref: ATP 7-100.3 (Aug '21), fig. 1-2. People's War.*

Second, the PLA further expanded its understanding of active defense. After watching nearly six months of force buildup on the Iraqi border during OPERATIONS DESERT SHIELD and DESERT STORM, the PLA concluded that the most efficient way to defend Chinese territory was to prevent potential opponents from building combat power in close proximity to Chinese borders. This in turn spurred the development and redevelopment of capabilities intended to defeat or interdict the buildup of combat power at extended ranges and across wide geographic areas. It also compelled development of capabilities designed to deny use of wide geographic areas to potential foes. Years later, in analyzing this approach, U.S. strategists denoted it antiaccess/area denial. This phrase was—and still is—an accurate depiction of the PLA approach to active defense in the Western Pacific.

Third, People's War in Conditions of Informationization encompasses the tacit admission that China may have to become involved in local or regional conflicts in order to maintain the regional or international economic order, protect the CPC, or otherwise support Chinese economic or political interests. This ostensibly contradicted Mao's initial philosophy of People's War, which is for the most part strictly a domestic defense-oriented military approach. This appreciation may represent fundamental change for the PLA, and it requires a vast increase in expeditionary and overseas sustainment capabilities. These requirements, in turn, demand significant increases in the PLA's military professionalization, equipment quality, and air and naval capabilities.

Local War is the PLA expression for a regional conventional conflict, likely close to Chinese territory. The PLA views this as the most likely situation calling for military force, and it has developed capabilities designed to excel in this environment. This conclusion makes sense, as every conflict the PLA has fought in its history was local, and all but the Chinese Civil War were limited. People's War in Conditions of Informationization, when applied to the Local War concept, is frequently referred to as Winning Informationized Local Wars.

Chinese military theory may be moving away from traditional Marxist-Leninist and Maoist theory as PLA leaders acknowledge the importance of technology, professionalization, and military hierarchy. The PLA, however, still identifies as staunchly Communist and views its modern theories as an evolution of People's War, not as a revision or repudiation. People's War in Conditions of Informationization is said to be a modern adaptation of The Art of War and People's War. It retains the concept of active defense as its centerpiece, and it considers deception and political willpower to be the most important—and most uniquely Chinese—elements of a successful military campaign.

The PLA anticipates Informationized Warfare evolving into Intelligentized Warfare in the relatively near future. Intelligentized Warfare incorporates numerous emerging technologies—including decentralized computing, data analytics, quantum computing, artificial intelligence, and unmanned or robotic systems—into the PLA's conceptual framework. Intelligentized Warfare seeks to increase the pace of future combat by effectively fusing information and streamlining decision-making, even in ambiguous or highly dynamic operating environments. Intelligentized Warfare also amplifies the nascent concepts embodied by the Military-Civil Fusion effort: many of the subsystems that create the backbone of an Intelligentized PLA are researched and developed initially in the civilian realm. Careful alignment of military and civilian efforts enables the synchronization of efforts and streamlines the fielding process for the PLA. Despite its focus on technology, Intelligentized Warfare remains informed by the People's War concepts: many Intelligentized Warfare initiatives are clearly shaped by the original People's War principles.



# III. China's Approach to Conflict

China considers three aspects in the country's view of conflict. They are Comprehensive National Power, deception, and the Three Warfares.

## A. Comprehensive National Power

China describes warfare as only one waypoint along a continuum. Military power is only one component of Comprehensive National Power (CNP), the Chinese method of ranking countries based on an assessment of all types of state influence. Hard power includes military capability and capacity, defense industry capability, intelligence capability, and related diplomatic actions such as threats and coercion. Soft power includes such things as economic power, peaceful diplomatic efforts, foreign development, global image, and international prestige. These two types of power combine to make up CNP. China views CNP as a vital measure of its global status. Ultimately, all forms of conflict—be they military, diplomatic, or other—must enhance China's CNP. CNP is viewed as a whole-of-government effort; the PLA is simply the military wing of CPC, so it makes sense to view military action as fundamentally political. The primary mission of the PLA is strategic deterrence, as illustrated in figure 1-3.

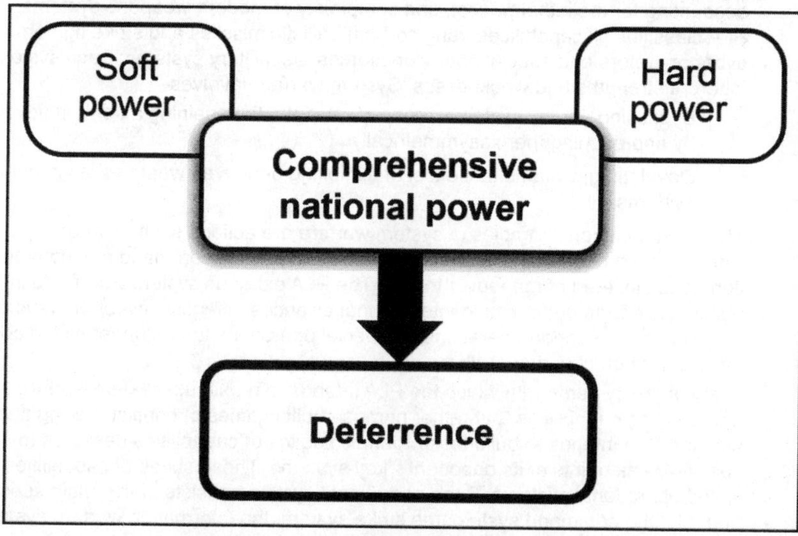


*Ref: ATP 7-100.3 (Aug '21), fig. 1-3. Comprehensive national power.*

## B. Deception

Deception plays a critical role in every part of the Chinese approach to conflict. The Chinese emphasis on deception can be traced to Sun Tzu, who believed that it was the basis for all warfare. PLA views on this topic differ considerably from those of most Western militaries. Instead of being a peripheral enabler, deception operations are seen as integral to every operation at all levels of war. Where U.S. Army operational planning uses the concept of a course of action— a scheme developed to accomplish a mission (JP 5-0)—PLA planners use stratagems. Rather than describing friendly operations, stratagems describe the enemy's mindset, focusing on how to achieve the desired perceptions by the opponent, and then prescribing ways to exploit this perception. Rather than focusing on defeating the opponent in direct conflict—as most Western militaries do—stratagems consider deception, trickery, and other indirect, perception-based efforts to be the most important elements of an operation. Deception is a fundamental aspect of the Chinese way of war, and applications of deception are considered a high priority.

# IV. PLA System Warfare

The PLA appears to recognize domains in the same way as the U.S. DOD, and it is actively seeking to enhance multi-domain capabilities and cross-domain integration. The organization applies all of its basic military philosophies and principles of active defense, deterrence, and deception to operations in all domains. The overarching PLA theoretical framework for its multi-domain effort is called system warfare. The system warfare concept seeks to identify critical or vulnerable system components, then degrade or destroy the effective use of larger systems through targeted attacks on these vulnerabilities. The primary principle of system warfare is the identification and isolation of critical or vulnerable subsystems or interdependence of these subsystems. The PLA believes that if key threat systems are rendered ineffective, the threat's ability and will to resist will crumble.

*See chap. 3, Tactical System Warfare for further discussion.*

System warfare is the most recent PLA effort to operationalize the principles of force concentration and asymmetric attack as outlined by Sun Tzu and Mao, while accounting for the lethality, cost, and complexity of modern weapons systems. The PLA classifies all capabilities, ranging from ballistic missiles and strike fighters to cyber operators and special operations forces, as military systems. Each system has inherent strengths and weaknesses. System warfare involves—

- Bypassing enemy systems' areas of strength, thus gaining a combat advantage by approaching them asymmetrically.
- Developing systems that excel at exploiting perceived weaknesses in enemy systems.

The most common examples of system warfare are actions such as targeting networks instead of shooters, sensors instead of aircraft, or command and communication nodes instead of maneuver forces. The PLA expands system warfare to include diplomatic efforts undermining international alliances, offensive cyber operations disabling air or seaport operations, or special operations forces undermining civilian morale through covert operations.

Many of the systems with which the PLA intends to prosecute system warfare are not standing capabilities, but rather purpose built in times of conflict. During times of war, the PLA intends to build task-organized suites of capabilities designed to strike specific weak points of its opponent's key systems. These suites of capabilities are called operational systems. Each operational system consists of five main subcomponents: the command system, the strike system, the information warfare system, the intelligence system, and the support system.

At the tactical level, system warfare centers largely on targeting high-value battlefield systems such as radars, command and communication nodes, and field artillery and

# C. The "Three Warfares" Concept *(See pp. 4-1 to 4-2 & 2-58.)*

*Ref: ATP 7-100.3, Chinese Tactics, (Aug '21), p. 1-14.*

China's strategic approach to conflict employs Three Warfares designed to support and reinforce the PLA's traditional military operations. Though these approaches are called warfares, these strategies—in Western thinking—fall somewhere between modern concepts such as information operations and historical concepts such as military operations other than war or effects-based operations. Despite the names, they are universally nonlethal: they do not involve direct combat operations. Instead, they are designed to pursue what Sun Tzu considered generalship in its highest form—victory without battle. If a battle must be fought, the Three Warfares are designed to unbalance, deceive, and coerce opponents in order to influence their perceptions. In a major change from the past, when political officers were mainly involved in rear area personnel functions, the Three Warfares make political officers and soldiers into nonlethal warfighters who provide essential support to combat units.

*See chap. 4, Information Operations, for further discussion.*

These Three Warfares are—

## Public Opinion Warfare

Public Opinion Warfare is referred to as huayuquan, which translates roughly as "the right to speak and be heard." To the Western mind this implies something along the lines of freedom of speech. Its meaning to the Chinese, however, is substantially different: it refers to the power to set the terms of a debate, discussion, or negotiation. In other words, it is China's high-level information campaign designed to set the terms of political discussion. China views this effort as influential not only on PLA operations, but also in support of Chinese economic interests worldwide. China views Public Opinion Warfare as capable of seizing the initiative in a conflict before any shots are fired by shaping public discourse, influencing political positions, and building international sympathy. Public Opinion Warfare operations are seen every day in the PLA's vast media system of newspapers, magazines, television, and internet sources that target both domestic and foreign audiences. Public Opinion Warfare supports the PLA's Psychological Warfare and Legal Warfare activities in peacetime and war.

## Psychological Warfare

Psychological Warfare is broadly similar to U.S. military information support operations in that it is intended to influence the behavior of a given audience. PLA Psychological Warfare seeks to integrate with conventional warfare and includes both offensive and defensive measures. The PLA views Psychological Warfare through the lens of Sun Tzu, emphasizing its multiplicative effect when coupled with comprehensive deception operations. Deception operations are critical to the PLA's entire warfighting approach, and Psychological Warfare represents the major information operations element of deception operations.

## Legal Warfare

Legal Warfare refers to setting the legal conditions for victory—both domestically and internationally. The U.S. does not have an equivalent concept, although State Department diplomatic and legal operations have roughly equivalent objectives. Legal Warfare seeks to unbalance potential opponents by using international or domestic laws to undermine their military operations, to seek legal validity for PLA operations worldwide, and to support Chinese interests through a valid legal framework. Legal Warfare has emerged with a particularly prominent role via the various Chinese political maneuverings in the Western Pacific, particularly those areas surrounding international waterways, disputed land masses, and economic rights of way. Legal Warfare is present at the tactical and operational levels of war. It guides how the PLA trains to treat prisoners of war, detainees, and civilians, and how it abides by international legal conventions, codes, and laws.

# V. Framework For PLA Operations

*Ref: ATP 7-100.3, Chinese Tactics, (Aug '21), pp. 1-14 to 1-16.*

The PLA operational framework is the lens through which the PLA develops capabilities, plans and executes operations, develops doctrine, and revises military philosophy and strategy. The operational framework is complex and comprehensive, running the entire scope of military operations from national strategy down to small-unit tactics. There is no direct U.S. equivalent, though most of the general ideas are replicated in various ways in U.S. doctrine. The PLA operational framework consists of five levels, ranging from the purely philosophical to the prescriptive and practical. These levels are—

## Military Thought

Military Thought represents the highest levels of PLA military thinking. It embodies China's long history as a land of philosophers, and it contains those works held in the highest esteem by the PLA. The Art of War and People's War are the two foundational documents and concepts for Military Thought, though there are many other works constantly being produced, debated, and revised. Military Thought provides the philosophical foundation for PLA operations, defines the culture and objectives of the PLA, and defines the PLA's relationship with the CPC. U.S. equivalents are military concepts and strategic directives, such as the National Security Strategy and the Capstone Concept for Joint Operations.

## Defense Theory and Defense Doctrine

Defense Theory and Defense Doctrine represent the PLA's strategic-level thinking. Defense Theory operationalizes the various principles laid out in Military Thought and other high-order documents. Defense Doctrine codifies the basic PLA approach to warfighting. Doctrine in this sense remains at the strategic level—the PLA does not appear to use the word "doctrine" to describe operational practices at operational or tactical levels of war. A U.S. equivalent to Defense Theory is the Air Force Future Operating Concept; an equivalent to Defense Doctrine is the U.S. Army in Multi-Domain Operations (formerly the Army Operating Concept). 1-64. Strategic Principles and Operational Principles align roughly with what the U.S. DOD refers to as doctrine; in other words, a series of situation-based discussions that serve as a guide to actions. These principles are intended to standardize language and understanding throughout the PLA. One important difference between Western doctrine and Chinese Strategic and Operational Principles is the importance that the PLA places on the interactions and relationship between the military and society. Western militaries are generally apolitical and consider themselves separate entities from civilian society, whereas the PLA codifies its place in Chinese society in these principles. In practice, this establishes the role of the PLA in shaping Chinese society and politics, explains the relationship between militias and the areas they serve, and describes the role of political officers in PLA units. A U.S. equivalent to Strategic Principles is JP 3-0, while an equivalent to Operational Principles is ADP 3-0.

## Combat Tactics and Regulations

Combat Tactics and Regulations address unit-level activities and tactical-level doctrine. Tactics and Regulations provide the basis for campaign operations, ensuring that all PLA units have a shared understanding of basic maneuvers, actions, and responses. Regulations in this context are not laws and rules as they are for the U.S. military; instead, they are similar to battle drills or other basic military actions. This mirrors the old naming convention for U.S. Army doctrine, as predecessors to ADP 3-0 and FM 100-5 were titled Field Service Regulations, Operations until the mid-1960s. This naming convention may also reflect greater centralization and top-down leadership. This approach may change in a similar way to the campaign model as the PLA moves toward a decentralized approach to leadership.

# Campaigns

Campaigns do not have a Western equivalent. A PLA campaign is a large-scale, operational-level effort that likely involves multiple services and a joint command, though it may also be conducted by units from a single service. Unlike Western theater-level operations that are generally objective-based, PLA campaigns are mission-based. PLA campaign types include—

- Joint blockade.
- Amphibious landing.
- Antiair raid.
- Mobile/maneuver warfare.
- Mountain offensive.
- Positional offensive.
- Positional defensive.
- Antiterrorism stability maintenance.
- Maritime group to destroy the enemy.
- Maritime interdiction.
- Offensive against coral reefs.
- Sea lane protection.
- Naval base defense.
- Air offensive.
- Airborne.
- Air defense.
- Conventional missile attack.

Each of these campaigns correlates to what U.S. forces would call a mission. In contrast, U.S. operations at this level tend to be largely based around an objective and a concept of operations. For example, the U.S.-led coalition operation, OPERATION IRAQI FREEDOM, had the objective of deposing the Ba'athist government and installing a stable democracy in Iraq, with the concept of operations being to defeat the Iraqi military, secure civil stability, and transition to a new government. The PLA's campaign approach suits its doctrine and philosophy well—it is more prescriptive, centralized, and narrow in scope, and consequently it requires less freedom of action from subordinates. It remains to be seen if this approach will be retained as the PLA moves toward more decentralized leadership and actions.

air defense systems, and it can include selective armored vehicles and critical logistics support means. Examples of tactical system warfare include using heavy rocket artillery to defeat or destroy enemy radars and artillery systems, EW to suppress or neutralize enemy command and communication networks, and deception operations to target enemy leadership's situational understanding and state of mind.

*See chap. 3, Tactical System Warfare for further discussion.*

The PLA's employment of system warfare supports the development of several traditional military strategies, such as preclusion, isolation, and sanctuary, throughout all domains and at all levels of war. Preclusion is achieved by keeping enemy commanders and forces off balance through asymmetric means, such as deception and information warfare, while simultaneously denying use of wide geographic areas through long-range reconnaissance-strike capabilities. Isolation is achieved by jamming or manipulating communications between units, employing psychological warfare to confuse and segregate enemy units from one another, then rapidly maneuvering to physically isolate them. Sanctuary is achieved through a mix of protection, defensive planning, information warfare, and deception operations. Sanctuary includes not only safety from physical attack, but safety from enemy information operations.

Chap 2

# III. PLA Joint Capabilities

Force Structure

*Ref: ATP 7-100.3, Chinese Tactics, (Aug '21), chap 3.*

The remainder of this chapter -- sections III(a) - III(f) -- describes People's Liberation Army (PLA) joint capabilities. The modern PLA is a sophisticated joint force, likely capable of conducting integrated operations across all domains. PLA joint operations are still in their relative infancy; the PLA has yet to employ a truly integrated joint force in combat, and many key joint integration enablers are still in the development stage. Nonetheless, PLA joint operations are likely to improve dramatically as integration technology advances.

## PLA Joint Forces

**A** **PLA Army (PLAA)**
*See pp. 2-25 to 2-38.*

**B** **PLA Navy (PLAN)**
*See pp. 2-39 to 2-44.*

**C** **PLA PLAN Marine Corps (PLANMC)**
*See pp. 2-45 to 2-46.*

**D** **PLA Air Force (PLAAF)**
*See pp. 2-47 to 2-52.*

**E** **PLA Rocket Force (PLARF)**
*See pp. 2-53 to 2-56.*

**F** **(Former) Strategic Support Force (SSF)**
*See pp. 2-57 to 2-58.*

*The proper name for China's military is the People's Liberation Army, or PLA. This organizational structure is comparable to the U.S. Department of Defense (DOD); it does not refer exclusively to ground forces, as does the U.S. Army. The PLA's land-based service is the People's Liberation Army Army (PLAA), the naval branch is the People's Liberation Army Navy (PLAN), and the aerial branch is called the People's Liberation Army Air Force (PLAAF). The PLA's strategic conventional and nuclear ballistic and cruise missile service is the People's Liberation Army Rocket Force (PLARF). The People's Liberation Army Strategic Support Force (PLASSF) is a new service-level organization that incorporates national-level cyber warfare, electronic warfare (EW), space, and other operational support elements under a single headquarters, and the People's Liberation Army Joint Logistics Support Force (PLAJLSF) is a new service-level sustainment organization.*

# Joint Logistic Support Force (JLSF)

*Ref: Department of Defense Annual Report to Congress: Military and Security Developments Involving the People's Republic of China 2024, pp. 72-73.*

## Key Takeaways

• The JLSF is concentrating improving joint strategic and campaign-level logistic efficiencies through training and integrating civilian products and services.

• The JLSF supports multimodal transportation methods to facilitate the movement of PLA forces and equipment for training.

• During wartime or mobilization, the JLSF probably will rely on its military representative offices (MROs) and dispatch centers to route military traffic.

## Service Roles and Missions

The PLA JLSF, established in 2016, supports integrated joint logistics for the PLA. The JLSF commander is Lt. Gen. Wang Liyan. The JLSF is directly subordinate to the CMC and is central to PRC efforts to build a joint, efficient "combat-oriented modern logistics system," which Beijing views as essential for modern warfare. The JLSF modernizes PLA strategic- and campaign-level logistics by overseeing theater-wide supply operations, establishing and coordinating support relationships among PLA service logistics elements, conducting joint logistics exercises with the PLA services, and integrating civilian logistics resources into military operations. JLSF exercises improve the PLA's ability to conduct joint logistics operations.

## Force Structure and Organization

The JLSF, headquartered at Wuhan Joint Logistics Support Base, is comprised of at least 100,000 personnel and supports integrated joint logistics for the PLA. It operates five joint logistic support centers (JLSCs) aligned with each theater command that are intended to streamline logistics support to the PLA. Under the JLSCs' control are units that provide materiel support to the PLA and joint logistics support BDEs, who focus exclusively on supporting mobile logistics during combat operations.

## Force Capabilities and Modernization

The JLSF provides the PLA with joint strategic- and campaign-level logistics, enabling large-scale operations. Elements of the JLSF who provide direct wartime support to the PLA fighting force are categorized as weapons and ammunition storage, warehousing, medical services, transport, fuel, engineering and construction management, reserve equipment, and procurement support. Since at least 2020, JLSCs have been incorporating a modular, mobile support team concept intended to provide logistics from rear depots to field depots during combat operations. These mobile support teams include logistics elements that provide ammunition, fuel, common supplies, vehicular maintenance, medical treatment, and food.

The JLSF integrates civilian resources and equipment into military operations and exercises, leveraging the PRC's civilian products, services, and transportation to improve resupply and move military personnel and equipment more rapidly. It does so primarily through MROs comprised of JLSF personnel who embed with civilian transportation companies (air, rail, road, and sea) to plan and manage military transportation. MRO personnel are vital to the JLSF's military-civilian integration efforts because they understand the technical loading procedures for various modes of transportation, can draft load plans, and provide expertise to reduce load times at points of embarkation. JLSF theater dispatch centers are the headquarters of regional MROs, coordinate PLA mobilization, and are tasked with improving the efficiency of transferring materials and forces between transportation modes. During wartime or mobilization, the JLSF probably will rely on its MROs and dispatch centers to route military traffic.

Chap 2

# A. Peoples Liberation Army Army (PLAA)

Force Structure

*Ref: Department of Defense Annual Report to Congress: Military and Security Developments Involving the People's Republic of China 2024, pp. 45-47.*

**Key Takeaways**

- The PLAA continues to modernize equipment and focus on combined arms and joint training to meet the goal of becoming a world-class military by 2049.
- The PLAA has continued to demonstrate long-range joint fires capability, such as during the April 2023 JOINT SWORD exercise in response to Taiwan's President Tsai Ing-wen's meeting with the U.S. Speaker of the House.

As the primary ground force in the PLA, the PLAA plays a key role in maintaining the PRC's national sovereignty, security, and development interests. The PLAA is tasked with transitioning from a regional defense to a global combat force, capable of multi-dimensional, trans-theater, multi-functional, sustained operations.

## I. Force Structure and Organization

The PLAA has approximately 965,000 active-duty personnel in its service. The PLAA is organized into five Theater Army Commands, the Xinjiang Military District, and the Tibet Military District. The PLAA has 13 group armies, which comprise multiple combined-arms BDEs that serve as the PLAA's primary maneuver force. The BDEs vary in size and composition. The PLAA delineates its combined-arms BDEs into three types: light (high-mobility, mountain, air assault, and motorized), medium (wheeled armored vehicles), and heavy (tracked armored vehicles and amphibious variants), with sizes ranging from approximately 4,500 to 5,000 personnel. Each GA controls six additional BDEs responsible for operational element functions: an artillery BDE, an air defense BDE, an army aviation (or air assault) BDE, a SOF BDE, an engineer and chemical defense BDE, and a sustainment BDE; however, some variations exist with at least one GA per theater separating its engineering and chemical defense BDEs into separate units. Although the PLAA has standardized its GAs, it does retain nonstandard divisions and BDEs that exist outside of the GAs. These units are typically located in areas the CCP considers sensitive, including Xinjiang, Tibet, Hong Kong, and Beijing. The PLAA commands border and coastal defense BDEs under the Theater Army Commands and regiments under the Xinjiang and Tibet Military Districts.

*See p. 2-27 for a map and overview of PLAA major ground units.*

## II. Capabilities and Modernization

To meet the stated ambitions to become a "world-class" military, the PLAA continues systems modernization and combined arms and joint training. However, it still employs a mix of modern and legacy military equipment. The PLAA continues focusing training on fighting as combined arms formations while adapting to the twice-yearly conscript induction change.

The PLAA's 15 SOF BDEs provide GA commanders with an organic unit capable of executing special operations. Typical PLAA SOF BDE missions include raids, harassment operations, target seizure and control, special reconnaissance, precision strike guidance, and rescue operations. PLAA Aviation and Air Assault units remained a focus of development in 2023. PLAA training events and reports in PRC media show that support to amphibious operations, multi-dimensional assaults, developing close air support tactics, and manned-unmanned teaming are now a standard part of training. PLAA air defense units in 2023 continued to improve their tactical air defense against low and slow threats like UAS and loitering munitions to meet evolving air defense requirements. Hybrid self-propelled air defense artillery systems, gun air defense artillery, small focused EW systems, and MANPADs form the core of the PRC's evolving solution to countering tactical UAS. The PLAA has continued to emphasize the fielding and application of modern EW capabilities designed to maximize a unit's combat effectiveness to detect, degrade, and disrupt adversary command and control (C2) communications and an adversary's ability to use its tactical ISR systems.



# III. Readiness

The PLAA continued to improve its methods and standards of training combined arms units. Training encompassed individual to collective soldier events integrating reconnaissance, infantry, artillery, armor, engineers, and signal units. In addition to continued PLAA deployments to the Indian border and Burma, the PLAA conducted multiple large-scale exercises in training areas throughout the country.

In April 2023, the PLA executed the JOINT SWORD exercise following Taiwan's President Tsai Ing-wen's meeting with the U.S. Speaker of the House. According to reports, the PCH-191 rocket launcher of the Army of the Eastern Theater Command (ETC) performed an exercise mission in the camp area/position of Xinshadi Village, Shishi City, Quanzhou, Fujian Province, and was loaded with 300 mm rockets. The PCH-191's use in the August 2022 U.S. CODEL visit and the 2023 JOINT SWORD exercise show the important role the system is expected to play in joint fires during a Taiwan campaign. In November of 2023, the PRC hosted the Peace and Friendship-2023 [Aman Youyi-2023] military exercises with five ASEAN countries. The multinational exercises were primarily focused on counterterrorism and anti-piracy.

# IV. PLAA Operational-Level Organizations

*Ref: ATP 7-100.3, Chinese Tactics, (Aug '21), pp. 2-6 to 2-18.*

The group army is the PLAA's basic operational-level organization. The group army structure is an evolution of the PLA's corps-based structure that comprised most of its history; it appears the group army is an attempt to retain most of the capabilities of the traditional corps, but with greater flexibility and ability to task-organize. Group armies are assigned to TCs; TC command authority is passed through TC PLAA headquarters to the group army headquarters. Group armies use the same dual-command structure as most other PLA units, employing both a military commander and a political commissar. Compared to geographic headquarters (MDs and LCs), it is likely that a group army's leadership is more heavily influenced by operational requirements, rather than local political relationships and responsibilities.

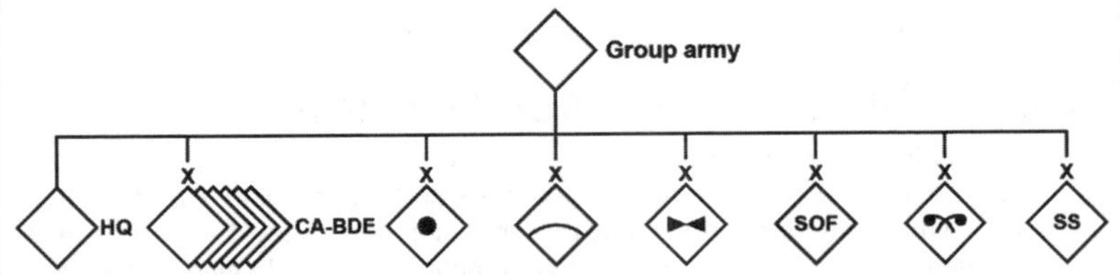


| | | | |
|---|---|---|---|
| CA-BDE | combined arms brigade | HQ | headquarters |
| SS | service support | SOF | special operations forces |

*Ref: ATP 7-100.3 (Aug '21), fig. 2-3. Group army structure (doctrinal).*

Following a complete overhaul and reorganization in 2017, each group army now directly commands 12 brigade-size organizations: six combined-arms brigades and

# PLA Army (PLAA)

*Ref: Department of Defense Annual Report to Congress: Military and Security Developments Involving the People's Republic of China 2024, p. 47.*

## Major Ground Units

*Note. The PLA understands rank in the same way that the U.S. military does, but it takes a fundamentally different view of grade. In the PLA, grade refers to one's duty position: platoon leader, brigade deputy leader, and so on. Hierarchy of grade is extremely important to the PLA—more so than rank. There are 15 different grades, and they are recognized across all of the PLA's different services. As such, a PLAA officer's grade can be directly compared with that of a PLAN officer, and seniority established between the two. In the U.S. military, grade refers to military pay, and duty position is clearly subordinate to rank when it comes to determining seniority.*

six support brigades of various types. Except for six legacy divisions, divisional headquarters have been eliminated, notionalized, or made strictly administrative. This overhaul seeks to eliminate excess command structure, reduce the number of general officers and associated staffs, and increase capabilities at tactical-level formations. The group army reorganization coincided with a massive drawdown in manpower across the PLA. This transition seeks to increase combined arms capability across the PLA through the application of improved technology and training, while simultaneously reducing PLA manpower and reducing quantities of obsolete equipment. A more standardized group army is a centerpiece of this effort.

The group army is likely not intended to be employed as an operational unit. Rather, it is the force pool from which operational systems are built as part of the wider system-warfare construct. Group army commanders facilitate the assembly of purpose-built operational systems, using their available force structure to create the command, maneuver, and support systems that execute operations in the group army's combat area. As such, concerns about the group army's large number of subordinate units—and the ability, or lack thereof, to control them—are not reflective of the PLAA's approach to building forces.

The main combat power of the notional group army consists of its six combined arms brigades (CA-BDEs). These brigades are supported by one artillery brigade; one air defense brigade; an aviation brigade; an SOF brigade; an engineer and chemical defense brigade; and a service support brigade, consisting of logistics, transportation, medical, repair, ammunition, communications, unmanned aircraft systems (UAS), and EW units. The 2017 reorganization placed a greater emphasis on system warfare capabilities at the group army level, providing a much more extensive suite of EW and cyber capabilities, long-range reconnaissance, and long-range fires under the direct control of group army commanders. Group army commanders can now support their assigned CA-BDEs with a significant suite of capabilities able to influence operations across all domains.

Group-army joint capabilities are limited, but they are expected to expand with ongoing reforms. TCs provide air, naval, and some SOF support, though group armies do typically contain an SOF brigade. The group army also appears to be largely dependent on higher commands—either the TC or PLAJLSF—for most of its logistics support. This emphasizes the lack of expeditionary capability throughout the PLA: sustainment within China's borders may be sufficient to support high-intensity operations, but current sustainment capabilities likely cannot support major combat operations overseas.

# V. PLAA Tactical-Level Organizations

The PLAA has completed a substantial period of reorganization with regard to its tactical unit structure. Traditionally, the PLAA was built around the division. Divisions were designed around the Soviet model. They were somewhat smaller and more homogenous than their U.S. counterparts, without the same combined arms and sustainment capabilities. In a transition similar to that undertaken by the U.S. Army, Ground Forces of the Russian Federation, and numerous other armies, the PLAA is in the process of "brigade-ization"—moving capabilities that used to reside at the division echelon to the brigade, with the goal of creating a more flexible force with a streamlined command structure, from group to army to brigade to battalion. This reorganization corresponds with the substantial drawdown in manpower ongoing throughout the PLA. PLAA divisions are now largely an administrative echelon—only six divisions remain operational, with none residing in group armies. Otherwise, frontline PLA units have moved entirely to a brigade-oriented structure.

*Unit descriptions in this section are notional, as described by the PLAA and supported by further intelligence analysis. Real-world units may vary widely in composition, organization, and equipment, though the PLAA is moving in the direction of force-wide standardization. It is important to note that PLAA units at battalion and below are designed to fight as structured, without the need for significant task organization.*

# CMC Command Authorities *(See pp. 2-8 to 2-9.)*

*Ref: ATP 7-100.3, Chinese Tactics, (Aug '21), pp. 2-5 to 2-6.*

CMC command authority is exercised operationally through theater commands (TCs). TCs are joint commands that exercise extensive—though not exclusive—command authority in their given region.

*See pp. 2-8 to 2-9 for further discussion of the theater commands.*

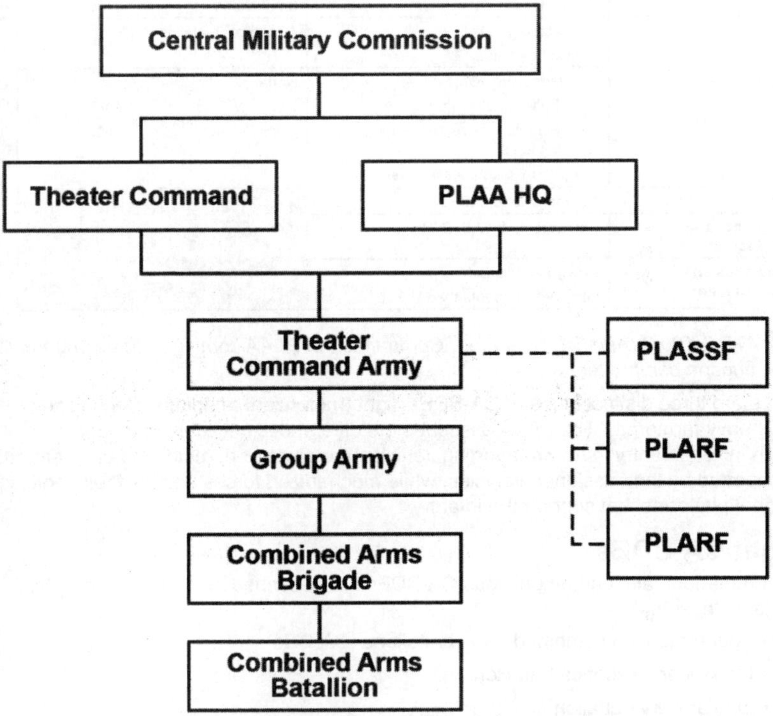


***Note.*** The theater command army has a duel reporting structure. In addition, support from other service-level organizations integrates at the theater command army level.

| | |
|---|---|
| HQ | headquarters |
| PLAA | People's Liberation Army Army |
| PLAJLSF | People's Liberation Army Joint Logistics Support Force |
| PLARF | People's Liberation Army Rocket Force |
| PLASSF | People's Liberation Army Strategic Support Force |
| ——— | command relationship |
| - - - - | support relationship |

*Ref: ATP 7-100.3 (Aug '21), fig. 2-2. PLAA chain of command, isolated.*

# A. Combined Arms Brigades (CA-BDE)

*Ref: ATP 7-100.3, Chinese Tactics, (Aug '21), pp. 2-7 to 2-9.*

The CA-BDE mixes different capabilities under a single headquarters: maneuver, fires, logistics, communications, engineer, and EW subordinate units are organically assigned to CA-BDE headquarters. CA-BDEs do not appear to have any organic joint capability, but they are developing joint capabilities to be able to control units from other services during training and operations.

| | *PLAA Combined Arms Brigade* | *U.S. Brigade Combat Team (BCT)* |
|---|---|---|
| **Maneuver** | 4–6 battalions | 3 battalions |
| **Artillery** | 1 howitzer battalion<br>1 rocket battalion | 1 howitzer battalion |
| **Air defense** | 3 batteries, including:<br>• Self-propelled guns.<br>• Man-portable air defense systems.<br>• Short-range missile systems. | None |
| **Reconnaissance** | Comparable to U.S. BCT | |
| **Engineer and protection** | Comparable to U.S. BCT | |
| **Logistics and sustainment** | Newly established<br>Likely less capable than U.S. BCT | |

*Ref: ATP 7-100.3 (Aug '21), table 2-1. Comparison of PLAA combined arms brigade to U.S. brigade combat team.*

There are three distinct types of CA-BDEs: light (motorized), medium (mechanized), and heavy (armored). The PLAA describes the differences between motorized and mechanized infantry in how supporting vehicles are employed: motorized units are only transported by their assigned vehicles, while mechanized forces employ their vehicles as combat platforms that support the infantry.

## Light CA-BDEs

Airborne, mountain, and amphibious CA-BDEs are described as light. The light CA-BDE contains these units—

- Four motorized combined arms battalions (CA-BNs).
- One reconnaissance battalion.
- One artillery battalion.
- One air defense battalion.
- One headquarters unit.
- One operational support battalion.
- One service support battalion.

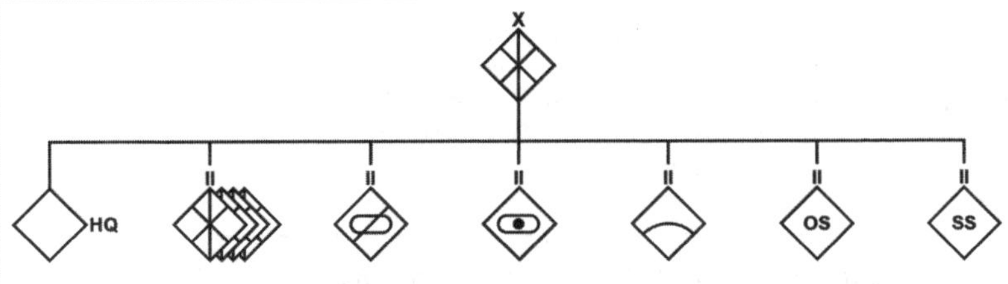


HQ headquarters OS operational support SS service support

*Ref: ATP 7-100.3 (Aug '21), fig. 2-4. Light combined arms brigade (doctrinal).*

## Medium CA-BDE

The medium CA-BDE contains these units—

- Four mechanized CA-BNs.
- One reconnaissance battalion.
- One artillery battalion.
- One air defense battalion.
- One headquarters unit.
- One operational support battalion.
- One service support battalion.

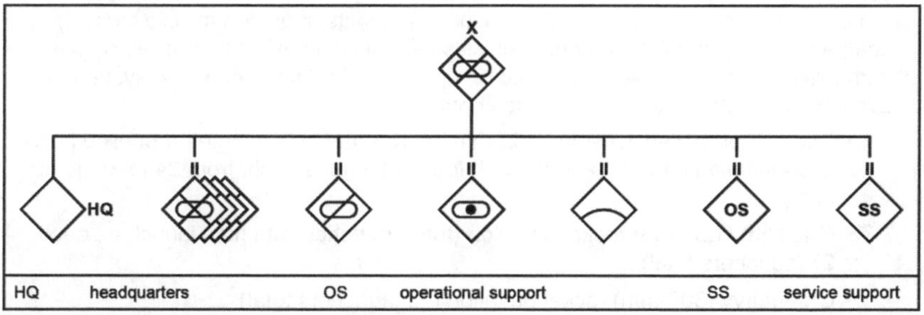


*Ref: ATP 7-100.3 (Aug '21), fig. 2-5. Medium combined arms brigade (doctrinal).*

## Heavy CA-BDE

The heavy CA-BDE contains these units—

- Four armored CA-BNs.
- One reconnaissance battalion.
- One artillery battalion.
- One air defense battalion.
- One headquarters unit.
- One operational support battalion.
- One service support battalion.

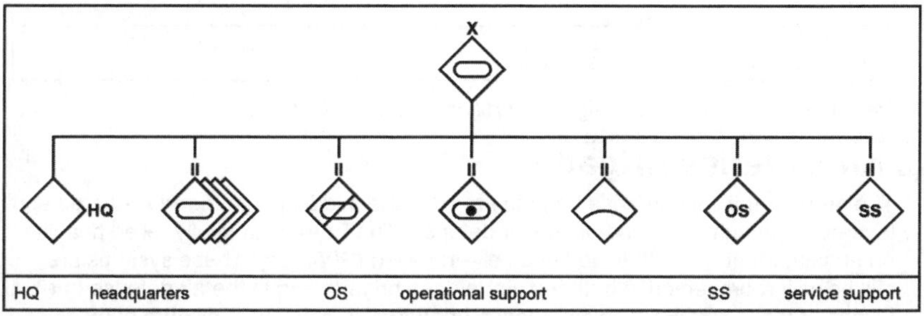


*Ref: ATP 7-100.3 (Aug '21), fig. 2-6. Heavy combined arms brigade (doctrinal).*

## B. Artillery Brigade

The artillery brigade in the group army employs a variety of towed guns, self-propelled guns (SPGs), light (122-mm) and heavy (300-mm) rocket artillery systems, and antitank and assault vehicles. These systems are employed to mass fires on critical targets, reinforce fires at lower echelons (chiefly the CA-BDE), deter or deny enemy actions, and offset enemy advantages in close combat. Artillery brigade assets may be employed in direct or general support of CA-BDEs. Artillery battalions include organic surveillance and target acquisition assets, including UAS, electronic intelligence (ELINT) systems, and traditional long-range visual forward-observation platforms. It is not clear how effectively the PLAA can task-organize fires; it traditionally preferred to centralize fires in order to maximize the effects of mass, but the movement toward smaller tactical formations and modularization requires that lower echelons be capable of employing effective fire support. In addition to its indirect fire capability, each artillery battalion includes an antitank guided missile (ATGM) company, employing light armored vehicles mounted with ATGMs. Artillery brigade composition varies significantly based on operational requirements and system availability. A notional artillery brigade contains—

- Two self-propelled 122-mm, 152-mm, or 155-mm towed or self-propelled howitzer battalions (three batteries with four to six guns each, from 24 to 36 guns total).
- One light (122-mm) rocket battalion (three batteries with nine launchers each, 27 launchers total).
- One heavy (300-mm) rocket battalion (12 launchers total).
- One target acquisition battery.
- One UAS company.
- One command battery.
- One support company.

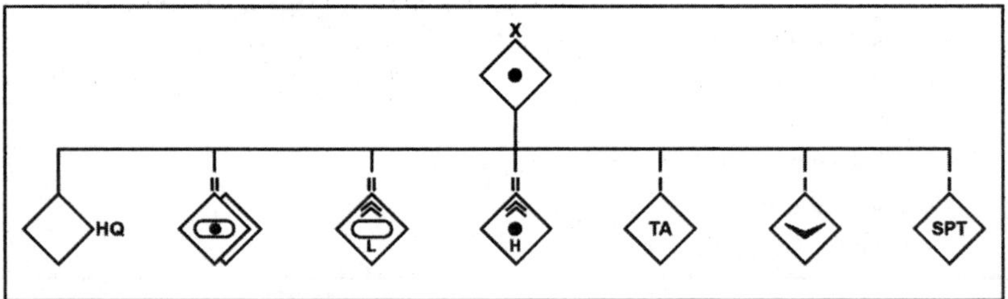


H heavy
HQ headquarters
L light
SPT support
TA target acuisition

*Ref: ATP 7-100.3 (Aug '21), fig. 2-7. Artillery brigade (doctrinal).*

## C. Air Defense Brigade

The group army's air defense brigade provides the middle tier in the PLA's tiered and layered approach to ground-based air defense. The PLAA historically relied heavily on antiaircraft guns, both towed and self-propelled (SPAAGs). These systems are still heavily represented in both the air defense brigade and in the air defense battalions of the CA-BDEs. The air defense brigade also employs a mixture of medium-range radar missile systems, short-range missile systems, and sensors designed to support engagement of aerial targets. It is not known to what extent these tactical air defense systems integrate with the much-larger theater-wide air defense networks employed by TCs, but it is likely that at least the medium-range surface-to-air missile (SAM) battalion has some sort of digital integration with a higher air defense echelon. As with artillery systems, air defense units may be employed in mass to defend critical assets from air attack, or they may be detached in direct or general support to subordinate CA-BDEs. Each brigade may also include an electronic air defense battalion, integrating EW assets with gun systems. As with artillery systems, air defense brigade composition varies significantly based on operational requirements and system availability. A notional air defense brigade contains—

- One medium-range SAM battalion (three batteries consisting of one radar section and three launcher sections per battery).
- Two short-range air defense (SHORAD) battalions (three batteries consisting of six to eight SPAAGs and four to six short-range SAM launchers per battery).
- One electronic air defense battalion (three batteries consisting of varied EW capabilities).
- One command battery.
- One support company.

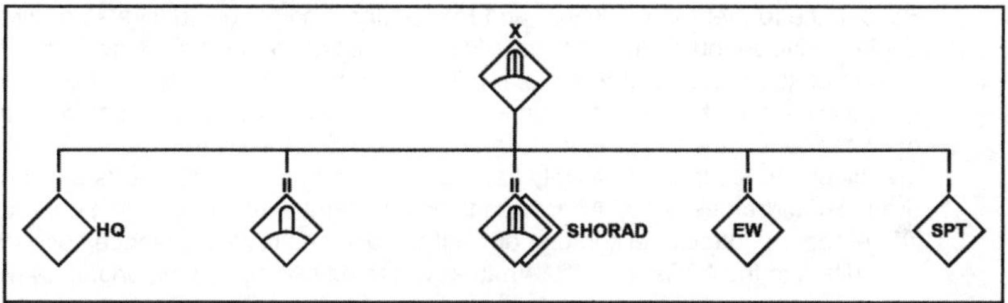


EW electronic warfare
SPT support
HQ headquarters
SHORAD short range air defense

*Ref: ATP 7-100.3 (Aug '21), fig. 2-8. Air defense brigade (doctrinal).*

## D. Engineer and Chemical Defense Brigade

The group army's engineer and chemical defense brigade is responsible for protection, mobility, countermobility, and chemical defense and obscuration functions within the group army's combat area. Historically, the PLA invested heavily in its engineer units due not only to its emphasis on mobility and countermobility, but also for contingency missions in support of disaster relief or other civil emergencies. This employment strategy is likely to continue. PLAA engineer brigades include mining, countermining, obstacle, bridging, smoke, chemical defense, and heavy maintenance units. These units provide much of the PLA's support to international peacekeeping and development missions.

## E. Service Support Brigade

The service support brigade includes logistics, transportation, medical, repair, ammunition, command and communication, UAS, and EW units. In addition to traditional combat support roles, the brigade is postured to perform signal support and EW missions. The service support brigade's signal element employs both traditional and network communications throughout the group army's combat area. Previously, communications were handled largely at the division, with minimal communications architecture being employed at lower echelons. With the PLAA's emphasis on information operations, signal capabilities were significantly enhanced at tactical echelons. The signal element has responsibility for providing network connectivity to PLAA units and protecting these communications networks from jamming, electromagnetic attack, and cyber intrusion. The EW regiment represents a serious investment in EW and cyber capability to be employed at the operational and tactical levels. It includes a jamming and electromagnetic attack section, a long-range electronic surveillance section, an electromagnetic protection section, a network operations section, and a communications operations section. It is unclear how the

PLA will allocate permissions and responsibilities to the EW regiment, though it can be safely assumed that rules of engagement will be more permissive than they are in Western militaries. It can also be assumed that the EW regiment will work closely with the artillery and air defense brigades in order to target long-range artillery fires in a construct similar to the Russian reconnaissance-fire system.

*Note. The PLAA describes signal and communications capabilities under the umbrella of command and communication.*

## F. Army Aviation Brigade

The army aviation brigade (AAB) provides the group army's rotary-wing aviation capability. The AAB's development is likely influenced by the U.S. Army's combat aviation brigade, though the AAB has far fewer airframes than does its U.S. counterpart. The AAB is built around medium-lift transport helicopters and scout and attack helicopters. The PLAA uses a number of different helicopter types: foreign systems from Russia, the former Soviet Union, and Europe, and indigenous designs are all present in significant numbers. The density of helicopters in the PLAA is relatively low as compared to the size of the rest of the force; the U.S. Army's ratio of helicopters to troops is over ten times that of the PLAA. Rotary-wing capability is a significant area of investment in the PLA's modernization strategy; it is likely that the AABs will be greatly expanded in the relatively near future, using mainly indigenous or license-produced airframes. The AAB is organized into eight battalions, mixing general-utility lift helicopters, attack helicopters, and light attack and reconnaissance helicopters. In addition to the AABs, the PLAA has also formed two air assault brigades, which include either two or three infantry battalions and up to six medium-lift helicopter battalions. AABs train frequently with both light CA-BDEs and SOF. A notional AAB organization contains—

- Four general-utility battalions (eight to 12 medium-lift or general-utility helicopters per battalion).
- Two attack battalions (eight attack helicopters per battalion).
- One reconnaissance battalion (eight reconnaissance or light attack helicopters).
- One headquarters unit.
- One aviation support battalion.

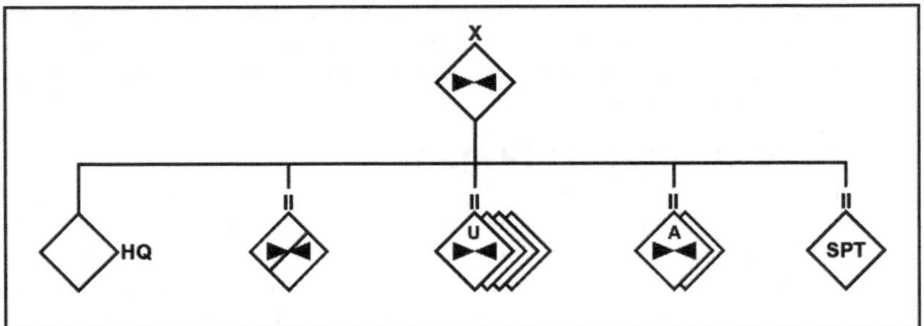


| | | | |
|---|---|---|---|
| A | attack | SPT | support |
| HQ | headquarters | U | utility |

*Ref: ATP 7-100.3 (Aug '21), fig. 2-9. Army aviation brigade (doctrinal).*

## G. Special Operations Forces Brigade

The SOF brigade provides the group army with an organic SOF capability. Group-army SOF brigade operations generally do not focus on training or interacting with foreign militaries; they instead focus on direct action—deep reconnaissance and commando operations in support of group army operations. Group army SOF brigades are highly specialized to operate in their specific theater. Mountainous theaters focus on alpine training, coastal theaters focus on amphibious operations, and urban theaters focus on urban operations. SOF brigades are equipped with organic surveillance UASs, likely to aid target acquisition in support of heavy artillery and rocket fires. Group army SOF brigades are more similar to U.S. Army Ranger light infantry and long-range reconnaissance units than traditional SOF; their primary mission is to support the group army's combined arms maneuver operations. SOF brigades habitually train with PLAA aviation units, but they do not enjoy the wide variety of fixed- and rotary-wing capabilities available to U.S. SOF units.



## H. Combined Arms Battalion

The CA-BN is a very new development in the PLAA organization. The CA-BN takes the basic combined arms approach used to build the CA-BDE and applies it to the battalion echelon. CA-BNs appear to only combine different maneuver elements along with organic short-range fires elements (assault guns and mortars), with the provision that CA-BDE headquarters can attach elements from other brigade organizations as required. The CA-BN is very similar to the U.S. battalion task force concept employed by mechanized and armored units since the World War II era, mixing company-level infantry and armor units to create a single combined arms command. Each CA-BN also houses an organic SHORAD capability in the form of man portable air defense systems (MANPADS). The PLAA breaks CA-BNs into three primary categories: light, medium, and heavy. Of note, the CA-BN appears to have only limited staff, which may affect its ability to function as the PLAA intends—as an independent unit.

A light infantry CA-BN contains—

- Three motorized infantry companies (10 light wheeled vehicles or APCs per company).
- One firepower company (six to nine rapid-fire 81-mm mortars, the form of man-portable air defense, and crew-served weapons).
- One headquarters unit.
- One service support company.

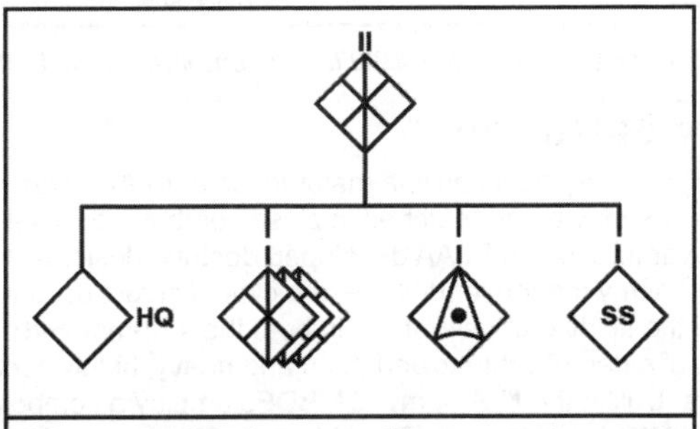


*Ref: ATP 7-100.3 (Aug '21), fig. 2-10. Light combined arms battalion (doctrinal).*

A medium CA-BN contains—

- Three mechanized infantry companies (10 wheeled or tracked IFVs per company).
- One assault gun company (14 wheeled 105-mm assault guns).
- One firepower company (six to nine rapid-fire 120-mm self-propelled mortars, MANPADS, and crew-served weapons).
- One headquarters unit.
- One service support company.

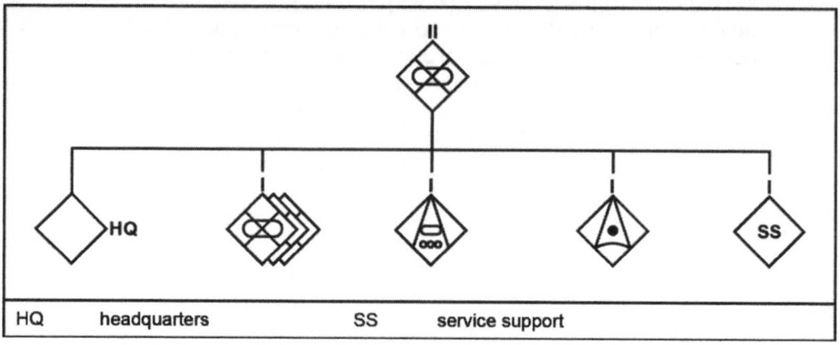


*Ref: ATP 7-100.3 (Aug '21), fig. 2-11. Medium combined arms battalion (doctrinal).*

A heavy CA-BN contains—

- Two tank companies (10-14 tanks per company).
- Two mechanized infantry company (10 IFVs per company).
- One firepower company (six to nine rapid-fire 120-mm self-propelled mortars, MANPADS, and crew-served weapons).
- One headquarters unit.
- One service support company.

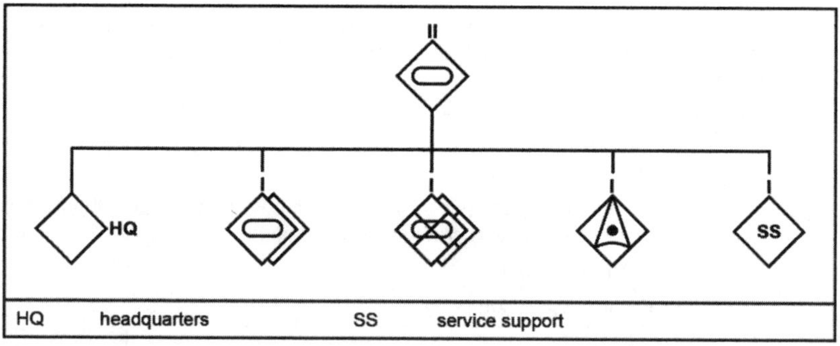


*Ref: ATP 7-100.3 (Aug '21), fig. 2-12. Heavy combined arms battalion (doctrinal).*

# I. Artillery Battalion

The CA-BDE's artillery battalion is a major focus of PLAA investment. Recognizing that PLAA forces may be outmatched in close combat with other industrialized nations' maneuver forces, the PLAA developed doctrine designed to defeat, suppress, or neutralize enemy maneuver units before close contact occurs. In practice, this necessitated the fielding of new SPGs in both the 122-mm and 155-mm class, along with a series of older 152-mm towed guns and heavy, highly capable rocket artillery platforms. Unlike the U.S. Army, CA-BDEs employ a composite of rocket and tube artillery at the brigade level. CA-BDEs can be reinforced with guns and rocket artillery from their respective TC's artillery brigade. Of note, the artillery battalion contains a robust antitank capability, consisting of light armored ATGM vehicles and several towed antitank guns. An artillery battalion contains—

- Two to three 122-mm or 155-mm SPG batteries (three platoons with three guns each, 18-27 guns total).
- One light (122-mm) rocket battery (three platoons with three launchers each, nine launchers total).
- One ATGM company (nine vehicle launchers or antitank guns).
- One command battery.
- One support company.

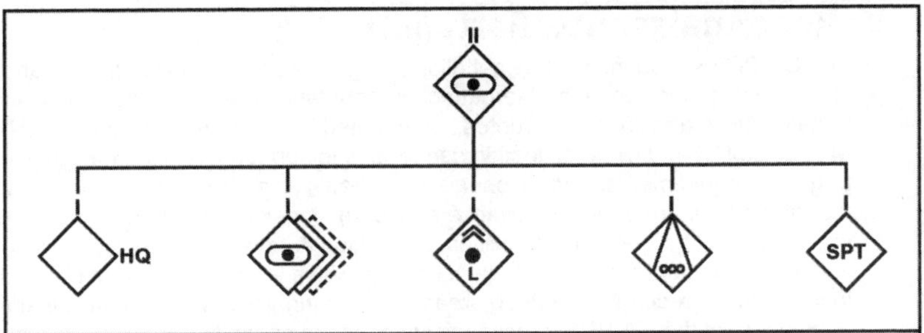


HQ headquarters L light SPT support

*Ref: ATP 7-100.3 (Aug '21), fig. 2-13. Artillery battalion (doctrinal).*

## J. Air Defense Battalion

The CA-BDE's air defense battalion provides low-altitude air defense in the CA-BDE's airspace. PLAA tactical air defense systems are generally mobile, lightweight, and within visual range. They consist of a mixture of guns and short-range missiles, including both vehicle mounted and MANPADS systems. The air defense battalion does not appear to have a robust long-range detection or early warning capability; integration with higher-level air defense units or PLAAF units is limited. Thus, the air defense battalion can only provide point, within-visual-range air defense against low-flying targets. The air defense battalion's primary target set is helicopters, unmanned aircraft, and low-flying aircraft. Fires are likely controlled locally through visual aircraft identification and weapons control statuses. Radar detection is limited to short-range organic radars mounted on weapons systems. The air defense battalion's systems are mobile, and they move closely with CA-BDE units while maneuvering. Gun systems are dual purpose and can engage surface targets if necessary. An air defense battalion contains—

- Three SPAAG batteries (six guns per battery, 18 guns total).
- One SHORAD battery (eight systems per battery and possibly one to two radar systems).
- One command battery.
- One support company.

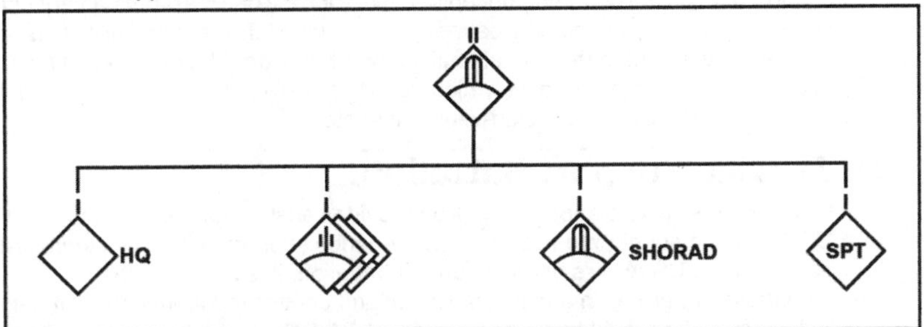


HQ headquarters SHORAD short-range air defense
SPT support

*Ref: ATP 7-100.3 (Aug '21), fig. 2-14. Air defense battalion (doctrinal).*

Force Structure

## K. Reconnaissance Battalion

The CA-BDE's reconnaissance battalion provides multi-domain reconnaissance, intelligence, surveillance, and target acquisition capabilities. The reconnaissance battalion uses a mixture of mounted, dismounted, UAS, human intelligence (HUMINT), ELINT, and cyber collection capabilities to support CA-BDE intelligence and targeting requirements. The reconnaissance battalion's structure is very similar to that of the U.S. Army reconnaissance squadron, consisting of three troops employing a mixture of collection capabilities. In addition to traditional scouts, the reconnaissance battalion may include an element of SOF-like troops, capable of conducting independent direct action in deep areas. The reconnaissance battalion is a critical component of the CA-BDE's reconnaissance-strike capability, providing much of the deep intelligence needed to support targeting for both the CA-BDE's organic surface-to-surface fires and any supporting or reinforcing fires from the artillery brigade. The PLAA places a very high priority on reconnaissance in general. In order to shape the mindset of one's opponents, one must know as much as possible about their disposition and intentions. It should also be noted that group armies do not have a dedicated ground reconnaissance unit as do U.S. corps; the group army's SOF brigade provides much of the ground intelligence in support of group army operations. A reconnaissance battalion contains—

- Two reconnaissance troops (mounted; six to 10 light armored vehicles per troop).
- One battlefield surveillance company (all-source intelligence organization).
- One UAS company (likely containing two to three aircraft).
- One headquarters unit.

## L. Operational Support Battalion

The CA-BDE's operational support battalion provides mobility, countermobility, and protection capabilities, along with an EW company, a command and communication company, a chemical defense company, and a security (military police) company. It is similar—though not identical—to the U.S. Army's special troops battalion. The operational support battalion likely operates in a highly decentralized manner, task-organizing support elements to other battalions within the CA-BDE. Operational support battalions are tailored to the needs of their parent unit; heavier CA-BDEs require heavier engineer support. One of the most important missions of the operational support battalion is supporting deception operations. The PLAA puts a very high priority on camouflage and concealment, and much of this responsibility falls to the engineering element of the operational support battalion. The operational support battalion also employs a relatively high density of vehicles, representing a significant portion of the CA-BDE's wheeled-vehicle inventory.

## M. Service Support Battalion

The CA-BDE's service support battalion provides sustainment support for the CA-BDE, including supply, medical, and maintenance support. It is likely structured similarly to the US Army's brigade support battalion. It is unclear if the service support battalion operates in a more traditional and centralized manner, or if it habitually task-organizes itself to support individual CA-BNs. It can be assumed that, in order to support the CA-BN concept, greater logistics decentralization must occur, but PLAA logistics infrastructure may not support this approach without significant augmentation.

# B. Peoples Liberation Army Navy (PLAN)



*Ref: ATP 7-100.3, Chinese Tactics, (Aug '21), pp. 3-3 to 3-5 and Office of the Secretary of Defense Annual Report to Congress: Military and Security Developments Involving the People's Republic of China 2021, pp. 48-57.*

The PLAN is the largest navy in the world with a battle force of over 370 platforms, including major surface combatants, submarines, ocean-going amphibious ships, mine warfare ships, aircraft carriers, and fleet auxiliaries. Notably, this figure does not include approximately 60 HOUBEI class patrol combatants that carry anti-ship cruise missiles (ASCM). The PLAN's overall battle force is expected to grow to 395 ships by 2025 and 435 ships by 2030. Much of this growth will be in major surface combatants. The PLAN's force structure consists of theater navies with subordinate submarine flotillas, surface ship flotillas, aviation units, and naval bases. The PLAN's Northern Theater Navy is subordinate to the NTC, the Eastern Theater Navy is subordinate to the ETC, and the Southern Theater Navy is subordinate to the STC.

Numerically, the PRC has the largest navy in the world with a battle force of over 370 ships and submarines, including over 140 major surface combatants. The PLAN is largely composed of modern multi-mission ships and submarines. As of mid-2024, the PLAN is outfitting the 9th and 10th RENHAI CG units. As many as 10 LUYANG III MOD DDGs are in various stages of construction, outfitting, or sea trials. The first two JIANGKAI III FFGs will conduct sea trials during summer 2024. The PRC will build more of these units as well as the smaller and less capable JIANGKAI II MOD FFG, at least two of which are under construction as of mid-2024. The PLAN is outfitting a fourth YUSHEN LHA which launched in late 2023 and will likely begin sea trials in the second half of 2024. The PLAN began construction on the new YULAN class (Type 076) LHA by early 2024, which is expected to be fitted with an electromagnetic catapult for UAVs. The PLAN's third aircraft carrier, Fujian, began its sea trials in the first half of 2024 and remains on track for a 2025 operational capability.

The PRC has long challenged foreign military activities in its territorial seas and EEZ in a manner inconsistent with customary international law in the UNCLOS. However, in recent years, the PLA has begun conducting the same types of military activities inside and outside the FIC in the EEZs of other countries, including the United States. This activity highlights the PRC's double standard in the application of its interpretation of international law. Examples include sending intelligence collection ships to collect on military exercises, such as the Rim of the Pacific exercises off Hawaii in 2014, 2018, and 2022, and the TALISMAN SABER exercises off Australia in 2017, 2019, and 2021, as well as operating near Alaska in 2017 and 2021. PRC intelligence collection ships operated near sensitive defense facilities off Australia's west coast in May 2022 and near Japan in July 2022. PRC survey ships are extremely active in the SCS, frequently operating in the claimed EEZs of other nations in the region, such as the Philippines, Vietnam, and Malaysia.

The PRC's 2019 defense white paper described the PLAN as adjusting to changes in the strategic requirements of near seas defense and far seas protection, noting that it was "speeding up the transition of its tasks from defense on the near seas to protection missions on the far seas..." Toward the PRC's goal of building a "strong and modernized navy force," the PLAN has replaced or updated its previous generations of platforms that had limited capabilities in favor of larger, modern multi-mission combatants. Today, the PLAN is largely composed of modern multi-role platforms featuring advanced anti-ship, anti-air, and anti-submarine weapons and sensors. The PLAN is emphasizing maritime joint operations and joint integration with the other branches of the PLA. This modernization aligns with the PRC's growing emphasis on the maritime domain and increasing demands for the PLAN to operate at greater distances from mainland China.



The PLAN organizes, mans, trains, and equips the PLA's naval and naval aviation forces as well as the PLANMC, which is subordinate to the PLAN. The PLAN continues to implement structural reforms that began in late 2015 and early 2016. The PLA-wide reforms removed the PLAN headquarters from conducting operations, which became the purview of the PLA's joint theater commands, focusing it instead on organizing, manning, training, and equipping naval forces.

## PLAN Roles and Missions

The PLAN continues to develop into a global force, gradually extending its operational reach beyond East Asia into a sustained ability to operate at increasingly longer ranges, including a continuous presence in the Gulf of Aden. The PLAN's latest surface and subsurface platforms enable combat operations beyond the reach of the PRC's land-based defenses. In particular, the PRC's growing force of aircraft carriers extend air defense coverage of deployed task groups beyond the range of land-based defenses, enabling operations farther from China's shore. The PLAN's emerging requirement for sea-based land-attack systems will enhance the PRC's ability to project power. Furthermore, the PLAN has a sizable force of highly capable logistical replenishment ships to support long-distance, long-duration deployments, including two new FUYU class fast combat support ships (AOEs) built specifically to support aircraft carrier and large-deck amphibious ship operations. The PLAN's expanding fleet of large modern amphibious warships enable it to conduct a wide range of expeditionary operations wherever PRC interests are threatened or in support of PRC participation in internationally sanctioned operations. The expansion of naval operations beyond the PRC's immediate region facilitates its NWMA and further legitimizes the PRC's growing global military posture, including at its base in Djibouti.

The PRC is in the beginning stages of operating its "multi-carrier force." The PRC continues to learn lessons from operating its first aircraft carrier, Liaoning. Its first domestically-built aircraft carrier, Shandong, was launched in 2017 and commissioned in December 2019. The PRC's next generation of carriers, the new FUJIAN class—the first of which it is outfitting—will have greater endurance and an electromagnetic catapult launch system, making them capable of launching various types of specialized fixed-wing aircraft for missions, such as airborne early warning (AEW) and EW. This will increase the striking power of a potential PLAN carrier battle group when deployed to areas beyond the PRC's immediate periphery. CV-18 Fujian launched in June 2022 and is expected to enter operational service in 2025.

The PLAN's newest amphibious ships, the YUSHEN LHA and YUZHAO amphibious transport docks (LPD), are modern platforms capable of regional and global expeditionary missions in support of wartime and non-war contingency operations singly or as part of capable and flexible task groups composed of multiple amphibious ships and surface combatants. An even larger expeditionary ship, the YULAN LHA (Type 076) is expected to join the fleet in the second half of the decade and provide additional sea control via catapult-launched UAVs.

The PLAN continues to build multiple new, large auxiliary ships to support force projection operations, including large logistic ships, such as the FUYU AOE, the large DAKAI and DAFENG-class submarine support ships, and specialized platforms for intelligence collection ships and ocean surveillance ships.

The PLAN's ability to perform missions beyond the FIC is modest but growing as it gains more experience operating in distant waters and acquires larger and more advanced platforms.

# PLA Navy (PLAN)

*Ref: Department of Defense Annual Report to Congress: Military and Security Developments Involving the People's Republic of China 2024, p. 52.*

## Major Naval Units

### Key Takeaways

- The PLAN is continuing construction of RENHAI class guided-missile cruisers (CG), LUYANG III MOD class guided-missile destroyers (DDG), and JIANGKAI II MOD and JIANGKAI III classes of guided-missile frigates as of the first half of 2024.
- A fourth YUSHEN (Type 075) class amphibious assault ship (LHA) is outfitting as of early 2024 and will begin sea trials in late summer. The PLAN began constructing the new YULAN class (Type 076) LHA in early 2024, which will be fitted with an electromagnetic catapult for UAVs.
- In 2022, the PLAN launched its third aircraft carrier, CV-18 Fujian, which began sea trials in mid-2024 and is expected to enter operational service in 2025.

# I. Submarines

The PRC continues to increase its inventory of conventional submarines capable of firing advanced ASCMs. Between the mid-1990s and mid-2000s, the PLAN purchased 12 Russian-built KILO class SS units, eight of which can launch ASCMs. PRC shipyards have delivered 13 SONG class SS units (Type 039) and 21 YUAN class diesel-electric air-independent propulsion attack submarines (SSP) (Type 039A/B). The PRC is expected to produce 25 or more YUAN class submarines by 2025. In late 2021, the PLAN retired the first two KILO class submarines (both non-ASCM capable) purchased from Russia in the 1990s.

Over the past 15 years, the PLAN has constructed 12 nuclear-powered submarines—two SHANG I class SSNs (Type 093), four SHANG II class SSNs (Type 093A), and six JIN class SSBNs (Type 094). Equipped with the CSS-N-14 (JL-2) submarine-launched ballistic missile (SLBM) (3,900 nm) or the CSS-N-20 (JL-3) SLBM (5,400 nm), the PLAN's six operational JIN class SSBNs represent the PRC's first credible sea-based nuclear deterrent. Each JIN class SSBN can carry up to 12 missiles. In 2019, Beijing displayed these missiles at the PRC's 70th founding anniversary parade. The PRC's next-generation Type 096 SSBN will reportedly be armed with follow-on longer range SLBM. The Type 096 will likely begin construction soon. Considering the projected 30-plus-year service life of the platforms, the PRC will operate its JIN and Type 096 SSBN fleets concurrently in the 2030s. This would align with Xi's 2018 directive for the SSBN force to achieve "stronger growth."

The PRC launched four SHANG III class (Type 093B) guided-missile nuclear-powered attack submarines (SSGN) between May 2022 and January 2023 and could have three hulls of this class operational by 2025. This new SHANG class variant enhances the PLAN's anti-surface warfare capability and could provide a clandestine land-attack option if equipped with LACMs. The PLAN is improving its anti-submarine warfare (ASW) capabilities through the development of its surface combatants and special mission aircraft but continues to lack a robust deep-water ASW capability.

# II. Surface Combatants

The PLAN remains engaged in a robust shipbuilding program for surface combatants. As of early 2024, the PLAN is building a new batch of CG, DDG, and two different classes of guided-missile frigates, including a larger, more capable variant: designated JIANGKAI III (054B). These assets will significantly upgrade the PLAN's air defense, anti-ship, and anti-submarine capabilities and will be critical as the PLAN expands its operations beyond the range of the PLA's shore-based air defense systems. By the end of 2023, the PLAN had commissioned its 40th JIANGKAI II class FFG and is building additional units with minor updates. The PLAN augmented its littoral warfare capabilities, especially for operations in the ECS and SCS, with the high-rate production of the JIANGDAO class corvettes (FFLs) (Type 056 and Type 056A). The PLAN commissioned the 72nd JIANGDAO in February 2021, completing the production run. The PLAN subsequently transferred the early flight Type 056 variants, likely 22 ships total, to the CCG in 2021, probably due to the early models' lack of towed-array sonar. The remaining JIANGDAOs (056A) are equipped with a towed-array sonar and are, thus, capable of contributing to ASW operations. The PRC continues to operate 60 HOUBEI (Type 022) missile boats, largely in a coastal defense role.

The PLAN has expanded its force of large surface combatants with two programs, the LUYANG III DDG and the RENHAI CG. By the end of 2023, the PRC had commissioned 25 LUYANG III DDGs—including 12 lengthened LUYANG III MOD DDGs—with additional hulls under construction. Both the standard LUYANG III and the LUYANG III MOD have a 64-cell multipurpose vertical launch system (VLS) capable of launching cruise missiles, SAMs, and anti-submarine missiles, and the

Mod variants can carry the new Z-20 anti-submarine helicopter. By mid-2024, eight RENHAI-class CGs were in commission in the PLAN with additional hulls under construction. The RENHAI has 112 VLS cells and can carry a large load out of weapons, including ASCMs, SAMs, torpedoes, and anti-submarine weapons along with likely LACMs and anti-ship ballistic missiles (ASBM) when those become operational. In early 2022, the PLAN released a video of RENHAI CG test launching an ASBM with a reported/estimated range of 540 nm. The new ship-launched ASBM can possibly be launched by the LUYANG III and LUYANG III MOD DDGs.



The PLAN continues to emphasize anti-surface warfare capabilities in its force developments. The PLAN's frigates and corvettes, as well as modernized older combatants, carry variants of the YJ-83/YJ-83J ASCM (135 nm) while newer surface combatants, such as the LUYANG II-class DDGs, are fitted with the YJ-62 (270 nm). The LUYANG III class DDGs and the RENHAI class CGs are fitted with a variant of the PRC's newest ASCM, the YJ-18A (290 nm). A few modernized destroyers have been retrofitted with the supersonic YJ-12A ASCM (270 nm), and the next-generation frigates may receive this missile. Eight of the PLAN's 12 KILO class SSs are equipped with the Russian built SS-N-27b ASCM (120 nm). The PRC's SONG-class SS, YUAN class SSP, and SHANG class SSN field the PLAN's newest domestic submarine-launched ASCM, the YJ-18, which constitutes an improvement over the SS-N-27b ASCM. It is possible the PRC is developing a launcher that can fit inside a standard commercial shipping container for covert employment of the YJ-18 aboard merchant ships.

The PLAN recognizes that long-range ASCMs require a robust, over-the-horizon (OTH) targeting capability to reach their full potential. To fill this capability gap, the PLA is investing in joint reconnaissance, surveillance, command, control, and communications systems at the strategic, operational, and tactical levels for high-fidelity targeting information for surface and subsurface launch platforms.

As the PLAN continues to transition into a global multi-mission force, the addition of land-attack capabilities to its modern array of anti-surface and anti-air capabilities is a logical next step. In the coming years, the PLAN will probably field LACMs on its newer cruisers and destroyers and the developmental SHANG III SSGN. The PLAN could retrofit its older surface combatants and submarines with land-attack capabilities. The addition of land-attack capabilities to the PLAN's surface combatants and submarines would provide the PLA with flexible long-range strike options, enabling the PRC to hold land targets at risk beyond the Indo-Pacific region.

# III. Amphibious Warfare Ships

The PRC's investment in LHA ships signals its intent to continue to develop its expeditionary warfare capabilities. In April 2021, the PRC commissioned the first YUSHEN class LHA (Type 075) followed by the commissioning of the second hull in December 2021. A third YUSHEN-class LHA was commissioned in October 2022 while the first hull achieved IOC in March 2022, and a fourth hull was outfitting during the second half of 2023 and will reach operational status during 2024. The YUSHEN class are highly capable, large-deck amphibious ships that will provide the PLAN with an all-aspect expeditionary capability, including the ability to carry a large number of landing craft, troops, armored vehicles, and helicopters. In addition, there are indications the PRC has begun construction on an even larger LHA, the YULAN (Type 076), at a shipyard in Shanghai. This ship will likely use a hybrid electric propulsion system and have electromagnetic catapults (EMALS) for launching large maritime UAVs for strike and ISR missions. In addition, the PLAN has eight large YUZHAO class LPDs (Type 071) in service. The YUZHAO class LPDs and YUSHEN class LHAs provide the PLA with greater capacity, endurance, and more flexibility for long-range operations than the PLAN's older landing ships, which it has reduced in number over the last decade with obsolete units being decommissioned. The YUSHEN and YUZHAO can each carry several of the new YUYI class air-cushion medium landing craft and the conventional YUBU class utility landing craft as well as a variety of helicopters, tanks, armored vehicles, and PLAN marines for long-distance expeditionary deployments.

# IV. Aircraft Carriers



The PLAN operates two moderate-capability "ski-jump" aircraft carriers based on an old Soviet design. While limited compared to U.S. aircraft carriers, the two KUZNETSOV variants (CV-16 and CV-17) provide the PRC with substantially more sea-based air power than any other country in the region. As previously noted, the PRC launched a third aircraft, CV-18 Fujian, in June 2022. The ship began sea trials in the first half of 2024. Fujian is larger than the ski-jump carriers and fitted with an electromagnetic catapult launch system. This design enables it to support additional fighter aircraft, fixed-wing early-warning aircraft, and more rapid flight operations, thus extending the reach and effectiveness of the PRC's carrier-based strike aircraft. CV-18 is expected to be operational in the first half of 2025, with additional carriers to follow.

The PLAN operates and is developing aircraft to operate from its carriers and combatants. In addition to the standard J-15 fighter that operates from PLAN carriers, a catapult-capable J-15 variant, is in development. The aircraft is testing from land-based steam and electromagnetic catapults. Two other J-15 variants are in development—the J-15S tandem-seat variant and the J-15D EW variant, which is equipped with wingtip electronic support measures/electronic intelligence gathering pods as well as several conformal antennas. The PRC is developing a carrier-capable variant of the fifth-generation J-31 fighter, known as the J-35, which conducted its first flight in 2021. Beyond fighter aircraft, the PRC is refining the design of a carrier-borne AEW aircraft, the KJ-600. A mockup of the aircraft, which appears externally similar to the E-2C/D Hawkeye, has existed for many years. Prototypes of the KJ-600 have been in flight testing since 2020. Beijing is developing the Z-20F helicopter for the PLAN intended for the RENHAI cruisers, LUYANG III MOD destroyers, and the JIANGKAI II MOD and JIANGKAI III frigate classes. The Z-20F is similar to the U.S. Navy's SH-60 and will provide significant improvements in ASW capabilities over the smaller Z-9 and Ka-28 helicopters the PLAN operates. The Z-20F will complement the larger Z-18Fs that operate from the PLAN's aircraft carriers.

Accompanying the manned fixed-wing and rotary aircraft will be UAVs. The PLAN has conducted sea trials on multiple surface combatants with vertical take-off and landing UAVs ranging in wingspan between 3.5 and 4 meters. These UAVs, used for ISR purposes, include the SD-40, CSC-005, S-100 CAMCOPTER, and AV-500 UAV systems.

Signifying its emphasis on developing future carrier-based aviation, the PLAN transferred several shore-based aircraft to the PLAAF. This includes all its H-6J bombers and JH-7 maritime strike aircraft. These aircraft were absorbed by the PLAAF and continue to remain available for maritime strike missions. PLAAF H-6Ks have been noted training for maritime strike missions since 2020.

The PRC operates fixed-wing, medium-to-large size UAVs from land bases. These UAVs specialize in ISR and include EO/infrared imaging, SIGINT capabilities, and synthetic aperture radar (SAR). Many of these UAVs can be operated in SATCOM mode, extending the datalink and operational ranges well past traditional line-of-sight (LOS) communications. Fixed-wing UAVs include the XIANGLONG high-altitude, long-endurance UAV, the BZK-005 medium-altitude, long-endurance UAV, and the ASN-209 medium-altitude, medium-endurance UAV.

# V. Auxiliary Ships

The PLAN continues to build many seagoing auxiliary and support ships, including hospital ships (AH), submarine salvage and rescue ships, and various other large auxiliaries for specialized missions. The PRC is expected to build additional fleet replenishment oilers soon to support its expanding long-duration combatant ship deployments.

Chap 2

# C. Peoples Liberation Army Navy Marine Corps (PLANMC)

Force Structure

*Ref: ATP 7-100.3, Chinese Tactics, (Aug '21), pp. 3-4 to 3-5 and Office of the Secretary of Defense Annual Report to Congress: Military and Security Developments Involving the People's Republic of China 2021, pp. 57-59.*

The People's Liberation Army Navy Marine Corps (PLANMC) roles and missions principally include defending PLA bases in mainland China, the SCS, and abroad; conducting amphibious operations to seize and defend small reef and island outposts; and conducting NWMA. Although the PLANMC has traditionally focused on its task to assault and defense of small islands in the SCS, more recently its focus has grown to include expeditionary operations beyond FIC. The PLANMC's roles under NWMA support protection of overseas interests, including resources, infrastructure, and citizens abroad. The missions for the PLANMC SOF BDEs are to conduct a myriad of diverse missions for the PLA, including infiltration, critical node seizure, C2 decapitation, maritime escort, and counterterrorism operations and providing precision strike guidance.

After taking possession of three PLAA BDEs in late 2022, the PLANMC now has 11 BDEs—eight maneuver, two SOF, and one aviation—and an approximate total force composition of 55,000 marines. With the assumption of the three PLAA BDEs, the PLANMC gained their legacy equipment, which the BDEs continued to train on in 2023. It is expected that the new BDEs will begin updating their inventory to match the other PLANMC BDEs in 2024. The addition of the three new BDEs places PLANMC elements in every theater except for the WTC. The PLANMC maintains a presence at the PRC's first overseas military base in Djibouti, extending the PRC's military reach and strategic influence in Africa and the Middle East. The assessed size of the PLANMC unit in Djibouti is a battalion and consists of approximately 400 marines. The PLANMC's presence in Djibouti provides the PRC with the ability to support a military response to contingencies affecting the PRC's investments and infrastructure in the region and the approximately 1 million PRC citizens in Africa and 500,000 in the Middle East, although the PLANMC in Djibouti has not been observed assisting in evacuation efforts to date. The PLANMC embarks a contingent of marines with the PLAN's Gulf of Aden counterpiracy-focused NETF that supports the PRC's trade interests. The PLANMC supports the PRC's military diplomacy. In 2023, the PLANMC trained with counterparts from Pakistan, South Africa, Russia, Thailand, Tanzania, and Saudi Arabia.

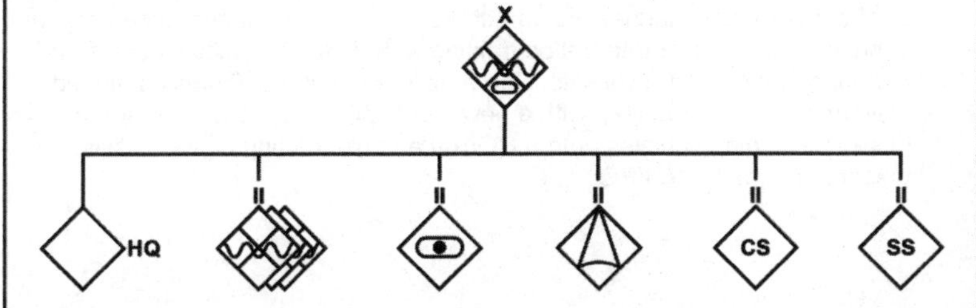


CS combat support HQ headquarters SS service support

*Ref: ATP 7-100.3 (Aug '21), fig. 3-2. PLANMC brigade (doctrinal).*

Force Structure

Each of the original six maneuver BDEs possess the following battalions: air assault, reconnaissance, air defense, artillery, operational support, and a service support battalion. The amphibious BDEs possess two amphibious mechanized infantry combined arms battalions and one light mechanized infantry combined arms battalion. The non-amphibious BDE possesses two light mechanized infantry combined arms battalions and likely a light high-mobility battalion.

The PLANMC SOF BDE has at least three special operations battalions and one reconnaissance battalion and consists of approximately 3,000 marines.

> In addition to the PLAN, China operates both the China Coast Guard (CCG)—which falls under the control of the People's Armed Police (PAP)—and the Maritime Militia under the China Militia. The CCG operates a large number of smaller surface vessels and has responsibility for coastal security, patrol, antipiracy, antismuggling, search and rescue, and maritime police operations. The Maritime Militia operates an exceedingly large number of small surface vessels, most of which are converted or co-opted civilian fishing or transport ships. While both the CCG and Maritime Militia have very traditional assigned roles, China aggressively employs both as part of a hybrid warfare approach during the competition phase of conflict. CCG and Maritime Militia vessels regularly and aggressively operate in international waters— and sometimes in the territorial waters of China's neighbors—ostensibly to protect Chinese fishing and shipping operations. This approach enables China to influence activity throughout the Western Pacific without involving the large surface combatants of the PLAN. In a time of war, the CCG and Maritime Militia will likely take on roles similar to the PAP and China Militia in ground operations: performing low-intensity or mundane maritime tasks in order to free up more-capable ships for more demanding operations.

## PLANMC Capabilities and Modernization

While the aviation BDE will likely continue to receive additional platforms over the next several years, the combined arms and SOF BDEs likely had completed their modernization efforts by the end of 2022. The only new piece of equipment the PLANMC received in 2023 was the PGZ-07 tracked self-propelled anti-aircraft gun (SPAAG). The PLANMC's efforts will likely shift to updating the new BDEs' table of organization and equipment in 2024.

## PLANMC Readiness

PLANMC training in 2023 largely resembled training observed during the past two years. The PLANMC continued to showcase its new PCL-181 vehicle-mounted, self-propelled gun-howitzer system. It revealed its new PGZ-07 SPAAG with live-fire training with ZTL-11 self-propelled anti-tank guns, PLZ-07B howitzer guns, and ZTQ-15 light tanks. The PLANMC highlighted its maturing aviation capability throughout the year. Pilots from the aviation BDE participated in training evolutions where they conducted troop transport for land and amphibious assault missions and delivered special operations marines via parachute. Amphibious marine forces continued to conduct routine driver integration training with PLAN amphibious ships and civilian roll-on/roll-off (RO/RO) vessels. Non-amphibious marine forces continued civilian-military integration training, with a new PLANMC BDE in the northern theater observed loading vehicles onto a RO/RO and conducting a sea crossing before unloading from the RO/RO.

Chap 2

# D. Peoples Liberation Army Air Force (PLAAF)

Force Structure

*Ref: Department of Defense Annual Report to Congress: Military and Security Developments Involving the People's Republic of China 2024, pp. 59-63 and ATP 7-100.3, Chinese Tactics, (Aug '21), pp. 3-1 to 3-3.*

The PLAAF and PLAN Aviation together constitute the largest aviation forces in the region and the third largest in the world, with over 3,150 total aircraft (not including trainer variants or UASs), of which approximately 2,400 are combat aircraft (including fighters, strategic bombers, tactical bombers, multi-mission tactical, and attack aircraft). The PLAAF's role is to serve as a comprehensive strategic air force capable of long-range airpower projection. The PRC's 2019 defense white paper described the PLAAF's missions and tasks as transitioning from territorial air defense to "offensive and defensive operations."

In 2021, General Chang Dingqiu assumed the post of PLAAF commander and continued to enact PLAAF reforms to improve the force's ability to accomplish joint warfighting tasks. The PLAAF is rapidly approaching technology typical of U.S. standards. The PLAAF has aviation, airborne, air defense, radar, electronic countermeasure, and communications forces organized into five Theater Command Air Forces, which command air bases, divisions, and BDEs. The PLAAF Airborne Corps (PLAAFAC) commands six airborne combined arms BDEs, a SOF BDE, an operational support BDE, an aviation transport BDE, a training base, and a new training BDE. The six combined arms BDEs consist of three airborne infantry, one air assault, one wheeled airborne mechanized (wheeled air droppable armored vehicles), and one tracked airborne mechanized (tracked air droppable armored vehicles). Each airborne combined arms BDE typically commands four combined arms battalions, an artillery battalion, a reconnaissance and pathfinder battalion, an operations support battalion, and a service support battalion, and possibly a transportation battalion.

The People's Liberation Army Air Force (PLAAF) is one of the world's largest air forces. Much like the rest of the PLA, the PLAAF is in the midst of a significant reorganization and modernization campaign, moving from employing massive numbers of 1960s-era aircraft with very limited capabilities to employing much smaller numbers of modern 4th-, 4.5th-, and 5th-generation multirole and fighter aircraft. The PLAAF enjoys a suite of capabilities largely comparable to modern Western air forces, including airborne early warning, aerial refueling, heavy and medium transport, and electronic warfare (EW), though relatively few airframes are available for these noncombat missions as compared to the United States Air Force (USAF). The PLAAF also operates a large and advanced ground-based air defense network employing medium-and long-range radar-guided missiles in an integrated, tiered, and layered deployment strategy.

For most of the PLAAF's history, its role has been defending Chinese airspace, with secondary importance given to strike and attack missions and little emphasis on close air support (CAS). This made sense for the period, considering the perceived threats to Chinese territory and the lack of available air-ground communications. As the PLAAF has modernized, greater emphasis has been placed on ground support, but these capabilities are developing and are likely limited in both availability and effectiveness. Although many PLAAF airframes are nominally multirole—meaning they are capable of conducting both air-to-air and air- to-ground missions—few air-to-ground-capable airframes are regularly employed in this mission type.

Force Structure

Most PLAAF combat aircraft date from the Cold War era. This status is changing rapidly, however, as newer and far more-capable types are being both purchased and indigenously produced. This creates a wide spectrum of capabilities within the PLAAF. Older aircraft are not capable of all-weather, around-the-clock operations, especially in ground-attack roles, and they employ mostly non-precision munitions with minimal targeting support. Newer aircraft employ precision-guided munitions, some with standoff capability, with targeting enabled by sophisticated onboard sensor suites. In general, more-capable aircraft are assigned to theaters with higher priority assets, while older aircraft are assigned to less important theaters. Nearly all PLAAF aircraft are either Russian-built, Russian-designed, or evolved from Russian designs. However, unique indigenous designs are starting to reach operational status. PLAAF combat aircraft capabilities are largely similar to those of their Russian equivalents, though the capabilities of the newest 5th-generation low- observable aircraft are not well known in the West.

PLAAF maintenance and logistics capabilities are immature and under-resourced, but they are likely to improve rapidly in the near future. As part of the PLAAF's modernization efforts, maintenance and logistics were relegated to a secondary role behind acquisitions; this in turn had negative impacts on pilot training and aircraft readiness. In addition, the PLAAF had to modernize the relationship between its maintenance and logistics backbone and China's state-run industry. Historically, repair and maintenance was largely outsourced to industry, but this relationship has become increasingly unwieldy in the modern age. The PLA's new expeditionary requirements place further strain on limited maintenance and logistics resources. Solutions to these capability gaps are currently in the conceptual and testing phase, but have yet to be widely implemented.

PLAAF units are organized into brigades, comparable to USAF groups. Brigades are assigned to PLAAF bases, which are the grade equivalent of the U.S. Army's corps and are generally at the disposal of the theater commander. Some high-value PLAAF assets, however, are controlled by national-level headquarters, including some of the newest fighter divisions and the airborne corps.

PLAAF doctrine employs a tiered and layered approach to defending Chinese airspace. Ground-based air defenses in the form of medium- and long-range radar-guided missiles combine with manned aircraft to deny use of airspace to enemy platforms of all types. The PLAAF in general is far less concerned with seizing or maintaining air superiority or air dominance than U.S. forces—it views simply denying the enemy use of airspace to be adequate. PLAAF forces do not have significant capabilities outside of Chinese airspace: excepting a few island airbases off the Chinese coast, neither basing nor aerial refueling capabilities are adequate to project power any significant distance past the Chinese coastline and territorial waters. Expanding power projection across wider areas is a focus of PLAAF training and acquisition. It is likely that the PLAAF can successfully deny use of airspace to any opponent wishing to operate over Chinese territory or ocean areas covered by its surface-to-air missile umbrella.

PLAAF ground-attack capabilities are immature but improving. Around two decades ago, PLAAF attack aircraft were limited to daytime visual attacks using only unguided munitions. Today, the PLAAF operates hundreds of advanced 4th-generation attack aircraft capable of all-weather operations using precision munitions. The primary ground-attack missions of the PLAAF are strike and interdiction. Both of these missions attack targets some distance away from friendly troops, and both generally support the strategic and operational levels of war. PLAAF CAS capabilities are limited due to a lack of training and integration, but they are seen as critical to developing the combined arms brigade (CA-BDE) as a true combined arms force. Efforts are currently underway to advance CAS capabilities. Gaps in CAS operations include a lack of trained forward air controllers, a lack of air-ground communications, and a lack of developed tactics and guidelines for CAS employment.

# PLA Air Force (PLAAF)

*Ref: Department of Defense Annual Report to Congress: Military and Security Developments Involving the People's Republic of China 2024, p. 60.*

## Major Aviation Units (w/ PLAN Aviation)

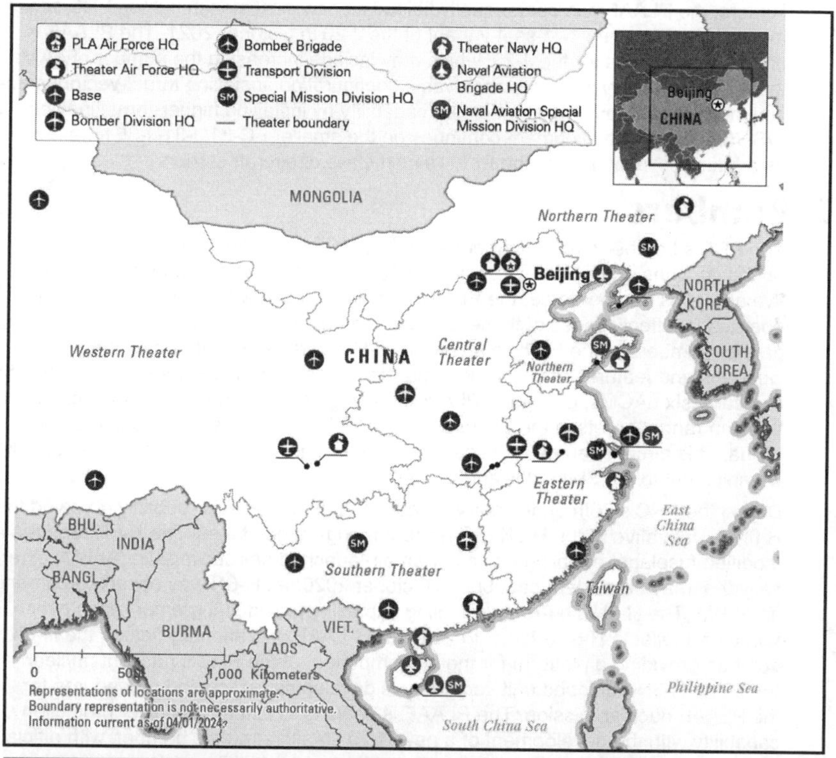


### Key Takeaways

- The PLAAF and PLAN Aviation together constitute the largest aviation force in the Indo-Pacific region.
- The PLAAF is improving technology standards. The PLAAF continues to modernize with the delivery of domestically built aircraft and a wide range of UASs.
- In October 2019, the PRC signaled the return of the airborne leg of its nuclear triad after the PLAAF publicly revealed the H-6N as its first nuclear-capable air-to-air refuellable bomber.
- In 2023, shifting missions for PLAN naval aviation triggered the transfer of significant portions of PLAN shore-based, fixed-wing combat aviation units, facilities, air defense, and radar units to the PLAAF. Given time, this shift will probably enable better C2 over the PRC's integrated air defense systems as well as the network of ground-based air domain awareness radars supporting the PRC's national integrated air defense system (IADS) network.

# I. Fighters

The PLAAF and PLAN Aviation continue to field greater numbers of fourth-generation aircraft (now more than 1,300 of 1,900 total fighters, not including trainers) and probably will become a majority fourth-generation force in the next several years. The PLAAF's J-16 is a fourth-generation plus multi-role fighter that will be capable of carrying the PLA's very long-range air-to-air missile, the PL-17. In 2023, over 225 J-16s were in service, with plans to produce more in the coming years. For fifth-generation fighters, the PLAAF has operationally fielded its new J-20 stealth fighter. PRC social media revealed a new two-seat variant of the J-20 in October 2021. The PLAAF is preparing upgrades for the J-20, which may include increasing the number of AAMs the fighter can carry in its low-observable configuration, installing thrust-vectoring engine nozzles, and adding super cruise capability by installing higher-thrust indigenous WS-15 engines. Development continues on the smaller FC-31/J-31/J-35 for export or as a future naval fighter for the PLAN's next class of aircraft carriers.

# II. Bombers

The PRC's bomber force is composed of H-6 Badger variants, domestically produced versions of the Soviet Tupolev Tu-16 (Badger) bomber. Despite the relative age of its bomber force, the PLAAF has worked to maintain and enhance the operational effectiveness of these aircraft. In recent years, the PRC has fielded greater numbers of the H-6K, a modernized H-6 variant that integrates standoff weapons and features more-efficient turbofan engines for extended-range. The H-6K can carry six LACMs, giving the PLA a long-range standoff precision strike capability that can range targets in the Second Island Chain from home airfields in mainland China. This aircraft carries six supersonic long-range YJ-12 ASCMs and can attack warships out to the Second Island Chain.

During the PRC's 70th anniversary parade in 2019, the PLAAF publicly revealed the H-6N, a derivative of the H-6K optimized for long-range strikes. The H-6N features a modified fuselage, enabling it to carry an air-launched ballistic missile (ALBM) externally that may be nuclear capable. In October 2020, an H-6N was observed carrying an ALBM. The H-6N's air-to-air refueling capability provides it greater reach over non-air refuellable H-6 variants. In 2020, the PLAAF operationally fielded the H-6N bomber, providing a platform for the air component of the PRC's nascent nuclear triad. The H-6N-equipped unit very likely is developing tactics and procedures for the PLAAF nuclear mission. The PLAAF is seeking to extend its power projection capability with the development of a new H-20 stealth strategic bomber, with official PRC state media stating that this new stealth bomber will have a nuclear mission in addition to filling conventional roles. The PLAAF is developing new medium- and long-range stealth bombers to strike regional and global targets. PLAAF leaders publicly announced the program in 2016; however, it may take more than a decade to develop this type of advanced bomber.

# III. Special Mission Aircraft

In 2019, the PLAAF publicly debuted its new Y-9 communications jamming/electronic countermeasures aircraft (GX-11). This aircraft can disrupt an adversary's battlespace awareness at long ranges. The PLA can conduct air-to-air refueling operations to extend the ranges of its fighter and bomber aircraft equipped with refueling probes using a small number of IL-78 Midas and a tanker variant of its Y-20 heavy-lift transport, Y-20U, which will improve the PLAAF's ability to operate beyond the FIC from bases in mainland China.

Production and deliveries of the KJ-500—the PRC's most advanced airborne early warning and control (AEW&C) aircraft—continued at a rapid pace, joining earlier KJ-2000 Mainring and KJ-200 Moth variants. These aircraft amplify the PLAAF's ability to detect, track, and target threats in varying conditions, in larger volumes, and at greater distances. It extends the range of the PLA's IADS network. Furthermore, the PRC has produced at least one KJ-500 with an aerial refueling probe, which will improve the aircraft's ability to provide persistent AEW&C coverage.

The PRC's aviation industry continues to advance with deliveries of its domestic Y-20 large transport aircraft and completion of the world's largest seaplane, the AG600. These transports will supplement and eventually replace the PRC's small fleet of strategic airlift assets, which, to date, consists of a limited number of Russian-made IL-76 aircraft. These large transports are intended to support airborne C2, logistics, paradrop, aerial refueling, and strategic reconnaissance operations as well as humanitarian assistance and disaster relief (HA/DR) missions.



# IV. Unmanned Aerial Vehicles (UAVs)

The PRC continues its comprehensive UAS modernization efforts, highlighted by the routine appearance of increasingly sophisticated systems across theater and echelon levels. The last three years have seen several key milestones, including the airshow display and operational appearance of the Xianglong jet-powered UAS as well as the unveiling of the supersonic WZ-8 UAS and a redesigned version of the GJ-11 stealth unmanned combat air vehicle. The PLA continues the maritime use of ISR UASs, featuring the venerable BZK-005 and the newer TW-328/TB001. The PRC is expanding the applications of large UASs by demonstrating uses, including disaster communications, anti-submarine roles, firefighting, and weather modification. Advanced small UASs are increasingly appearing in military and civilian applications, with PRC industry remaining a key exporter of UASs and components of all sizes.

In addition to maturing their capabilities, the PRC is signaling its efforts in next generation capabilities. Air and trade shows are displaying growing numbers of autonomous and teaming systems, including for combat applications. In these concepts, PRC developers are demonstrating an interest in additional growth beyond ISR and EW into air-to-air and air-to-ground combat, with substantial development efforts to produce swarming capability for operational applications. PRC researchers have disclosed the development of a future multi-domain kill-web designed to target penetrating counterair by coordinating across aircraft, sensors, and missiles.

# V. Air and Missile Defense

The PLAAF possesses one of the largest forces of advanced long-range SAM systems in the world, composed of Russian-sourced SA-20 (S-300) battalions and domestically produced CSA-9 (HQ-9) and follow-on HQ-9b battalions. To improve its strategic long-range air defenses, in 2019, the PRC acquired the SA-21 (S-400) SAM system from Russia. The PRC is fielding its indigenous CH-AB-02 (HQ-19), which has a ballistic missile defense (BMD) capability. PRC researchers are interested in developing SAMs with a range similar to the DF-17 to provide ultra-long-range air defense using space-based ISR to support in-flight target updates. The PRC is developing kinetic-kill vehicle technology to field a mid-course interceptor, which will form the upper layer of a multi-tiered missile defense. The PLA tested a land-based, mid-course interceptor on February 4, 2021.

# VI. PLAAF Airborne Corps

*Editor's note: The following discussion of the PLAAF Airborne Corps is carried forward as referenced previously from the Office of the Secretary of Defense Annual Report to Congress: Military and Security Developments Involving the People's Republic of China 2021.*

The PLA's airborne corps is assigned to the PLAAF, and it is the PLA's strategic airborne unit. It comprises most of the PLA's Rapid Reaction Unit, a cohort of light, strategically mobile ground forces that can provide a significant military presence anywhere in China in a very short period of time.

Force Structure

Though called a corps, the airborne corps mimics the group army in composition, though with fewer support assets. The PLAAF Airborne Corps commands six identified airborne combined-arms brigades (three light motorized, two mechanized, and one air assault), a special operations brigade, an operational support brigade, an aviation transport brigade, a training base, and a new training brigade.

An airborne corps includes—

- Four airborne infantry brigades (three airborne infantry battalions and one towed light howitzer battalion per brigade).
- One airborne infantry (mechanized) brigade (three mechanized infantry battalions and one self-propelled howitzer battalion per brigade).
- One air assault brigade (three air assault battalions and one towed light howitzer battalion per brigade).
- One service support brigade.
- One special operations forces brigade.
- One air transport brigade (fixed- and rotary-wing transport capability).
- One headquarters unit.

The PLAAF has the air transport capacity to deploy either two light brigades simultaneously or half of the mechanized brigade. These units are employed in much the same way as the U.S. Army employs the 82nd Airborne Division: at least one brigade-size unit on very short-term deployment orders, with other units to follow if required.

2020 was the 70th anniversary of the founding of PLA airborne forces. PRC media reflected on the evolution of PLA airborne into a joint-operations armed branch with diversified capabilities—e.g. from a simple airborne infantry force into an airborne combined-arms force. Modern-day airborne units are expected to perform integrated air-land missions in multiple strategic directions, in various battlefield environments, with diverse services and arms. Significant new equipment development in recent years enables airborne units to achieve these effects and accomplish these missions.

Airborne units conducted significant joint training events in 2020. Airborne units conducted training and operations in a maritime environment, and responded directly to requests for assistance from a PLA Navy unit. In at least two separate instances, PLAAF Airborne units conducted inter-service training exercises with PLAA units, which reportedly included force-on-force maneuvers. The airborne unit specifically trained on infantry assault, reconnaissance, and fire strike, as well as gathering data on equipment, and improving leadership. Airborne units increased integration with other PLAAF units, including the first heavy equipment airdrop from the Y-20 in a high plateau environment. Airborne units also worked with military and civilian logistics (airlines, railways, etc.) to move equipment long distances to perform subsequent airborne operations thousands of kilometers from their home base—sometimes in just a few hours. Airborne units worked on nighttime offense and defense operations, including battalion-sized, full-equipment night jumps. Airborne units experimented with using UAS for continuous airdropping operations, dispersed replenishment, fast-roping replenishment, and wounded personnel evacuation.

Throughout the summer of 2020, airborne forces assisted in supporting the national, regional, and local responses to emergencies—specifically the COVID-19 outbreak and widespread flooding. PRC media claimed the units quickly transitioned from disaster response to combat training, including the integration of new fighting vehicles.

In August 2020, PLA airborne troops participated in the Airborne Platoon contest at the International Army Games in Russia. The PLA contingent used Russian infantry vehicles for the first time and placed first in various events, according to PRC media.

Equipment upgrades for airborne forces in 2020 include a new assault rifle and a new wheeled armored vehicle.

# E. Peoples Liberation Army Rocket Force (PLARF)



*Ref: ATP 7-100.3, Chinese Tactics, (Aug '21), pp. 3-5 to 3-6 and Department of Defense Annual Report to Congress: Military and Security Developments Involving the People's Republic of China 2024 pp. 63-67.*

The People's Liberation Army Rocket Force (PLARF) is the branch responsible for operating China's strategic missile and rocket forces and a substantial portion of the country's long-range tactical missiles. It is the largest missile force in the world, operating well over 1,000 short-range, medium-range, and intercontinental ballistic missiles (ICBMs) and over 300 long-range cruise missiles. There is not a U.S. equivalent to the PLARF. U.S. ICBMs are operated by the USAF, a very limited quantity of short- range ballistic missiles (SRBMs) are operated by the U.S. Army, and both the U.S. Navy and USAF operate cruise missiles.

The PLARF is descended from the Second Artillery Corps, China's original nuclear missile unit. It is a discrete military branch similar to PLAAF or PLAN. The PLARF operates using the same base-brigade-battalion structure as the PLAAF. PLARF bases are corps-grade, and each base operates multiple PLARF brigades. It is unclear exactly how the PLARF command interacts with theater commands (TCs), but it is known that each TC houses a PLARF contingent designed to integrate ballistic missile fires into its operations. PLARF missiles are considered strategic assets, and authority for their use is likely retained at the national level. It is also likely that high-value strategic missile assets, such as ICBMs and long-range ballistic missiles, are to be employed only by national authorities, while SRBM brigades may be attached to TCs in support of their operations.

## Key Takeaways

- The PLARF is advancing its long-term modernization plans to enhance its "strategic deterrence" capabilities.
- The PRC probably completed the construction of its three new solid-propellant silo fields in 2022, which will cumulatively contain at least 300 new intercontinental ballistic missiles (ICBM) silos and has loaded at least some ICBMs into these silos.
- The PRC is developing new ICBMs that will significantly improve its nuclear-capable missile forces and require increased nuclear warhead production, partially due to the introduction of multiple independently targetable reentry vehicle (MIRV) capabilities.
- The PRC may be exploring development of conventionally-armed intercontinental range missile systems. If developed and fielded, such capabilities would enable the PRC to threaten conventional strikes against targets in CONUS, Hawaii, and Alaska.

The PLARF operates most of China's nuclear arsenal through a fleet of 60-70 ICBMs. China professes a "no-first-use" nuclear policy and maintains the nuclear missile fleet in a deterrent capacity— though there is an ongoing internal dialogue in China about possible situations were first-use may be necessary. The country at present does not possess an immediate second-strike nuclear capability, and it has relatively few nuclear warheads. Its nuclear strategy can be described as a minimal

# I. PLARF Missile Capabilities

*Ref: Department of Defense Annual Report to Congress: Military and Security Developments Involving the People's Republic of China 2024, pp. 63-67.*

## Conventional

The PLARF fields a variety of conventional mobile ground-launched, short-, medium-, and intermediate-range ballistic missiles and ground-launched cruise missiles (GLCMs). The PLARF's ground-based missile forces complement the air- and sea-based precision strike capabilities of the PLAAF and PLAN. The PLARF's conventional missile forces includes the CSS-6 (DF-15) short-range ballistic missile (SRBM) (range 725 km–850 km), the CSS-7 (DF-11) SRBM (600 km), the CSS-11 (DF-16) SRBM (more than 700 km), land-attack and anti-ship variants of the CSS-5 (DF-21) medium-range ballistic missile (MRBM) (approximately 1,500 km), the hypersonic glide vehicle (HGV) capable CSS-22 (DF-17) MRBM, the CSS-18 (DF-26) IRBM (3,000 km–4,000 km), and the CJ-10 (DH-10) and CJ-100 (DF-100) GLCM (approximately 1,500 km–2, 000 km). The PLARF's conventionally-armed CSS-5 Mod 5 (DF-21D) ASBM variant gives the PLA the capability to conduct long-range precision strikes  against ships, including aircraft carriers, out to the Western Pacific from mainland China. The DF-21D has a range exceeding 1,500 km, is fitted with a maneuverable reentry vehicle, and is reportedly capable of rapidly reloading in the field.

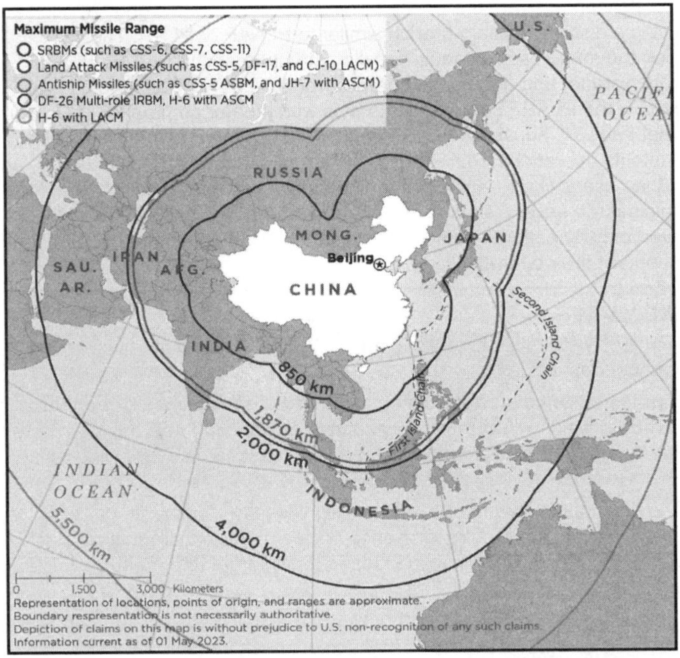


*Conventional strike ranges (Annual Report to Congress: Military and Security Developments Involving the People's Republic of China 2024, p. 67.)*

The PLARF is developing and testing new variants of theater-range missiles and developing capabilities and methods to counter adversary BMD systems. The DF-17 passed several tests and is deployed operationally. In 2020, a PRC-based military expert described the primary purpose of the DF-17 as striking foreign military bases and fleets in the Western Pacific. The PLARF continues to expand its DF-17 inventory and this weapon's HGV payload is designed to evade adversary radars and BMDs. The PRC may be exploring development of conventionally-armed intercontinental range missile systems. If developed and fielded, such capabilities would enable the PRC to threaten conventional strikes against targets in CONUS, Hawaii, and Alaska.

| THE PRC'S ROCKET FORCE | | | |
|---|---|---|---|
| **System** | **Launchers** | **Missiles** | **Estimated Range** |
| ICBM | 550 | 400 | >5,500 km |
| IRBM | 250 | 500 | 3,000 km–5,500 km |
| MRBM | 300 | 1,300 | 1,000 km–3,000 km |
| SRBM | 300 | 900 | 300 km–1,000 km |
| GLCM | 150 | 400 | >1,500 km |



## Nuclear

The PLARF continues to grow its inventory of DF-26 IRBMs, which it first revealed in 2015 and fielded in 2016. As the PLARF expands its DF-26 inventory, it is phasing out DF-21 equipment. The PLA no longer fields any dedicated BDEs of the dual nuclear-conventional capable DF-21Cs. The multi-role DF-26 is designed to rapidly swap conventional and nuclear warheads and is capable of conducting precision land-attack and anti-ship strikes in the Western Pacific, the Indian Ocean, and the SCS from mainland China. In 2020, the PRC fired ASBMs against a moving target in the SCS.

The PLARF is developing and testing new variants of theater-range missiles and developing capabilities and methods to counter adversary BMD systems. The DF-17 passed several tests and is deployed operationally. In 2020, a PRC-based military expert described the primary purpose of the DF-17 as striking foreign military bases and fleets in the Western Pacific. The PLARF continues to expand its DF-17 inventory and this weapon's HGV payload is designed to evade adversary radars and BMDs. The PRC may be exploring development of conventionally-armed intercontinental range missile systems. If developed and fielded, such capabilities would enable the PRC to threaten conventional strikes against targets in CONUS, Hawaii, and Alaska.

The PLARF is developing more survivable ICBMs to significantly improve its nuclear-capable missile forces. The PRC probably will continue to grow the number of launchers at most ICBM units over the next several years. The PRC's ICBM arsenal consists of approximately 400 ICBMs, including fixed and mobile launchers capable of launching unitary and multiple reentry vehicles. The PRC's fixed ICBMs consist of multiple variants of the CSS-4 (DF-5) class missiles, one of which can carry up to five MIRVs and a silo-based CSS-10-class missile. The solid-fueled, road-mobile CSS-10 (DF-31) class and CSS-20 (DF-41) ICBMs complement this force. The CSS-10 Mod 2 (DF-31A), with a range exceeding 11,000 km, can reach most CONUS locations. The DF-41 ICBM has been operationally deployed. In the 2019 parade, commentators noted that two BDEs existed for the system. The PRC appears to be considering additional DF-41 launch options, including rail-mobile and silo-basing.

The PLA may have deployed a "long-range" DF-27 ballistic missile to the PLARF and has a HGV payload option as well as conventional land-attack, conventional antiship, and nuclear capabilities. Official PRC military writings indicate this range class spans 5,000 km–8,000 km, designating the DF-27 as an IRBM or ICBM. PRC media indicates that it can potentially range as far as Alaska and Hawaii.

Force Structure

deterrence approach, possessing only the nuclear capability necessary to deter a nuclear attack. Future modernization accompanied by an expansion of the nuclear force, however, is a clear possibility.

In contrast to its relatively small nuclear force, the PLARF's conventional ballistic missile force is the world's largest and among the world's most technologically advanced and most capable. The PLA employs ballistic missiles as its primary precision deep-strike capability. Ballistic missiles target high-value assets, including air and seaports, supply depots, and command and communication nodes. Ballistic missiles represent a significant element of the Chinese antiaccess strategy. More-advanced ballistic missiles are designed specifically to engage hardened or mobile high-value assets, such as aircraft carriers and antiballistic missile systems. Chinese ballistic missile capabilities represent one of the PLA's strongest investments in system warfare, as ballistic missiles asymmetrically destroy or neutralize assets that traditionally required force-on-force methods to effectively attack. China is not a signatory to the Intermediate-Range Nuclear Forces Treaty, and thus it is free to develop short- and medium-range missiles of all types. Due to the country's strict no-first-use nuclear policy and its substantial investment in ballistic missile capabilities, it is unlikely that China will ever voluntarily downgrade its conventional missile-strike capability.

While many PLARF ballistic missiles are strategic assets, hundreds of SRBMs may be used to target high-value tactical assets or in support of tactical-level missions. PLARF ballistic-missile launchers are mobile, and their facilities hardened, making it difficult to target or neutralize these systems prior to launch. The PLARF also operates much of China's land-based cruise-missile fleet. Unlike the United States, PLA cruise missiles are seen as complementary to the ballistic missile force, rather than the other way around.

PLARF missile tactics are advanced and mature, taking full advantage of the suite of capabilities available. PLARF units will work along with other national assets, such as cyber warfare, EW, and air and naval forces, to neutralize or destroy antiballistic missile and air defense systems using a system warfare, combined-arms approach. PLARF ballistic missiles are accurate and precise, integrated with advanced surveillance and targeting capabilities, equipped with advanced penetration aids and countermeasures, and can be employed in sophisticated structured attacks. The PLA will seek to deplete expensive and rare antiballistic interceptors; defeat or destroy air defense radars; and either sink, damage, or threaten antiair surface naval vessels, giving it freedom of maneuver when employing its missile force.

# II. PLARF Readiness

The PLARF's growing arsenals of advanced and longer-range conventional missile systems improves its readiness for long-range precision strikes on U.S. and allied forces and bases throughout much of the Indo-Pacific. The PLARF routinely practices live-fire strikes on mock airfields, bunkers, aircraft, and ships, indicating that the PLARF is improving its readiness for several counter intervention strike contingencies. In April 2023, the PLARF participated in the JOINT SWORD joint exercise, in coordination with land-based aircraft and the PLAN operational CV-17 Shandong carrier group. The exercise involved encircling Taiwan and testing the PLA's joint and coordinated strike capabilities in responding to perceived threats from U.S. warships. In August 2023, the PLARF participated in a joint exercise with the PLAN and PLAAF, testing the PLA's ability to conduct joint operations in vessel-aircraft integration and ASW.

# F. The Former Strategic Support Force (SSF)



*Ref: Department of Defense Annual Report to Congress: Military and Security Developments Involving the People's Republic of China 2024, pp. 67-71 and ATP 7-100.3, Chinese Tactics, (Aug '21), pp. 3-6 to 3-7. (See also p. 4-1 to 4-2.)*

As of April 19, 2024, Beijing announced the dissolution of the SSF and the alignment of subordinate forces directly under the CMC. Beijing established of a new Information Support Force under the CMC—commanded by Lieutenant General Bi Yi, the former deputy commander of the SSF—to coordinate the military's networks and communications systems. The updated PLA organizational structure features four theater-grade services—the PLAA, PLAN, PLAAF, and PLARF—and four deputy-theater grade forces or service "arms," comprising the Aerospace Force (ASF), Cyberspace Force (CSF), ISF, and the JLSF.

On April 19, 2024, the PRC announced the establishment of a new PLA force, the **Information Support Force (ISF)**, and the subsequent dissolution of the SSF. The SSF had been established in 2015 to centralize the PLA's strategic space, cyberspace, electronic, and psychological warfare missions and capabilities. The SSF reported to the CMC and supported the entire PLA with its capabilities. The PRC's 2019 defense white paper described the SSF's modernization goals as "seeking to achieve big development strides in key areas and accelerate the integrated development of new-type combat forces, so as to build a strong and modernized strategic support force."

## Key Takeaways

• The CMC did not release a detailed explanation of why it dissolved the SSF. It had created the SSF in December 2015 to serve as a theater command-level organization to centralize the PLA's strategic space, cyberspace, electronic, information, communications, and psychological warfare missions and capabilities. The CMC's decision to dissolve the SSF after only 8 years reveals compelling concerns over its contribution to joint operational effectiveness as well as severely inefficient management and leadership.

• The CSF, or the Network Systems Department (NSD) of the former SSF, is responsible for information warfare with an integrated mission set that includes cyberspace warfare, technical reconnaissance, EW, and psychological warfare. The ASF, or the Space Systems Department (SSD) of the former SSF, is responsible for military space and counterspace operations. The Information Communication Base (ICB) of the former SSF was responsible for PLA communications networks and network defense.

• The PRC continues to develop counterspace capabilities—including direct-ascent anti-satellite missiles, co-orbital satellites, EW, and directed-energy systems—to contest or deny another nation's access to and operations in the space domain.

• The PRC's space enterprise continues to mature rapidly and Beijing has devoted significant resources to growing all aspects of its space program, from military space applications to civil and commercial applications, such as profit-generating launches, scientific endeavors, and space exploration. The SSF worked with civilian organizations, such as universities and research organization, to integrate civilian support to military efforts.

With the dissolution of the SSF, two formerly subordinate deputy theater command-level departments—the ASF and CSF—realigned directly under the CMC. The ASF, previously also known as the former SSF SSD, is responsible for military space and counterspace operations, and the CSF, previously also known as the former SSF NSD, is responsible for IO, which includes technical reconnaissance, EW, cyberspace warfare, and psychological operations. Also subordinated under the CMC, the ISF coordinates the development and application of PLA network information systems and provides communications support to the PLA.



With the SSF's disbandment, the ASF and CSF resubordinated directly under the CMC, along with the new ISF. The PLA identified an updated organizational structure with the ASF, CSF, ISF, and JLSF serving as the PLA's four service "arms." "Arms" are not a new concept for the PLA—before it became the PLARF, the Second Artillery Force was referred to as an "arm" of the PLA rather than a service.

# I. Cyberspace Force (CSF)

The CSF, formerly the SSF NSD, is responsible for information warfare with a mission set that includes cyberspace warfare, technical reconnaissance, EW, and psychological warfare. The CSF operates five theater-aligned technical reconnaissance bases, several signals intelligence bureaus, and several research institutes. The CSF provides intelligence support to the theater commands by leveraging a diverse suite of ground-based technical collection assets to provide a common operating picture to geographically dispersed operational units.

### The "Three Warfares" Concept *(See pp. 4-1 to 4-2 and 2-58.)*

The CSF performs missions and tasks associated with the PLA's concept of "Three Warfares," which comprises psychological warfare, public opinion warfare, and legal warfare. This department is the only publicly known organization in the PLA that performs psychological warfare operations. Refer to Special Topic: PRC Views of Information and Information Dominance for more on information warfare, including psychological warfare.

# II. Aerospace Force (ASF)

The ASF, formerly the SSF SSD, is responsible for nearly all PLA space operations, including space launch and support; space surveillance; space information support; space telemetry, tracking, and control; and space warfare. The PRC seeks to enhance the PLA's C2 for joint operations and establish a real-time surveillance, reconnaissance, and warning system. It is increasing the number and capabilities of its space systems, including communications and intelligence satellites as well as the BeiDou navigation satellite system.

The ASF operates at least eight bases, including those whose core missions are the launch, tracking, R&D, and operation of the satellites vital to the PRC's overhead C4ISR architecture. The ASF operates tracking, telemetry, and command (TT&C) stations in multiple locations worldwide to guide space missions around the Earth as well as in cislunar and deep space. The ASF operates Yuanwang space support ships that track satellite and ICBM launches.

The PRC has five launch sites: four land-based and one sea-based. The ASF's China Launch and Tracking Control operates all four fixed launch sites in the PRC, in addition to Yuanwang space support ships, two major satellite control centers—Xian Satellite Control Center and the Beijing Aerospace Control Center—and the PLA TT&C system for all PRC satellites.

# Chap 3 Tactical System Warfare

*Ref: ATP 7-100.3, Chinese Tactics, (Aug '21), chap 4.*

This chapter provides an overview of system warfare as it is applied at tactical echelons. System warfare is the basic theoretical construct the People's Liberation Army (PLA) uses to build capabilities at all echelons. People's Liberation Army Army (PLAA) units historically did not often task-organize, especially at tactical echelons; as such, the system warfare concept is relatively immature at these levels. System warfare requires a modular approach and substantial task organization in order to be successful. It also requires enhanced cooperation between sister and adjacent units, a skill that is similarly underdeveloped in PLAA formations. Nonetheless, it appears that the PLAA is deeply committed to the system warfare methodology, and it will likely govern PLAA tactical operations for the foreseeable future.



## I. Tactical System Warfare Principles

Land warfare traditionally involves imposing one's will upon an uncooperative opponent by the use of direct lethal attack or threat of attack. Maneuver is the process by which one puts oneself in an advantageous position to conduct an attack. Thus, ground combat components traditionally maneuver with the objective of physically isolating their opponents, then defeating them in detail; creating numerical or firepower advantages at key positions on the battlefield; or deceiving opponents into believing their position to be hopeless or otherwise indefensible. The PLAA still views land warfare through this lens, but it sees attacks that do not directly target one's opponent with the threat of physical danger as equal to or more important than direct lethal attacks on the modern battlefield. For example, an opponent may be isolated by disabling network or communications connectivity, rather than by being physically surrounded and cut off. An opponent may believe its position untenable by having vulnerable nodes of its systems rendered ineffective, rather than having the entire system destroyed. An opponent may view continued resistance as futile—not because of the direct threat of physical force—but because it has been deceived into thinking its situation is hopeless. In other words, system warfare takes the basic principles outlined by Sun Tzu and Mao and applies them to conflict utilizing modern, high-technology weapons systems that are manned and operated by professional soldiers.

The PLA describes system warfare using several different official names, including system destruction warfare and system confrontation. All of these terms refer to the same basic concept: a type of conflict wherein systems clash with one another in an attempt to neutralize, destroy, or offset key capabilities and thus grant one side a decisive advantage. System warfare differs from traditional Western military thinking in that it does not necessarily consider the human doing the fighting as the most important element of a combat system. Humans are viewed as subcomponents of a system of systems, to be assessed and targeted much as any other subcomponent. System warfare represents Military Thought supporting the Defense Theory and Defense Doctrine concepts within People's War in Conditions of Informationization. (See chapter 1 for more information on Defense Theory and Defense Doctrine.) The backbone of the system of systems is the network; the importance of this backbone in-turn gave rise to the concept of network-centric warfare.

The system warfare concept consists of two basic ideas: creating purpose-built operational systems that combine key capabilities under a single command, and the use of these operational systems to asymmetrically target and exploit vulnerable components of an opponent's system. If done effectively, this method will render the opponent's key systems ineffective or otherwise unable to function. The PLA believes that by effectively destroying, isolating, neutralizing, or offsetting key capabilities, the enemy's will or ability to resist will degrade and victory will be achieved. Building operational systems is similar to creating task forces, but broader in scope, attempting to create a comprehensive suite of capabilities under a single command. An operational system consists of a number of groups—subordinate entities that are custom-built for a specific mission, task, or purpose. Virtually every battlefield function is represented by one or more groups; this publication discusses only the most significant groups as employed by tactical-level units.

At the tactical level, operational groups are also referred to as combat teams or combat groups. While the PLAA stresses a modular approach to building operational systems, it also recognizes that the less radical the reorganization, the more cohesive a unit will be. The combined arms battalion (CA-BN) structure is designed to reflect this. It is the basic building block of the tactical operational system, and it is intended to be employed in something close to its organic or peacetime configuration. Conversely, the combined arms brigade (CA-BDE) is intended to be easily augmented or task-organized as conditions dictate, flexibly employing a variety of subordinate CA-BNs, supporting battalions, or other nonorganic capabilities.



# II. Planning and Organizing PLAA Operations

The PLAA places great emphasis on planning. A meticulous approach to operations underpins the PLAA throughout its history, and this tradition remains in place today. Though the PLAA seeks to gradually move to an increasingly decentralized leadership philosophy, careful planning at all echelons will remain a basic principle.

Movement toward greater decentralization, modular combined arms units, widespread downsizing, and the creation of new headquarters at the national and theater command levels have significantly complicated the PLAA's employment of command and support relationships. The PLAA was traditionally a strongly centralized and hierarchical force, relying on a mixture of discipline and obedience to overcome shortages in combat technology and firepower. The evolved PLAA recognizes that greater decentralization requires a more sophisticated understanding of command and support relationships; improved professional military education for commanders, staffs, and noncommissioned officers; continued PLAA doctrinal adjustments as new equipment and technologies enter the force; and a task-oriented approach that underpins the challenging requirement of building operational systems.

## A. PLAA Planning Process

The PLAA planning process is broadly similar to the U.S. Army's military decision-making process. The primary outputs of the planning process are the objectives and scheme of maneuver, along with the structure of the operational system being used to conduct the operation.

*See following pages (pp 3-4 to 3-5) for an overview and further discussion.*

## B. PLAA Control Measures

The PLAA employs zones to define the combat area—a designated area assigned to a military, paramilitary, or security force in which combat or security operations occur—for both offensive and defensive actions. These are made up of several secondary zones or areas; each has specific characteristics and is designed to be occupied by one or more groups. Control measures are established for each echelon and are typically nested within the combat area for the higher echelon unit.

*See pp 3-6 to 3-7 for an overview and further discussion.*

# C. PLAA Command Post Operations

*Ref: ATP 7-100.3, Chinese Tactics, (Aug '21), pp. 4-4.*

The PLAA defines a command post as a temporary command structure formed around a commander and associated staff. Up to four command posts are established in support of an operational system: a base command post, an advance command post, a rear command post, and a reserve command post. Command posts are led by a commander and manned by various command groups. Certain command posts—most likely those more forward on the battlefield—may be designed as mobile command posts, able to conduct command post operations while rapidly emplacing and displacing, thus making it more challenging for the enemy to detect and target. The number and type of command posts are situationally dependent, not prescribed.

## Base Command Post

The base command post (also called the main command post) is the primary command structure for the operational system, and it is the center for executing combat command throughout the operational area. The base command post must be well protected, well concealed, and roughly central within the operational area. The base command post houses all command groups: command and communication, reconnaissance and intelligence, firepower coordination, electronic warfare (EW) and cyber warfare, engineering, battlefield management, and political work. The commander operates out of the base command post, along with the chief of staff and all primary staff officers.

## Advance Command Post

The advance command post (also called the forward command post) is a forward-based structure designed to enhance command and communication in a key direction of the battle. Ideally, it is situated near the main offensive or defensive effort, and it is well concealed and protected. The advance command post is typically led by the deputy commander and staff, and it houses a command and communication group, a reconnaissance and intelligence group, and a firepower coordination group.

## Rear Command Post

The rear command post (or alternate command post) is responsible for logistics and equipment support, along with rear area security. It consists of a combat support and service support group headquarters, a political work group, and security personnel. The commander of the combat support or service support group is likely the commander of the rear command post. The rear command post serves as the primary backup command post in case the base command post is destroyed, neutralized, or otherwise compromised. Depending on the situation, the deputy commander may choose to operate from the rear command post.

## Reserve Command Post

The reserve command post is a backup structure in case one of the other three command posts is compromised, damaged, destroyed, or otherwise neutralized. It is necessarily a smaller and less-capable structure than the base command post, but it must be able to conduct all base command post operations. The reserve command post may also serve as an interim command post in case more forward command posts are temporarily unavailable due to movement or enemy action.

# PLAA Planning Process

*Ref: ATP 7-100.3, Chinese Tactics, (Aug '21), pp. 4-2 to 4-4.*

The PLAA planning process is broadly similar to the U.S. Army's military decision-making process. The primary outputs of the planning process are the objectives and scheme of maneuver, along with the structure of the operational system being used to conduct the operation. The operational system may include multiple subordinate operational systems in addition to specialized supporting systems.

The PLAA planning process consists of five steps, each with distinct outputs. Subordinate units conduct planning in sequence with their higher echelon headquarters and adjust their plan according to inputs received from it.



## Step 1: Assess the Situation

During this step, the commander assesses four critical components of the upcoming confrontation: terrain, enemy forces, friendly forces, and other considerations, such as civilian presence and political elements. The outputs from this step include an early outline of the desired friendly operational systems, a comprehensive report on enemy strength and disposition enabled by reconnaissance and intelligence support, and a thorough analysis of all other factors in and around the battlefield that may influence operations for either side.

## Step 2: Make Decisions

This step requires commanders to make several key decisions that determine the direction of the rest of the planning process and the operation at large. Commanders must establish the purpose and objective of the operation, the general scheme of maneuver to be employed to achieve this purpose and objective, the basic structure of the operational systems under their command, and key offensive and defensive points of interest. The output of this step allows subordinate commanders to begin their own planning process, staffs to begin building operational systems, and command posts to deploy in order to establish command systems. Reconnaissance operations should also commence at this point in order to support the intelligence requirements for the wider operation.

## Step 3: Issue Orders

This step consists of two main phases. First, commanders conceptualize the battle to their staffs, accounting for enemy strengths and disposition, friendly forces, and the higher echelon commanders' intent. The staff then creates an order that clearly lays out the commander's concept in such a way that subordinate units can easily understand it. Second, the order is issued to subordinate units, either verbally or in writing. Orders are ideally issued to subordinate commanders in groups, allowing subordinates to ask questions and collaborate with one another prior to conducting their own planning process.

## Step 4: Organize Coordination

This step refines the orders issued in step 3. Staffs conduct planning that synchronizes purpose between subordinate units, ensures that sister units' schemes of maneuver are integrated, provides useful predictions about how the battle will be conducted, clarifies issues of confusion with subordinate commanders, and ensures that shared resources and enablers—such as communications and network support—are deconflicted.

## Step 5: Organize Support

This step organizes external capabilities that support the operation. Examples include combat support, logistics support, equipment support, and political support. Specific capabilities include firepower and artillery support, information warfare, sustainment, casualty evacuation, and medical support. This step requires staffs to interact with external enablers; higher echelon staffs may provide assistance in building this part of the plan. Commanders must account for available resources, mission requirements, and friendly forces when developing the support plan.

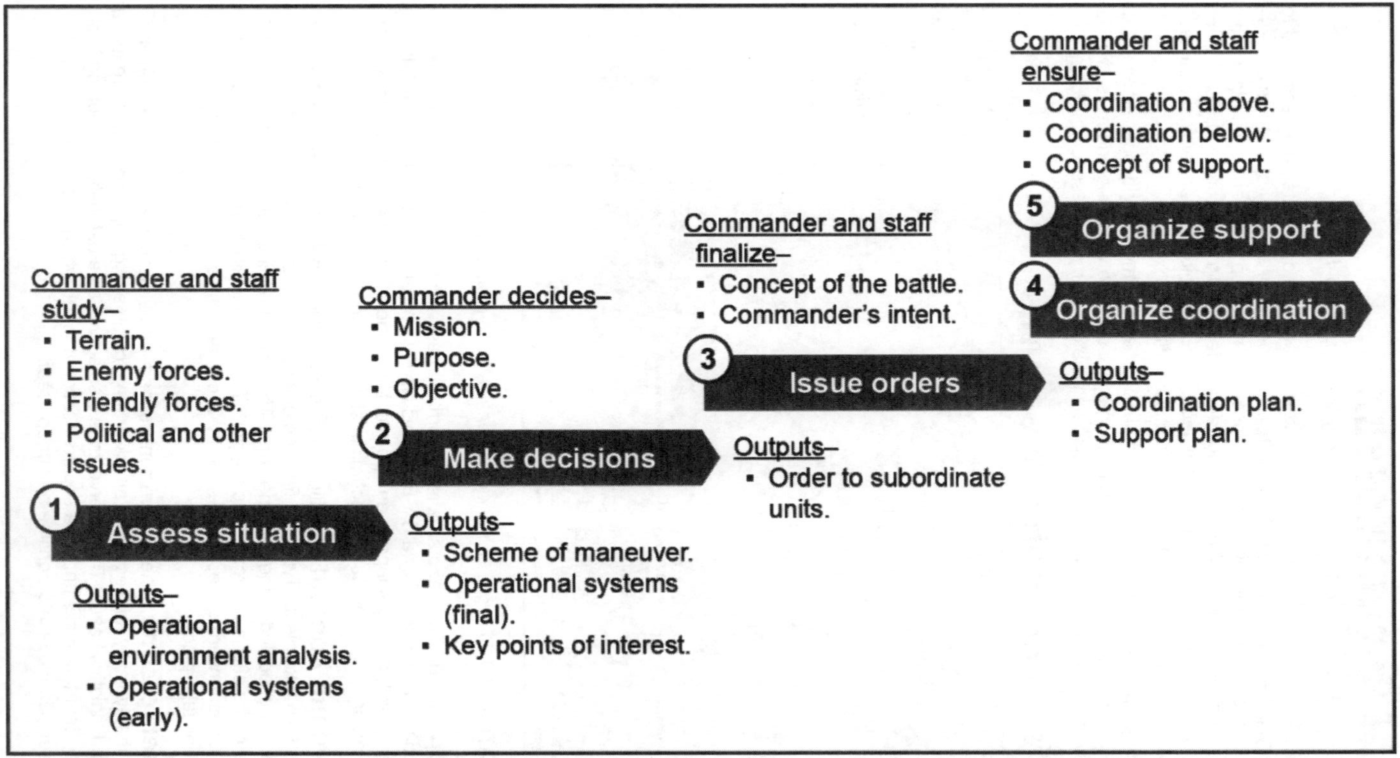


*Ref: ATP 7-100.3 (Aug '21), fig. 4-1. PLAA planning process.*

# PLAA Control Measures

*Ref: ATP 7-100.3, Chinese Tactics, (Aug '21), pp. 4-5 to 4-7.*

The PLAA employs zones to define the combat area—a designated area assigned to a military, paramilitary, or security force in which combat or security operations occur—for both offensive and defensive actions. These are made up of several secondary zones or areas; each has specific characteristics and is designed to be occupied by one or more groups. Control measures are established for each echelon and are typically nested within the combat area for the higher echelon unit.

Security zones are established throughout a zone any time security operations are conducted. Annihilation zones—specific geographic areas to which the enemy is to be lured or driven and then destroyed—may also be created in any of the secondary zones.

System Warfare

## Offensive Control Measures

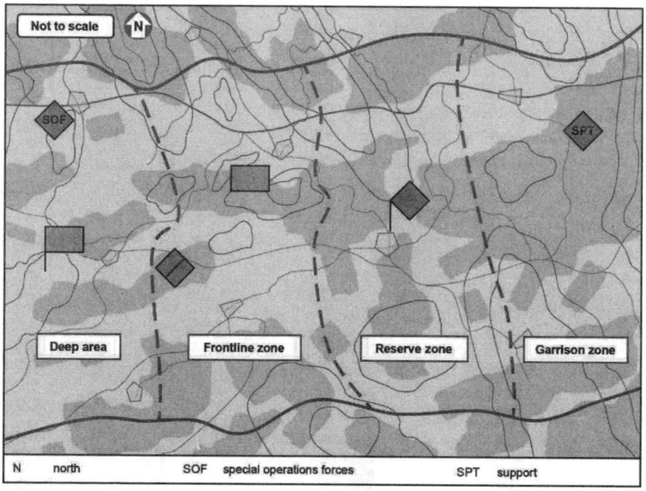


*The offensive zone typically consists of four secondary zones or areas: the deep area, the frontline zone, the reserve zone, and the garrison zone.*

*Ref: ATP 7-100.3 (Aug '21), fig, 4-2. Offensive zone (conceptual).*

**Deep Area.** The deep area is the territory past which a unit's organic sensors and weapons can operate. For a CA-BDE, this typically means the area past which its rocket artillery and targeting support can operate. Units operating in deep areas are usually more autonomous and can expect minimal support from their parent unit.

**Frontline Zone.** The frontline zone is the territory in which the main offensive action is to take place. Typically, first-line objectives and the enemy's main defensive position are located in the frontline zone. The entire frontline zone should be within the range of the offensive group's fire support. The frontline zone typically contains a security zone on its forward edge, where security, reconnaissance, and counterreconnaissance activities take place.

**Reserve Zone.** The reserve zone is the territory just to the rear of the frontline zone that typically houses depth attack groups, command groups, firepower groups, and forward logistics bases. The reserve zone also serves as a defensive bastion against enemy counterattacks and as a secure location through which follow-on forces and supplies can move into frontline and deep areas.

**Garrison Zone.** The garrison zone consists of rear areas not actively occupied by the offensive group. Supporting capabilities such as logistics, EW, and long-range artillery reside here. Garrison zones typically contain one or more security zones that surround key positions such as bases, supply routes, or command posts. The People's Armed Police (PAP) may take on much of the security load in garrison zones in order to free up PLAA forces for more-intense duties.

# Defensive Control Measures

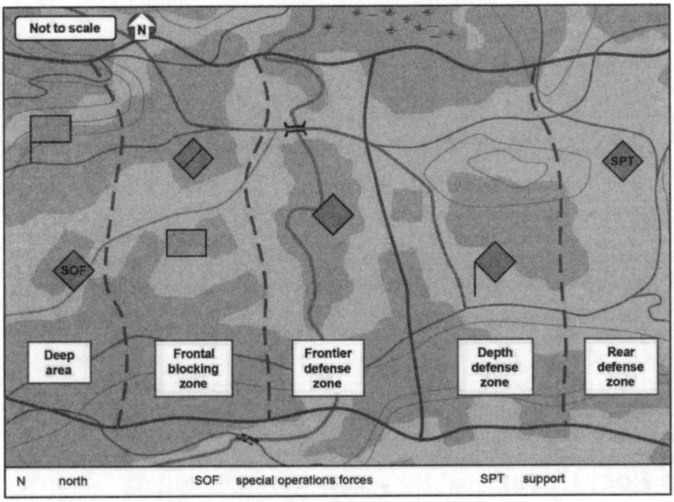


*The defensive zone typically consists of five secondary zones or areas: the deep zone, the frontal blocking zone, the frontier defense zone, the depth defense zone, and the rear defense zone.*

*Ref: ATP 7-100.3 (Aug '21), fig. 4-3. Defensive zone (conceptual).*

**Deep Area.** The deep area is defined in the same way as the deep area in an offensive zone. Usually it is the area in which independent units conduct reconnaissance and disruption activities prior to heavy enemy contact.

**Frontal Blocking Zone.** The frontal blocking zone is the forward-most area occupied by the defensive group. Screening, reconnaissance, and counterreconnaissance missions are the primary focus in this zone. Groups in the frontal blocking zone are usually used to disrupt, canalize, and slow down enemy assault forces in preparation for a future decisive counterattack.

**Frontier Defense Zone.** The frontier defense zone is the zone that typically contains the main line of defense, consisting primarily of the frontier defense group. The bulk of the defensive group is deployed here, making best use of terrain to slow, disrupt, and degrade enemy units as they conduct their attack. Most fortifications and other defensive engineering works are focused in the frontier defense zone.

**Depth Defense Zone.** The depth defense zone is the zone that houses the depth defense group, and it may contain the firepower strike or coordination group or parts of the command group. The depth defense zone is designed to facilitate the commitment of the depth defense group to defeat the enemy's main effort.

**Rear Defense Zone.** The rear defense zone is the zone that houses the combat reserve group, long-range firepower, command posts, and forward logistics support. A secondary defensive line is often established in the rear defense zone as a fallback position in case forward zones are overrun.

# III. Tactical System Warfare Methodology

Building the operational system is one of the primary outputs of the PLAA's planning process. Commanders are given general guidelines on how to assemble the groups that comprise the operational system, and then they build the operational system to suit the mission and situation. Using the commander's purpose, intended scheme of maneuver, and analysis of enemy forces and other battlefield conditions, groups are formed using the available force pool, weighting specific groups with the appropriate levels of firepower, mobility, and protection. In tactical ground operational systems, groups can be organized into five general categories: command groups, offensive groups, defensive groups, firepower strike groups, and support groups. Not all operational systems will have all groups; most will consist of command groups supported by one or more additional groups.

*Note. PLA literature uses a very wide variety of names for these groups; its approach appears to be unstandardized and still somewhat in development. This list represents a cross-section of different group names and mission descriptions and is not exhaustive.*



## A. Command Groups

Command groups are the components of the operational system that enable leadership to control the system, make rapid and effective command decisions, inform subordinate commanders, and coordinate capabilities. They are built using a variety of different force pools—from command posts and staffs, to reconnaissance and intelligence units, to network support units. Some functions, such as engineer and chemical support, are prescribed in both the command group and a support group; it is unclear how the decision to place these units is made. The following paragraphs describe several of the most important command groups.

### Command Post Group

The command post group establishes and operates the operational system's command posts. It is also responsible for protecting command posts from direct assault and concealing them from enemy surveillance and targeting. In an operational system designed around a CA-BDE, the CA-BDE's command post provides most of the manpower and equipment for the command post group.

### Command and Communication Group

The command and communication group establishes the communications and network architecture necessary to support the operational system. The PLAA places a high priority on information systems that are adaptable and reliable, enabling them to operate in a variety of battlefield conditions or when under electromagnetic or network attack by enemy forces. Communications systems are automated wherever possible, enabling critical information to be rapidly distilled and then passed to the appropriate consumer. The PLAA prefers to use a top-down and centralized approach to its information systems, ensuring interoperability across disparate units. The command and communication group is also responsible for establishing and monitoring the control measures that control information systems within its combat area— functions such as assigning radio frequencies, deconflicting electronic emitters, and ensuring network security. Finally, the command and communication group is responsible for protection of the operational system's network backbone. In an operational system designed around a CA-BDE, the CA-BDE's communications company within the service support battalion likely forms the basis for the command and communication group.

## Reconnaissance and Intelligence Group

The reconnaissance and intelligence group is comprised of reconnaissance forces and the operational system's intelligence personnel, task-organized under a single group structure. This structure enables close integration of intelligence and reconnaissance, outlined in detail in chapter 6. The reconnaissance and intelligence group is responsible for developing the reconnaissance plan, conducting reconnaissance in support of the operational system, and disseminating relevant intelligence to all members of the operational system. In an operational system designed around a CA-BDE, the CA-BDE's reconnaissance battalion and intelligence headquarters section form the core of the reconnaissance and intelligence group.

## Firepower Coordination Group

The firepower coordination group, also called the firepower coordination center, is responsible for integrating firepower capabilities of all types into a coherent, synchronized plan in support of the operational system. Firepower capabilities include, but are not limited to, tube and rocket artillery, ballistic and cruise missiles, fixed- and rotary-wing aircraft, network, electromagnetic, and information attack. The firepower coordination group is enabled by the firepower coordination system, an automated or semiautomated command and communication network that distributes targeting and system data across a variety of integrated firepower systems and seeks to streamline targeting and fire control. The firepower coordination group is closely linked with all of the firepower groups.



## Electronic and Network Warfare Group

The electronic and network warfare group is comprised of electronic countermeasure forces, network warfare forces, and other associated personnel. Its missions include electronic reconnaissance, interference, sabotage, targeting, network defense, and cyber intrusions against vulnerable enemy networks.

## Battlefield Management Group

The battlefield management group is responsible for managing personnel and administrative functions in support of the operational system. It consists of the personnel staff officer, supporting staff, and support teams.

## Political Work Group

The political work group is responsible for political support to the operational system. This includes managing the political messaging of the commander, ensuring synergy of purpose for all troops in the operational system, conducting propaganda operations in support of friendly operations and against enemy operations, and supporting psychological warfare operations. The political work group is led by the political officer and manned by supporting staff.

## B. Offensive Groups

Offensive groups are the elements of the operational system that are responsible for fixing, assaulting, and annihilating enemy forces. Offensive groups are deployed in such a way as to ensure substantial numerical and firepower advantage over the enemy: PLAA guidelines suggest offensive groups seek a four-to-one advantage in maneuver forces, a five-to-one to seven-to-one advantage in artillery firepower, and three PLAA antitank systems for each anticipated enemy armor system. An offensive group deploys in a focused manner, establishing the depth and three-dimensional organization necessary to overwhelm the enemy when an assault commences. Offensive groups are integrated with firepower strike groups in order to achieve synchronization between maneuver and fires capabilities. Offensive tactics and techniques are discussed in greater detail in chapter 7.



### Advance Group

The advance group is composed of troops that take advanced positions ahead of a primary attack, conducting a mission roughly similar to the advance guard. The group provides security for the main body, screens against enemy troops, conducts counterreconnaissance, occupies terrain favorable for the main body, and initiates contact with the enemy main body. After completing these tasks, the advance group typically assumes a security or reserve role.

### Frontline Attack Group

The frontline attack group consists of those troops whose mission is to conduct the initial assault against a hardened enemy target, with the goal of achieving a breach or other small penetration. The frontline attack group includes infantry units, armor units, firepower units—likely in the form of mortars supported by artillery—and air defense, antitank, engineer, chemical weapons defense, and electronic countermeasure units. The group's primary task is to break through and capture enemy first-line positions. The frontline attack group is typically subdivided into assault teams and firepower teams, and it seeks to support each assault team with a firepower team. The frontline attack group attempts to concentrate combat power on the narrowest possible front, then launch a well-supported attack. Armored units likely form the core of this main thrust, supported by infantry. Once the assault is successful, the frontline attack group transitions to a security-or vigilance-role, providing flank and rear security for the depth attack and thrust maneuvering groups as they commence their follow-on missions.

### Depth Attack Group

The depth attack group (or in-depth attack group) consists of those troops that advance deep into the enemy position once the initial breach is achieved. The depth attack group likely consists of the best-available armored forces, supported by mechanized infantry, air defense, antitank, engineer, chemical weapon defense, and electronic countermeasure units. The tasks of the depth attack group include breaking into the depth of the enemy's defense, seizing critical terrain or other targets deep in enemy territory, annihilating the enemy's in-depth positions, and occupying positions favorable for defense against the enemy's counterattack. The depth attack group is staged in order to ensure that the penetration does not lose momentum as forces tire, face challenging terrain, or meet enemy resistance.

### Thrust Maneuvering Group

The thrust maneuvering group consists of those troops that exploit the advantages created by the depth attack group. The thrust maneuvering group may be a highly mobile armored force or an air assault force, enabled by light artillery, antitank, air defense, engineer, and EW units. The thrust maneuvering group continues the attack against enemy deep positions, targeting command nodes, supply areas, and key terrain. Its most important missions, however, are to cut off enemy retrograde routes—enabling the annihilation of enemy units—and to spoil or disrupt counterattacks as the enemy attempts to reinforce its units that are caught in the annihilation zone(s).

### Combat Reserve Group

The combat reserve group consists of those troops that remain in the operational system's rear area, with missions to reinforce the frontline attack group or depth attack group if necessary, spoil enemy counterattacks that threaten other offensive groups or the rear area, and deal with any other situations that may arise in the operational system's combat area. The combat reserve group likely consists of either recently used armored units on a rest cycle or lighter mobile units that are able to respond rapidly to developing situations. The combat reserve group is augmented by antitank and engineer units.

## C. Defensive Groups



Defensive groups are those groups charged with defending friendly forces, systems, or key terrain from enemy attack. Defensive groups conduct either a position-based or mobile defense, with the intent of blunting the enemy's attack, attriting enemy forces, and enabling friendly counterattack and transition to the offense. Defensive groups are integrated with firepower strike groups in order to enhance their combat power and defeat superior enemy forces.

### Cover Group

The cover group is made of those troops assigned to a cover mission in support of the operational system's main body. The group's primary tasks are to conduct counterreconnaissance, defend stubbornly in the face of an enemy attack, screen the main body, and cover the main defensive line's deployment and disposition. After the cover mission is complete, the cover group may withdraw to deeper areas and conduct vigilance or security operations, serve as the combat reserve group, or continue to operate as part of the main defensive line. If possible, a component of the cover group will attempt to remain behind the enemy's main advance in order to conduct operations behind enemy lines. The cover group consists of reconnaissance or light armored units enabled by light artillery, antitank, antiair, and EW units.

### Frontier Defense Group

The frontier defense group consists of those troops assigned to the main line of defense. Frontier defense groups are further divided into a main defensive direction group and secondary defensive direction groups, depending on the enemy's predicted course of action. The main defensive direction group is weighted more heavily with numbers and firepower; secondary defensive direction groups are accordingly less powerful, proportional to the commander's assessment of the enemy's approach. The primary tasks of the frontier defense group are to hold the defensive line, blunt the enemy's offensive attack, inflict heavy casualties upon the enemy, buy the commander decision space, and inform the commander about how best to commit reserves and counterattack forces. In a position-based defense, the frontier defense group is likely much larger, as a position-based defense requires the defender to hold a fortified position with minimal maneuver. The frontier defense group consists of infantry and armor, augmented by artillery, antitank, engineer, antichemical weapon, and electronic countermeasure units. It likely consists of a higher proportion of infantry compared to other defensive groups.

### Depth Defense Group

The depth defense group consists of those troops assigned to defend deep areas of the defense. Its primary mission is to conduct counterattacks against enemy penetrations; reinforce weak areas of the frontier defense group's lines; defeat aerial incursions into the rear defense area; and encircle, isolate, and assault any enemy forces operating in rear areas. The depth defense group must be mobile, and it likely contains a high proportion of armored and mechanized forces. In a mobile defensive operation, the depth defense group is the decisive component. Commitment of the depth defense group is the decisive point in a defensive battle. The depth defense group is augmented by artillery and antitank units.

## Combat Reserve Group

The combat reserve group is composed of those troops that are retained in depth areas, with a mission to maintain security and reinforce the main defensive line, if necessary. The group may also conduct counterattacks against enemy penetrations. The combat reserve group likely consists of light, mobile troops or troops on a rest cycle following previous action.



## Firepower Strike Groups

Also called fire assault groups, these groups provide the bulk of the operational system's firepower. Firepower strike groups include a variety of heavy weapons, such as artillery systems, mortar systems, air defense systems, and antitank systems, and they include electromagnetic attack and psychological attack systems. These disparate systems are integrated into the operational system by the firepower coordination group, and they can further be subdivided into firepower teams charged with supporting small subordinate elements of the operational system.

## Artillery Group

The artillery group consists of the indirect fire elements of the firepower strike group. These include 122-mm and 155-mm tube artillery systems, light and heavy multiple launch rocket systems, and short-range ballistic missile systems. The PLAA's firepower is seen as the backbone of its operations, and the artillery group provides most of the tactical-level firepower to the operational system commander. The artillery group is positioned in deep areas, and its primary missions are to—

- Support scouting and security operations.
- Disrupt or pre-empt advancing or unfolding (pre-attack positioning) enemy troops.
- Support defensive groups.
- Disrupt or defeat enemy flanking and encirclement attempts.
- Disrupt enemy penetrations and counterattacks.
- Suppress enemy artillery.
- Strike at key targets, such as enemy command posts and supply areas.

Broadly speaking, fire support missions are conducted by tube artillery units, while annihilation, suppression, and counterfire missions are conducted by rocket artillery units. Missile units, if available, conduct precision deep strike missions. The artillery group is built around artillery battalions combined in such a way as to achieve a combined arms effect.

## Air Firepower Strike Group

The air firepower strike group (also called the aerial firepower assault team) consists of the operational system's rotary-wing airframes that are capable of ground-attack operations. The PLAA employs its helicopters as airborne fighting vehicles, closely attaching them to ground units and seeking to maximize their unique capabilities against vulnerable parts of enemy formations. The air firepower strike group is employed in a variety of ways: as a highly mobile reserve, attacking enemy penetrations and supporting friendly counterattacks; as mobile artillery, targeting enemy artillery systems and other weapons that were not destroyed by the artillery group; as an antitank unit targeting enemy tanks and other armored vehicles; as a deep penetration force targeting enemy command, EW, and sustainment systems; as reconnaissance and surveillance platforms; and as security and vigilance platforms.

### Antitank Group

The antitank group consists of those troops specially equipped and tasked with targeting and destroying enemy armored vehicles. The antitank group consists of a combination of mounted antitank guided missile launchers, dismounted antitank guided missile teams, and antitank artillery—mainly in the form of assault guns. The antitank group is positioned near the main direction of attack or defense, and it is employed as a rapid, mobile reserve whenever enemy armored forces are encountered. If armored targets are not available, the antitank group can serve as one or more firepower teams, augmenting assault teams as they carry out their missions. The PLAA employs a high density of antitank weapons throughout its formations, and employment of antitank groups is a key planning consideration.

### Mobile Artillery Group

The mobile artillery group is comprised of highly mobile artillery and mortar systems that provide direct, close-in fire support to offensive and defensive groups of all types. Missions for mobile artillery groups include suppressing or destroying enemy positions, delivering obscurants or illumination, conducting short-range counterfires, and disrupting enemy reinforcement or counterattack maneuvers. The mobile artillery group is viewed as vulnerable to counterfire and direct fire, and rapid movement is considered key to its survival. The mobile artillery group is likely built around either 82-mm or 120-mm rapid-fire mortars or mobile 122-mm self-propelled howitzers.

### Air Defense Group

The air defense group is comprised of most of the operational system's air defense weapons systems, plus associated sensors and command nodes. At the tactical level, this consists of short-range air defense (SHORAD) and man-portable air defense systems (MANPADS), plus a variety of towed and self-propelled antiaircraft guns (SPAAGs). The air defense group's primary mission is to deny the use of airspace over the battlefield by enemy aircraft through a mixture of deterrence and lethal engagements of in-range aircraft. The air defense group may also conduct air ambushes, intended to surprise and destroy enemy aircraft. Per Chinese doctrine, the group should be employed using a defense-in-depth approach, with deployments weighted according to the value of the assets defended.

## D. Support Groups

Support groups are responsible for the operational system's support and sustainment functions and rear area security functions. At the CA-BDE level, these are built primarily using the CA-BDE's two support battalions: combat support and service support.

### Combat Support Group

The combat support group consists of a variety of combat support capabilities. At tactical echelons, most of this group is comprised of the CA-BDE's combat support battalion. Capabilities include engineering, mobility, protection, anti-chemical weapons, and communications. Special emphasis is given to obstacles: there are discrete barrier set-up and removal teams. The combat support group also provides deception and camouflage support and smokescreen generation.

### Rear Area Support Group

The rear area support group is comprised of most of the operational system's supply and logistics capabilities. The rear area support group is divided into two subgroups: logistics support and equipment support. The logistics support group conducts the supply mission, while the equipment support group performs maintenance.

### Psychological Warfare Group

The psychological warfare group is comprised of specially trained soldiers that provide support to frontline units. The group's mission consists of three tasks: present a clear picture of the enemy's psychological situation, conduct psychological attacks on the enemy in support of the operational system's operations, and protect friendly troops from enemy psychological attack. The psychological warfare group also works carefully with the political group to build consensus and morale among friendly troops.

## IV. Tactical-Level Notional Operational System



This section describes one possible construction of a tactical-level operational system, also called a combat group. Since operational systems are, by definition, ad hoc and task oriented, there is not one single or correct way to assemble them. Each operational system is different due to battlefield characteristics, the nature of the enemy, and the preferences of the commander. This section is intended to demonstrate the building-block nature of the operational system and how a system of systems is created.

The example combat group is assembled by a group army, using a medium CA-BDE and a heavy CA-BDE as the two main force providers. The group army's artillery, air defense, aviation, and special operations forces (SOF) brigades provide additional forces to the combat group. The group is assembled in order to conduct an attack against a fortified enemy. The group army command and staff build a single, coherent combat group, minimizing ad hoc or unnecessary reorganization and maintaining the cohesion of subordinate battalions wherever possible.

The force pool that the combat group is built from consists of all the organic units of the two CA-BDEs, plus augmentation from the group army. Available forces to build the combat group, *with units in italics being reinforcements from the group army,* include—

- Four medium CA-BNs.
- Four heavy CA-BNs.
- Two reconnaissance battalions.
- *One unmanned aircraft system (UAS) reconnaissance section.*
- *One SOF company.*
- *One scout helicopter company.*
- One UAS targeting section.
- Two artillery battalions.
- *One heavy multiple rocket launcher (MRL) battery.*
- *One light MRL battery.*
- *One attack helicopter company.*
- *One medium lift helicopter company.*
- Two air defense artillery battalions, consisting of three SPAAG batteries and one SHORAD battery each.
- *One heavy SPAAG or missile battery.*
- *One medium-range surface-to-air missile (SAM) battery.*
- *One electronic air defense battery.*

# Operational System Force Pool (Example)

*Ref: ATP 7-100.3 (Aug '21), fig. 4-4. Operational system force pool (example).*

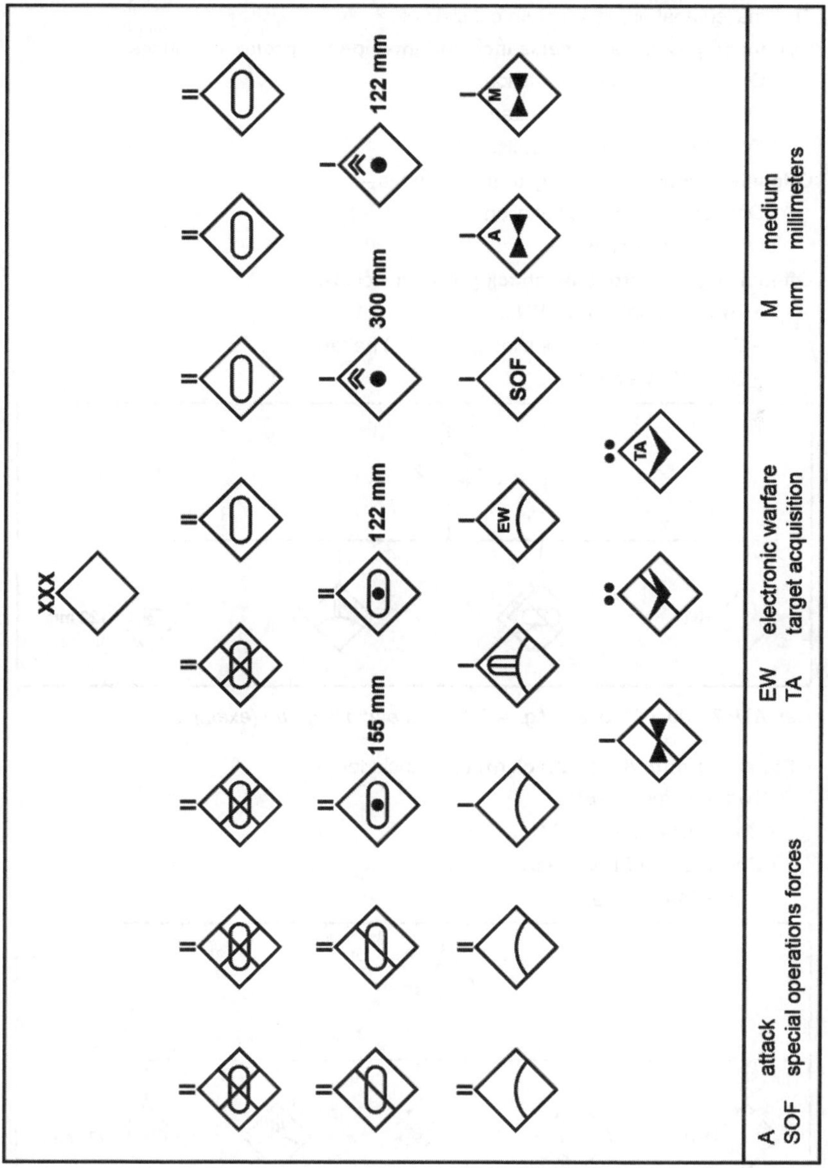


System Warfare

# Final Combat Group Structure (Example)

*Ref: ATP 7-100.3, (Aug '21), pp. 4-14 to 4-16.*



The final combat group structure consists of—

**Command group: reconnaissance and intelligence group.** It includes—

- One reconnaissance battalion.
- One scout helicopter company.
- One UAS section (reconnaissance).

**Offensive group: advance group.** It includes—

- One reconnaissance battalion.
- One SOF company.

**Offensive group: frontline attack group**. It includes—

- Two mechanized CA-BNs.
- One 122-mm self-propelled gun (SPG) battery.
- One SPAAG battery.

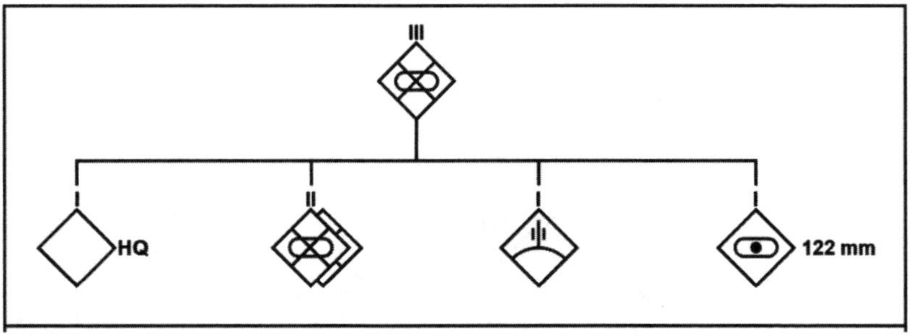


*Ref: ATP 7-100.3 (Aug '21), fig. 4-5. Frontline attack group (example).*

**Offensive group: depth attack group.** It includes—

- Two armored CA-BNs.
- One mechanized CA-BN.
- One 122-mm SPG battery.
- One SPAAG battery.

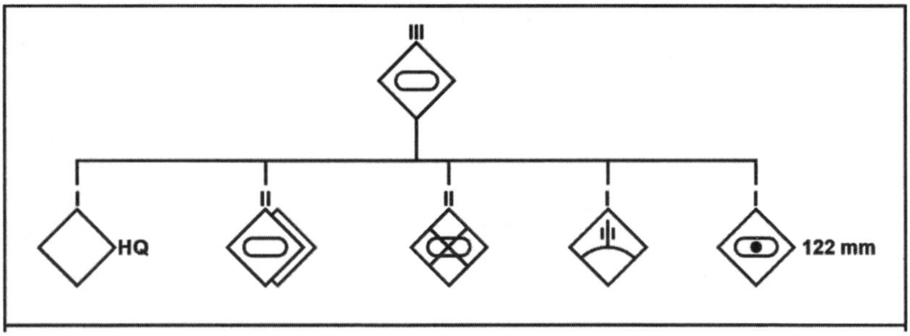


*Ref: ATP 7-100.3 (Aug '21), fig. 4-6. Depth attack group (example).*

**Offensive group: thrust maneuvering group.** It includes—
- Two armored CA-BNs.
- One 122-mm SPG battery.
- One SPAAG battery.

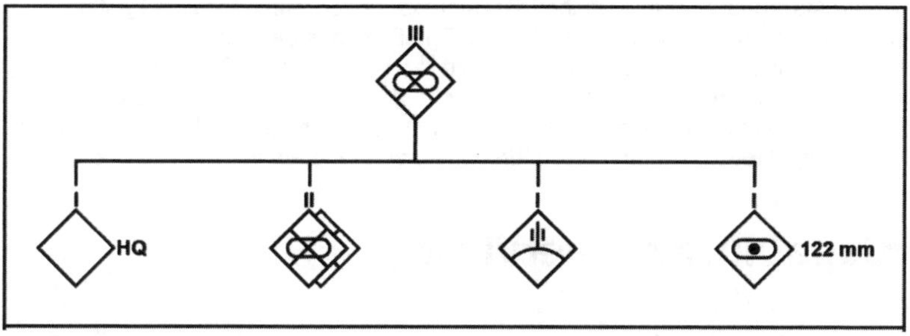


*Ref: ATP 7-100.3 (Aug '21), fig. 4-7. Thrust maneuvering group (example).*

**Offensive group: reserve group.** It includes—
- One mechanized CA-BN.
- One SPAAG battery.

**Firepower strike group: artillery group.** It includes—
- One 155-mm SPG battery.
- One heavy MRL battery.
- One light MRL battery.
- One UAS section (targeting).

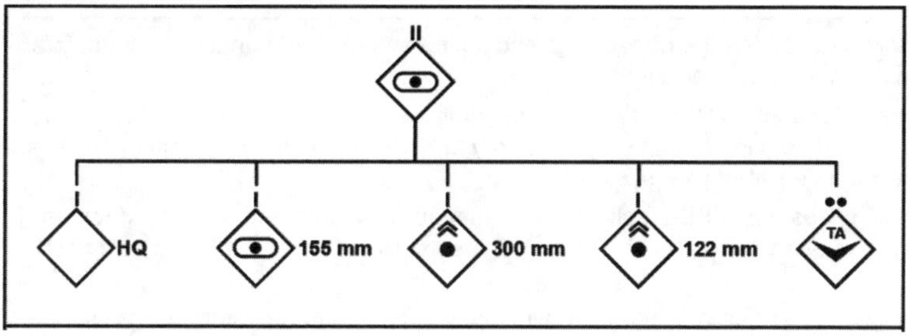


*Ref: ATP 7-100.3 (Aug '21), fig. 4-8. Artillery group (example).*

**Firepower strike group: air group.** It includes—
- One attack helicopter company.
- One medium lift helicopter company.

**Firepower strike group: mobile artillery group.** It includes two 155-mm SPG batteries.

**Firepower strike group: air defense group.** It includes—
- One medium-range surface-to-air missile battery.
- Two SHORAD batteries.
- Two SPAAG batteries.
- One heavy SPAAG or missile battery.
- One electronic air defense battery.

# PLA Capabilities

*Ref: ATP 7-100.3, (Aug '21).*

The PRC is a near-peer competitor to the United States, with significant political and strategic interests worldwide, though the PRC's primary geographic focus is Eastern and Southeastern Asia and the Western Pacific. The PRC has developed significant capabilities in all domains: land, air, maritime, space, and cyberspace. Capabilities in specific operational environments include a number of different unit and system combinations able to challenge the combat power of U.S. and allied forces. The PLA represents a significant competitor to U.S. and allied forces, especially in the Western Pacific.



## PLA Capabilities Overview



Appendix A describes the capabilities and limitations of PLA maneuver forces: infantry, armor, marine, and airborne.

Appendix B describes PLA indirect fire capabilities, including mortars, tube artillery, and rocket artillery in support of group army and tactical operations; this includes an overview of integrated fires systems.

Appendix C describes PLA tactical air defense forces controlled by its land forces, including an overview of integrated air defense systems up to the group army echelon.

Appendix D describes how the PLA aviation assets—primarily rotary-wing manned aircraft and unmanned aircraft—support tactical operations.

Appendix E describes the use of engineers and chemical defense forces in support of tactical operations.

Appendix F describes the PLA's communications and network architecture at the tactical level.

Appendix G describes special operations forces both at tactical and national/strategic echelons.

Chap 4

# I. China's Influence Operations

*Ref: Department of Defense Annual Report to Congress: Military and Security Developments Involving the People's Republic of China 2024, pp. 37 to 39.*

**Key Takeaways**

- The creation of the (*former) PLA Strategic Support Force (SSF) in 2015 reflected the CCP's understandings of cyber operations as the primary means for psychological manipulation. Even as the SSF was dissolved in 2024 in favor of support forces across the PLA's other military departments, psychological warfare remains prominent in CCP thinking of the military's evolving roles and responsibilities. *(*See pp. 2-57 to 2-58.)*
- The PLA concept of cognitive domain operations (CDO) combines psychological warfare with cyber operations to shape adversary behavior and decision-making. *(See p. 4-2.)*
- The PLA probably intends to use CDO as an asymmetric capability to deter U.S. or third-party entry into a future conflict or as an offensive capability to shape perceptions or polarize a society.



## I. The "Three Warfares" Concept *(See pp. 2-19 and 2-58.)*

The PLA views controlling the information spectrum in the modern battlespace as a critical enabler of information dominance early in a conflict. Beginning in the early 2000s, as part of the PRC's overall influence operations, the PLA began developing the "Three Warfares" concept, which calls for the coordinated use of public opinion warfare, psychological warfare, and legal warfare.

- Public opinion warfare creates and disseminates information to guide an adversary's public opinion and gain support from domestic and foreign audiences.
- Psychological warfare uses propaganda, deception, and coercion to induce pressure and affect the behavior of the target audience.
- Legal warfare uses domestic and international laws to shape narratives that advance PRC interests and undermine those of an adversary.

The PLA likely seeks to combine digital influence activities with the "Three Warfares" concept to demoralize adversaries and influence domestic and foreign audiences, creating an environment advantageous to the PRC. Since at least the mid-2010s, the PLA has been incorporating the concept of cognitive warfare or CDO into PLA frameworks for conducting influence operations. While the concept of cognitive warfare appears to be a PLA specific, overall PRC influence operations reflect a whole-of-government approach to shaping the information environment.

From the PRC's perspective, all nations—especially the United States—that use digital narratives to undermine the CCP's authoritarian system and employ offensive influence operations. Hence, the PRC considers its influence operations that counter this perceived subversion as defensive to protect the party and the military.

Another primary goal of the PRC's influence operations is to maintain domestic stability and protect CCP rule. Domestically, the CCP uses influence activities to protect its image with the public and garner popular support for the military. Internationally, Beijing aims to create an information environment favorable to the PRC and its strategic foreign policy objectives.

# II. PRC Cognitive Domain Operations (CDO)

*Ref: Department of Defense Annual Report to Congress: Military and Security Developments Involving the People's Republic of China 2024, pp. 38 to 39.*

The creation of the PLA SSF in 2015 reflected that the CCP understands cyber operations as the primary means for psychological manipulation. **As the PLA seeks to expand the reach of its influence operations around the world and seize information dominance on the battlefield, it is researching and developing what it believes to be the next evolution of psychological warfare, which it calls CDO.** CDO blends previous PRC concepts, such as public opinion guidance and psychological warfare, with modern internet technologies and communication platforms and is designed to achieve strategic national security goals by affecting a target's cognition to change the target's behavior and decision-making. The PLA has recognized the importance of incorporating emerging technologies, such as AI, big data, brain science, and neuroscience into CDO as the PLA perceives that these technologies will lead to profound changes in the ability to subvert human cognition.

The goal of CDO is to achieve what the PLA refers to as "mind dominance," which the PLA defines as the use of information to influence public opinion to affect change in a nation's social system, likely to create an environment favorable to the PRC and reduce civilian and military resistance to PLA actions. The PLA probably intends to use CDO as an asymmetric capability to deter U.S. or third-party entry into a potential conflict, or as an offensive capability to shape perceptions or polarize a society. Authoritative PLA documents describe one aspect of deterrence as the ability to bring about psychological pressure and fear on an opponent and force them to surrender. PLA articles on CDO state that seizing mind dominance in the cognitive domain and subduing the enemy without fighting is the highest realm of warfare.



## Synthetic Media in CDO

An aspect of cognitive warfare and CDO calls for the use of emerging technologies—such as AI and big data—to enhance operations, including the production and use of deepfakes. Deepfakes are a type of synthetic media where a person in an image or video is swapped with another person's likeness. PLA researchers have been interested in developing this technology since 2011, when National University of Defense Technology researchers proposed using audio-visual technology to imitate voices of foreign senior political and military leadership to mislead adversaries and shape their decision-making process. The researchers identified what they called "voice information synthesis technology" as a key enabling technology to accomplish this, indicating the development of underlying technologies used in deepfakes. In 2018, PLA researchers found that the PLA needed to improve other AI techniques, such as deep learning and natural language processing, to employ voice information synthesis technology to spread content online. In 2020, elements of the PLA had reportedly created a deepfake to mislead the U.S. public. The PLA has identified that using deepfakes presents the PLA with a low cost and quick-to-develop capability to shape the information environment with a low threshold for use in operations.

In 2023, PLA publications argued that cutting-edge generative AI applications, such as large language models, can enhance CDO by increasing the efficiency of creating synthetic media, including deepfakes. Leading generative AI technologies have greater authenticity and require less human input than previous AI technologies used for deepfake creation. PRC military researchers have complained that the PLA lacks the necessary staff with adequate foreign-language skills and cross-cultural understanding for authentic content generation. Leading generative AI technologies offer a potential technical solution to overcome this deficiency. Numerous PRC institutions, including leading technology companies, such as Baidu, Alibaba, and Huawei, are developing generative AI technologies for capabilities such as text, image, audio, and video creation.

Chap 4

# II. Tactical Information Operations (IO)

*Ref: ATP 7-100.3, Chinese Tactics, (Aug '21), chap 5.*

This chapter outlines the characteristics and principles that govern People's Liberation Army (PLA) and People's Liberation Army Army (PLAA) information operations at tactical echelons. In keeping with the teachings of Sun Tzu, the PLA considers information operations to be at least as important—if not more important— than maneuver or firepower. Deception, trickery, and concealment are to be employed extensively throughout the information operations campaign in order to manipulate the enemy commander's state of mind, the morale of enemy troops, and the enemy's understanding of the battlefield to the PLAA's advantage.

*Note. The concept of information operations will be used in this document consistent with the Chinese expression xinxi zuozhan. This is similar to the U.S. DOD term information operations, though it is not as inclusive. Chinese information operations include information warfare, concealment, deception (general efforts to mislead an opponent), and trickery (specific plans targeted at a particular opponent). The Chinese expression for information warfare, xinxi zhan, refers to direct, specific offensive and defensive actions, such as EW and cyber warfare, and is not analogous to the U.S. concept of information warfare that pertains to an opposing force.*



## I. Overview of PLA Information Operations

Information operations (IO) are comprised of the wide variety of lethal and nonlethal capabilities with the intent to influence the information environment. This includes information attack and defense, psychological warfare, and deception, trickery, and concealment efforts. The PLA considers IO to be a constant and ongoing effort before, during, and after conflict. Information warfare (IW)—also known as information confrontation—is a subcomponent of IO, and it consists of specific offensive and defensive information actions that are in direct support of military operations. Along with maneuver, firepower, special operations, and psychological warfare, IW is conducted as part of a campaign, usually referred to as campaign information warfare. This is defined by the Chinese as a series of integrated operational battlefield activities that target enemy sensors, information channels, information processing, and decision making. The PLA views IO as a zero-sum activity: they involve both attack and defense, and any information advantages gained in one area may be lost in another. As such, both information attack and defense are considered critical.

---

# II. Principles & Characteristics of IO

*Ref: ATP 7-100.3, Chinese Tactics, (Aug '21), pp. 5-3 to 5-4.*

## Principles

Four principles describe the PLA's concept of IO: actively attack, target nodes, achieve synthesis, and protect tightly.

### A. Actively Attack

Attack is the primary method by which initiative is gained in an IO campaign. The PLA stresses aggressive and integrated attack methods early in the conflict as the best way to ensure information superiority. Further, in an information war it is harder to steal or regain the initiative once it is lost, so greater emphasis is placed on seizing information superiority early in the campaign. Three specific characteristics of IO contribute to the focus on early action: its clandestine nature and ease of concealment; the low requirements on manpower and material resources, making operational sustainment relatively easy; and the inherent vulnerabilities of information systems due to their reliance on the electromagnetic spectrum. A strong initial attack takes advantage of these characteristics, but it requires the conditions to be set long before active hostilities commence.

### B. Target Nodes

The Chinese define a node as a critical component of an information system that either provides a capability or links other nodes. In this context it refers primarily to sensors, information processing centers, and the network backbone that enables them. Nodal attack attempts to identify, isolate, and target these objects, and it is the centerpiece of the IO campaign. Destroying or neutralizing nodes has the highest efficiency of any IW operation: if a critical node is destroyed or neutralized, then all systems reliant on it become either degraded or disabled. Nodal attack embodies the basic principle of targeting points of weakness rather than strength and, if applied appropriately, it can concurrently weaken strong points of the enemy system through isolation and confusion. It is best employed in an integrated manner along with psychological attack in order to maximize confusion and extend the effects of nodal destruction.

### C. Achieve Synthesis

Synthesis in the context of an IO campaign implies a variety of different measures all intended to carefully align, synchronize, and coordinate different IW efforts. This includes coordinating information attack and defense to maximize the efficiency of both; assuring deconfliction between different efforts so that one effort does not interfere with another; ensuring that all efforts are unified in their movement towards a clear objective; and focusing on building mutually supportive efforts that in turn create a combined arms effect.

Adjustment is an analogue to synthesis: it refers to the act of assessing and changing operations quickly and without the need for extensive planning. IO are inherently fluid; changes are rapid and often unpredictable. Agility in operations is vital to ensuring IW efforts are targeting the right objects and achieving desired effects. The primary focus of adjustment is to evaluate different friendly systems and their respective targets, then deconflict actions between them.

### D. Protect Tightly

Attack is viewed as the most important basic requirement in campaign IW. The corollary to this axiom is that one must blunt the enemy's attacks in order to effectively gain information superiority. The PLA views information defense as fundamentally more difficult than information attack due to the broad reliance upon—and disparity in—information systems. Because of the wide variety of different points of vulnerability, comprehensive information defense is a practical impossibility—so building systems with resiliency and

Information Operations

redundancy is critical. While information defense typically takes the form of passive defense, active defense also has a significant role. This may be described as information counterattack: the use of information attack capabilities to strike at and undermine enemy information attack efforts.

# Characteristics of Information Operations

Chinese IO have four primary characteristics: universal permeation, high target value, the importance of integration and synthesis, and the linkage between attack and defense.

## A. Universal Permeation

IO permeate all domains and are throughout all campaigns. Due to the widespread proliferation of advanced weapons systems, reliance on communications network technology, and information management systems, IO either govern or affect every aspect of modern conflict. Without information superiority, the Chinese believe that they cannot achieve dominance in any other domain, be it land, sea, air, or space. If the information battle is lost, initiative for the entire campaign cannot be achieved, and campaign activities will be fundamentally less efficient or otherwise blunted. IO are an ongoing and continuous process, while IW will likely begin before formal hostilities, making it a precursor to or first shot in active conflict. The conditions to achieve information superiority must be set prior to the commencement of hostilities, must actively support PLA forces during open conflict, and must be maintained through security and stability operations once open conflict has concluded.

## B. High Target Value

Information systems are high-value targets. The PLA defines information systems very broadly, including all sensors, information management systems, communications network systems, and decision-making centers at all echelons. System warfare theory suggests that targeting these subcomponents is often an optimal way to undermine the rest of the system, and it is in close keeping with the PLA's warfighting principle of avoiding the enemy's strengths and attacking its weaknesses. Information systems are considered to be the senses and brain of a modern force, and destroying or neutralizing them is an effective way of neutralizing or defeating a stronger enemy. The PLA refers to this approach as paralysis attack or decapitation attack. Attacking enemy information systems is a high-priority mission, while defending friendly information systems is of similar importance. The PLA prescribes a mix of capabilities, both lethal and nonlethal, when targeting enemy information systems.

## C. Importance of Integration and Synthesis

IO require a high level of integration and synthesis of a wide variety of systems, personnel, and functions. Effective integration of these varying components is referred to as synthetic quality, and it is one of the unique aspects of IO. Similarly, tactics such as deception, feints, or camouflage must be integrated into the wider information campaign in order to maximize their effectiveness.

## D. Linkage Between Attack and Defense

Information attack and defense must be closely linked. Information attack and defense are viewed as mutually supporting; information attack is most effective when friendly information and information systems are carefully protected, while information defense is enhanced when enemy IO are disrupted by offensive actions. IW is a zero-sum game, and any information superiority gained cannot be undermined by information lost in a different way. Initiative in the information battle is much the same as it is for maneuver or firepower: the side that effectively seizes initiative and then successfully maintains it enjoys a fundamental advantage throughout the conflict. Initiative cannot be either seized or maintained without effective integration of both attack and defense.

The objective of PLA IO are to gain information superiority or information dominance. Information superiority requires one to deprive the enemy of information, disrupt the enemy's ability to control or manipulate information of all types, and ensure friendly forces enjoy freedom of maneuver in the information domain. The PLA considers information superiority to be the new high ground of warfare in much the same way as the air domain during the mid-20th century or the land domain prior to that. Moreover, achieving information superiority is thought to have a very low cost-to-effectiveness ratio, achieving significant effects with relatively minimal expenditure in either lives or resources.

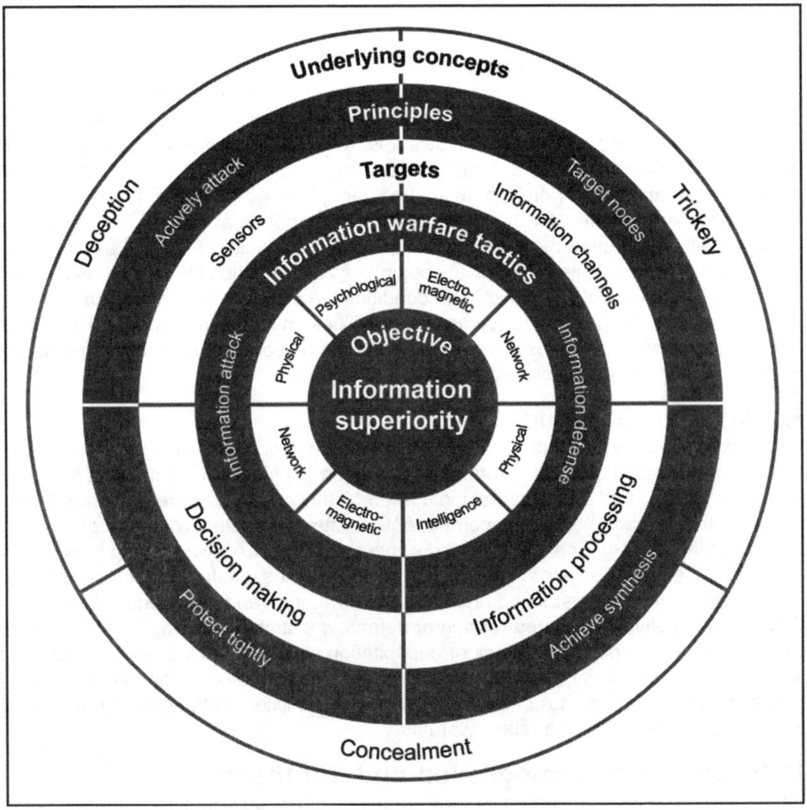


*Ref: ATP 7-100.3 (Aug '21), fig. 5-1. Chinese information operations.*

# III. Information Warfare Tactics

The PLA breaks IW techniques into two broad categories: information attack and information defense, with each of these containing several subcategories.

## A. Information Attack

The PLA defines information attack as any IW activity intended to weaken or deprive the enemy of control of information. Information attack is the primary means by which information warfare is won, and it is the key to achieving information superiority. There are four subcategories of information attack: electromagnetic attack, network attack, psychological attack, and physical attack. Of note, psychological attack is considered both a form of information attack and its own unique campaign; it is unclear what the relationship is between the two.

## Electromagnetic Attack

Electromagnetic attack encompasses those activities that manipulate the electromagnetic spectrum to jam, suppress, deceive, or neutralize enemy information systems. There are three subcategories of electromagnetic attack: electronic reconnaissance, electromagnetic suppression, and electromagnetic deception. Electronic reconnaissance attempts to use electromagnetic and other reconnaissance platforms to collect intelligence on enemy information operations platforms, enabling more effective electromagnetic attack operations. Electromagnetic suppression uses electronic warfare capabilities to jam or suppress enemy sensors or communications, softening or neutralizing enemy information systems. Electromagnetic deception manipulates enemy information operations systems—ideally, without enemy knowledge—in support of friendly operations. The PLAA views electromagnetic attack as the centerpiece of most IW operations; this belief is broadly represented in the increasing concentration of IW capabilities at tactical levels, including ground-based and aerial jammers and electronic reconnaissance platforms.

## Network Attack

Network attack includes those activities that target enemy computer information systems, software, hardware, and their associated networks. Network attack differs from electromagnetic attack in that it is conducted digitally and through networks, rather than through the electromagnetic spectrum. There are two major types of network attack: computer virus attack, which makes use of preprogrammed viruses, and hacker invasion, which uses active code experts working in real time. The aim of both methods is to penetrate enemy information systems. The two techniques can be mutually supportive, and should be closely coordinated. The PLAA broadly prefers clandestine or concealed network attack, wherein the enemy's information is manipulated without its knowledge. This is viewed as the most efficient form of information attack, allowing for more aggressive activities that remain beneath the threshold of active conflict. Concealed network attack operations in peacetime focus on understanding the enemy's operational architecture, disposition, and possible weaknesses, enabling more aggressive network attacks when hostilities commence. Concealed network attacks also focus on acquiring and exfiltrating critical information, particularly that surrounding the defense industrial base. The PLAA seeks to employ network attack even at tactical echelons in order to manipulate enemy situational understanding and trick the enemy into behavior conducive to PLAA tactical operations.

## Physical Attack

Physical attack comprises those activities that target enemy IO systems with direct physical damage or destruction. At tactical echelons, this primarily entails attacking command and communication centers, network nodes, and sensors. There are three subcategories of physical attack: force strength, which is essentially using maneuver forces to penetrate enemy defenses and destroy the IO systems at close range; firepower assault, which is the use of fires systems of all types—particularly artillery—to target and destroy IO systems at extended ranges; and energy weapons, which involves the use of highly specialized weapons systems to destroy enemy IO systems. The latter includes both current (such as antiradiation missiles) and future (laser or microwave) weapons systems. The PLAA emphasizes firepower assault as the most effective method of physical attack, and firepower assaults against IO systems are a key element of tactical system warfare.

## Psychological Attack

Psychological attack is composed of those activities that target the enemy's mindset and morale. While categorized as a form of information attack, psychological attack is also a component of psychological warfare.

## B. Information Defense

The PLA defines information defense as all those operational activities that protect friendly IW systems from enemy IO. The PLA views IW as a zero-sum activity: any successes gained from information attack are eroded by losses due to inadequate or ineffective information defense. Every potential offensive tactic has a corresponding defensive responsibility. Information attack and defense are viewed as mutually supporting, and they should be carefully integrated.

### Electromagnetic Protection

Electromagnetic protection contains those measures put into place to resist enemy electromagnetic attack. There are two main modes of electromagnetic protection: counterelectronic reconnaissance and counterelectronic jamming. Counterelectronic reconnaissance is the use of both active and passive means to prevent enemy collection on friendly IW systems. This includes active suppression of enemy collection systems, concealment of friendly electromagnetic signals, the use of decoy or spoof signals to confuse enemy collection, and physical targeting of enemy collection systems. Counterelectronic jamming consists of those systems and techniques that either eliminate or weaken the effects of enemy jamming. This includes hardening IW systems, using of more powerful or more resistant emitters, and carefully monitoring information to weed out disinformation planted by the enemy.

*Note. The Chinese concepts of electromagnetic attack and electromagnetic defense comprise what both China and the U.S. refer to as electronic warfare (EW).*



### Network Protection

Network protection encompasses those measures put into place to resist enemy network attack. There are two main types of network protection: computer virus defense and hacker defense. Each seeks to proactively protect friendly hardware, software, and networks from both overt and covert enemy intrusion. Network defense is both an active and passive activity. Passive defense seeks to prevent, disrupt, or delay intrusion, while active defense seeks to identify and stop intrusion after it occurs. Due to the extensive use of computer networks, along with the wide variety of hardware and software in use, network defense is seen as a highly difficult—though highly important—tactical task.

*Note. The Chinese concepts of network attack and network defense comprise what both China and the U.S. refer to as cyber warfare.*

### Physical Protection

Physical protection consists of those activities that defend IW systems from direct lethal attack by enemy forces. There are two categories of physical protection: force strength and firepower, and quality protection. Force strength and firepower incorporates tactical activities that preclude, preempt, neutralize, or defeat enemy direct lethal attack against IW systems. They can be broadly grouped into either defensive and security activities, such as tactical ground defense, or firepower activities, such as counterfire. Quality protection comprises those activities that harden or hide IW systems, making them more difficult to neutralize or destroy. This includes both engineering measures—such as building physical barriers, using underground facilities, or camouflage and concealment—and survivability measures, such as adding armor. The PLAA is very concerned about the vulnerability of IW systems to direct attack—particularly by firepower and energy weapons—and it emphasizes survivability against such attacks.

### Intelligence Protection

Intelligence protection is IW's counterintelligence component. This includes measures and activities that defeat enemy reconnaissance, surveillance, and espionage, concealing friendly operations and capabilities. Intelligence protection in the form of disinformation can be integrated with information attack—particularly psychological attack—in order to effectively deceive the enemy and trick it into behaving in a preferred way.

# III. PLA Psychological Warfare

*Ref: ATP 7-100.3, Chinese Tactics, (Aug '21), chap 5.*

## I. Overview of PLA Psychological Warfare

Chinese psychological warfare encompasses those IO activities wherein a combatant employs information and media in order to target human thought, emotion, and spirit. At tactical echelons, the objective of psychological warfare is to create a psychological condition favorable to friendly forces and unfavorable to the enemy, reducing enemy morale, and consequently, the enemy's will to resist. The PLA views psychological warfare as the operational element of the fundamental reason for conflict: a contest of wills. As such, the PLA notably prioritizes psychological operations. Psychological warfare is considered a fourth operational mode, in addition to land, air, and sea warfare. An effective psychological warfare campaign is considered to be the best possible trade-off, paying a small price in lives and material for as big a victory as possible.

*Note. The PLA appears to categorize domains in much the same way as does the United States, but its categorization of psychological warfare as an operational mode is unique. It is unclear how operational modes differ from domains.*

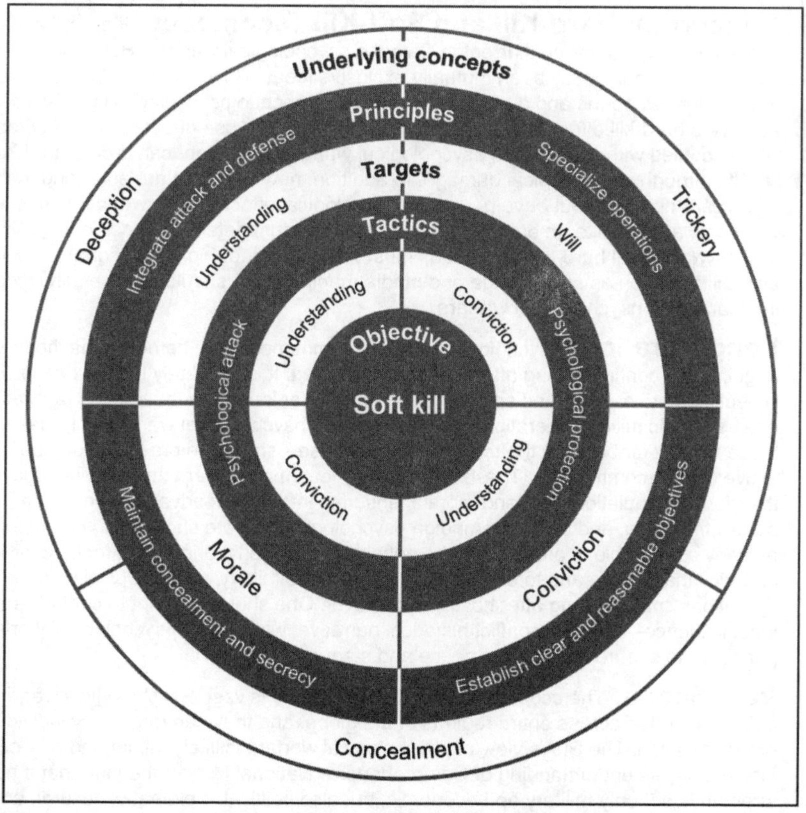


*Ref: ATP 7-100.3 (Aug '21), fig. 5-2. Chinese psychological warfare.*

# II. Characteristics & Principles of Psychological Warfare

*Ref: ATP 7-100.3, Chinese Tactics, (Aug '21), pp. 5-7 to 5-10.*

## Characteristics of Psychological Warfare

**A Central Role.** Psychological warfare has a central role in modern conflict. The informationized battlefield transitioned psychological warfare from a peripheral or auxiliary military activity to one of the most important operational forms. Due to modern media and social integration, psychological warfare efforts can now target the enemy anywhere and at any time. This includes soldiers on the battlefield, in rear areas, or on the home front; civilians, both enemy and others across the world; and decision makers, including those at the highest level. Psychological warfare must be proportional to the conflict; however, a full-scale deep psychological warfare campaign should not be employed in a localized or limited conflict, for instance. Every military activity has some value psychologically and politically, and these second-order effects must be considered carefully by commanders. A campaign's success is no longer measured by casualties, material loss, or territory; instead, it is measured by its overall psychological effect on enemy forces, friendly forces, and neutral or outlying parties around the globe. The ultimate goal of warfare of all types is winning the contest of wills, and this does not always mean that the side winning the tactical battle wins the war.

**Concurrent Hard-Kill and Soft-Kill Techniques.** Hard-kill and soft-kill techniques are concurrent in Chinese psychological warfare. Hard kill and soft kill have traditionally been mutually exclusive ideas. However, the PLA believes that modern weapons and a highly integrated approach to psychological warfare can achieve a hard-kill effect through soft-kill capabilities. The use of overwhelming firepower, when coupled with an effective psychological warfare campaign, can create hard-kill results through psychological damage. In addition, new weapons are emerging that are technically nonlethal but have powerful psychological effects: sonic weapons, microwave weapons, and the like. In order to achieve the maximum deterrent effect, both physical and psychological hard-kill capabilities must be considered. More-traditional soft-kill capabilities, such as propaganda and media manipulation, should also be fully integrated into campaign psychological warfare.

**Long Duration.** Psychological warfare commences long before formal hostilities begin, and it continues long after they are concluded. It is ultimately tied to a nation's collective morale, unity, and spirit, and so it lasts a far longer period than do tactical or even strategic military operations. Subtle or covert psychological warfare activities—those that fall far beneath the threshold of response—should be practiced long before active conflict commences. These will naturally lean more toward the defense, buoying the friendly population's will and morale, ensuring information advantage in media across the globe, and so on. Campaign psychological warfare should also commence as early as possible in order to set the right information environment; to fool, deceive, and trick the enemy; and to erode the enemy's will. Similarly, psychological warfare operations continue long after hostilities are over. One should attempt to control—or at least influence—the post-conflict historical narrative, while using psychological warfare principles to ensure long-lasting peace and stability.

**Vast Scope.** The scope of psychological warfare is vast. Psychological warfare activities stretch across entire regions of the globe, and they can target any individual or organization. The PLA's view of psychological warfare is likely influenced by—or influences—its understanding of Comprehensive National Power: the idea that a nation's strength is not only military and economic, but also political, diplomatic, cultural, and religious in nature. Psychological warfare activities can influence or target anything within any of these fields, and they render the battlefield wholly nonlinear, making coordinated attack and defense not only complicated, but perhaps one and the same. Achieving psychological superiority in conflict establishes a baseline advantage for one's side that permeates to the lowest tactical echelons, and it can enable a technologically or tactically inferior force to prevail in the face of otherwise insurmountable odds.

# Principles of Psychological Warfare

**Integrate Psychological Attack and Protection.** The PLA views campaign psychological warfare through much the same lens as any other combat campaign; in this sense, it includes both offensive and defensive activities. Further, one should target an enemy's weaknesses while attempting to fool or deceive it into attacking one's strengths. Just as with maneuver warfare, success in psychological warfare centers on seizing the initiative, and this is accomplished through attack. Maintaining the initiative is largely a defensive effort. Though different missions, attack and protection are fundamentally linked, and they are mutually supporting. One key difference in campaign psychological warfare is that most psychological protection (or defense) is passive in nature. For example, one cannot hope to prevent an enemy from attempting to influence a population through the global media structure; instead, one must prepare the population to resist these efforts. This process is ongoing and cannot be successfully implemented on a short timeline. Another key idea is that of the psychological counterattack: a targeted effort designed to blunt, disrupt, or marginalize enemy psychological attacks. While broadly similar to other forms of counterattack, the PLA designs psychological counterattacks to target wide audiences and offset enemy psychological attacks on a broad scale, rather than targeting specific enemy operations.

**Specialize Operations.** The idea of a specialized psychological operations section is relatively new in both the PLA and the PLAA. In the past, psychological operations generally fell to political officers and nonspecialized troops. Following the model of the U.S. military, the PLAA now builds specialized psychological warfare groups at a variety of different echelons, ensuring integrated support for ground operations

**Establish Clear and Reasonable Objectives.** The wide scope of psychological warfare naturally lends itself to making campaign psychological warfare objectives extremely broad, perhaps excessively so. Planners must carefully assess operational needs and then craft specific, achievable psychological warfare goals. This is particularly true for information and disinformation campaigns. While it is now possible to promulgate a message to a large number of people with relative ease, messages must be at least plausible in order to be of any use. Indeed, messaging viewed as fanciful or clearly false may lead the target to dismiss further messages and sabotage follow-on psychological warfare efforts. The most effective psychological warfare campaigns are those that are focused and that present their targets with believable messages that roughly align with their existing perspectives and biases.

**Maintain Concealment and Secrecy.** Deception is ultimately at the heart of nearly every psychological warfare operation. Psychological warfare is one of the only forms of warfare wherein awareness of an enemy's presence decreases the overall effectiveness of the effort. Concealment is at the core of all deception efforts. The source and delivery mechanism for psychological warfare actions should remain opaque to the intended recipient whenever possible. Deception operations are at the core of a number of other PLA and PLAA campaigns, and nearly all of these efforts can be enhanced by effective psychological warfare operations. Secrecy can be considered in much the same vein as concealment: a fundamental component of a successful psychological warfare campaign. Since the main point of psychological warfare is to misinform, confuse, fool, or trick the enemy, any breaches in operational security can conclusively undermine psychological warfare efforts by allowing the enemy to know what is true and what is false.

The PLA integrates psychological warfare into all elements of its operations. Psychological warfare is one of the main reasons why the PLA employs political officers in most of its organizations, and the PLA considers itself to have a fundamental advantage in psychological warfare against nearly any opponent due to its political unity of purpose.

Psychological warfare seeks to achieve the soft kill: the use of nonlethal coercive means to impose one's will upon an uncooperative opponent. This contrasts with hard-kill techniques, which involve the use of physical force to achieve this coercion. Psychological warfare is thought to have uniquely powerful soft-kill capabilities: a well-constructed psychological campaign is the only thing that can achieve Sun Tzu's definition of supreme excellence—subduing one's opponent without fighting.

# III. Psychological Warfare Tactics

The PLA divides campaign psychological warfare into attack and protection, similar to IW. These two aspects result in discrete but mutually supporting operations, and they should be tightly integrated in order to achieve a combined arms effect.

## A. Psychological Attack

Psychological attack contains those activities that target an enemy's situational understanding, conviction, will, and morale. The goal of psychological attack is to control the enemy's awareness and fighting spirit such that its combat power is weakened. The PLAA views this as a supporting effort to information attack, although psychological attack is necessarily broader and less technical in nature. There are two primary methods by which psychological attack is carried out: attacking psychology of understanding and attacking psychology of conviction. Psychology of understanding attacks target the enemy's situational understanding, and they can include everything from camouflage and concealment, to feint attacks or demonstrations, to disinformation efforts. The general objective of these attacks is to lead the enemy into an information trap, a situation wherein enemy leaders and planners make erroneous, faulty, or inefficient decisions due to faulty, insufficient, or misleading information. This can be described as getting inside of the enemy's decision-making cycle, and it is considered one of the most difficult and valuable operational arts in all of warfare.

Deception is the activity at the heart of psychology of conviction attacks. Practically every PLAA operation places as high a premium on deceiving the enemy as it does maneuver or firepower. Attacks against psychology of conviction target the enemy's fighting spirit, will, and morale. At political echelons, this involves broad campaigns to convince or coerce enemy civilians and politicians to abandon support for the conflict. At tactical levels, the PLAA takes a notably Marxist approach.

## B. Psychological Protection

Psychological protection encompasses those activities that attempt to blunt, minimize, or neutralize enemy psychological warfare activities. Psychological protection is divided into two subtypes in much the same way as psychological attack: protecting psychology of understanding and protecting psychology of conviction.

Protecting psychology of understanding involves attempting to defeat enemy deception or concealment activities or to minimize enemy disinformation efforts. The objective is to give commanders a clear and accurate picture of the battlefield, enabling quick and accurate decision making, and keeping the enemy out of the friendly decision cycle.

Protecting psychology of conviction includes those activities that maintain friendly cohesion, morale, and will to fight. It includes defeating broad enemy propaganda efforts or more-direct attacks against friendly soldiers. It also includes political education efforts and other measures designed to enhance esprit de corps, and mental strength and resiliency measures put in place to protect and enhance the individual soldier's psychological well-being.

Chap 5

# I. Reconnaissance Actions

*Ref: ATP 7-100.3, Chinese Tactics, (Aug '21), pp. 6-1 to 6-12.*

The People's Liberation Army Army (PLAA) places a high priority on its reconnaissance and security operations. These are integral to creating and sustaining situational awareness, and they are seen as essential to developing an advantage over one's opponent. System warfare and the application of combat systems are underpinned at all levels by extensive reconnaissance. PLAA maneuver units place a high priority on mounted and dismounted reconnaissance: these units are seen as prestigious, and their soldiers and leaders are highly regarded. Ground reconnaissance remains the chief method by which tactical units gain intelligence on their enemies, but air, space, and electronic reconnaissance are all integrated to varying degrees, even at tactical echelons.

PLAA security operations are linked to ground reconnaissance. Their primary intent is to protect main-body formations, give commanders information and decision space, and enable effective risk assessment. The People's Armed Police (PAP) also conduct combat and area security operations, both domestically and in support of PLAA operations. In combat, the PLAA and PAP likely cooperate and coordinate to ensure security throughout the combat area.

## I. Reconnaissance Fundamentals

Recognizing the changes brought by the informationized battlefield, the PLAA invested substantial time and resources in developing its reconnaissance capabilities. Throughout the Cold War, PLAA units had only light mounted scouts and unreliable manned aircraft to conduct reconnaissance. Despite experiences in the Korean War and the basic tenets of People's War that underpinned the need for effective reconnaissance operations, the PLAA approach during this era was generally to conduct a massive reconnaissance effort using primitive methods and obsolete equipment. The post-reform PLAA, however, places heavy emphasis on advanced reconnaissance methods: more-advanced surveillance platforms, advanced techniques and better training, mechanization and motorization, and intelligence integration. Modern PLAA reconnaissance doctrine is designed to support a sophisticated mechanized force and the system warfare approach.

PLAA reconnaissance operations obtain information, develop situational understanding, support decision making and actions, and protect key elements of combat power. Reconnaissance and security operations are seen as continuous. Their most important role is providing early and accurate warning of adversary or enemy actions and intent, information about enemy dispositions and mindset, and decision space for commanders at all echelons. The PLAA puts a high priority on reconnaissance operations of all types, considering a commander's situational awareness to be a key factor in being able to detect and exploit enemy weaknesses. Reconnaissance and security missions are typically interwoven with combined arms mission tasks. Reconnaissance is considered an ongoing task that is conducted before, during, and after operations. PLAA leaders match reconnaissance capabilities with missions, seeking to ensure that reconnaissance efforts support coordinated intelligence collection efforts.

*The PLAA recognizes four key fundamental principles when conducting reconnaissance at all echelons: continuous, command-directed, performed through action, and agile.* *See p. 5-3.*

Recon & Security

# II. Reconnaissance at Echelons Above Brigade

The PLA views reconnaissance as a national-level mission, and it employs national assets in support of reconnaissance operations at all echelons. Examples of national-level reconnaissance assets include satellite surveillance, cyber and electronic surveillance, national information and disinformation campaigns, and national-level special operations forces (SOF) operations. Collectively, these resources focus either on strategic-level targets or on deep areas of the battlefield. Tactical commanders may benefit obliquely from surveillance on strategic-level targets, but they are reliant on these assets' contribution to deep reconnaissance.

Deep reconnaissance is loosely defined as reconnaissance operations focusing on areas behind that which can be serviced by a given unit's organic weapons systems. For the PLAA combined arms brigade (CA-BDE), this means the areas past the ranges of its own tube and rocket artillery, beginning at somewhere between 35 kilometers (km) and 100 km, and extending through the remainder of the theater of operations. This area typically contains enemy command posts, supply areas, air and sea ports, reinforcement route and staging areas, and long-range fires.

PLA deep reconnaissance operations consist of a mix of long-range persistent aerial and ground sensors using a variety of manned and unmanned systems, long-endurance SOF and similar light infantry units, and space-based systems. A significant proportion of deep reconnaissance assets are housed in the theater command's reconnaissance and intelligence brigade, while others are national-level assets. Some deep reconnaissance operations are focused on supporting national- and strategic-level assets, while some play an important role in support of tactical operations. Deep reconnaissance capabilities in support of the latter provide targeting for long-range fires, supply commanders with imagery analysis and signals intelligence (SIGINT), and collect on the morale, disposition, and cohesion of large enemy formations. Deep reconnaissance assets are deployed using a combined arms approach, where the strengths of one system offset limitations of another. Collection efforts that cannot be processed in real time, such as imagery analysis, are cross-referenced with collection efforts from SOF and clandestine personnel or information warfare capabilities to create a more holistic intelligence picture than would be possible with a lone system.

Deep reconnaissance operations feed the PLAA's military planning process during all phases. PLAA commanders may request or be allocated deep reconnaissance assets based on mission requirements. Much of the collection performed in support of tactical echelons is performed by manned and unmanned aircraft, with dismounted ground forces in support. Intelligence is collected in a cycle, wherein collection assets are paired with areas of interest as part of a wider collection plan.

# Reconnaissance Principles

*Ref: ATP 7-100.3, Chinese Tactics, (Aug '21), pp. 6-2.*

Reconnaissance and security missions are often linked. Reconnaissance can be seen as a proactive task, seeking to develop situational understanding through action, while security is a reactive task, done primarily to preclude enemy actions and enable effective transition to offensive operations. PLAA leaders build reconnaissance intelligence systems to support each mission, task-organizing reconnaissance capabilities that are coordinated with the requirements of the unit's main effort. Dedicated security missions may be assigned to assault groups, combat reserve groups, or defensive groups of all types. These units are designated to screen and protect critical assets from enemy actions. Local security is the responsibility of every unit at all times.

The PLAA recognizes four key fundamental principles when conducting reconnaissance at all echelons: continuous, command-directed, performed through action, and agile.

## Continuous

Reconnaissance is continuous. Though a reconnaissance mission may be in support of a specific mission or objective, it must commence before and continue after its supported mission. Collection plans and updated reconnaissance objectives ensure an uninterrupted flow of information to the commander, enabling appropriate reactions to enemy activities and changes on the battlefield.

## Command-Directed

Reconnaissance is command directed. Reconnaissance efforts must be focused on those objectives that support the unit's mission. Reconnaissance forces must be given clear objectives, rules for engagement and disengagement, and follow-on missions. Commanders must ensure that reconnaissance formations are deployed to support established collection efforts.

## Performed Through Action

Reconnaissance is best performed through action. PLAA forces place great emphasis on direct action in reconnaissance—engaging enemy forces in order to develop situations, fix enemy positions, and force enemy commanders into early decisions. Reconnaissance through action also includes the use of airborne sensors to trigger enemy responses and reconnaissance by fire to reveal enemy positions and systems. Direct action also achieves a disruption effect against the enemy, reducing its cohesion and increasing uncertainty.

## Agile

Reconnaissance must be agile. The People's Liberation Army (PLA) is universally struggling with a transition from a strong top-down command approach toward a more decentralized methodology. In previous generations, PLAA reconnaissance efforts—particularly patrols—were strictly controlled by commanders, with patrolling elements expected to stick steadfastly to predetermined routes and observation objectives. The reformed PLAA is placing great emphasis on a more flexible approach to patrolling, allowing the patrol leader discretion in supporting the commander's objectives.

# III. Reconnaissance at the CA-BDE

*Ref: ATP 7-100.3, Chinese Tactics, (Aug '21), pp. 6-3 to 6-5.*

Reconnaissance at the CA-BDE is fully integrated with the general intelligence efforts at higher echelons. Intelligence officers are charged with developing an intelligence plan that supports the objective of the commander. Reconnaissance operations are a major component of this plan, especially at tactical echelons. The PLAA employs a simple and robust intelligence cycle designed to integrate all-source intelligence assets with unit operations. PLAA intelligence is intended to be very specific, with discrete questions driving discrete answers.

## Collection Principles

Five principles guide PLAA intelligence collection efforts: mission-supporting, target-oriented, well-planned, well-hidden, and selective.

### Mission-supporting

Perhaps more than any other soldier, the reconnaissance soldier must understand the commander's mission and how the reconnaissance effort fits into it. This includes not only understanding the design of the mission, but also its intent.

### Target-oriented

Intelligence efforts must focus on targets—those objects or activities that best inform commanders and enable their decision-making process. Accuracy is inherent in this principle; inaccurate intelligence is at best useless, or at worst, dangerous.

### Well-planned

Reconnaissance operations must be well conceived and carefully thought out. This includes identifying objectives, schemes of maneuver, and criteria for engagement, disengagement, and reporting. Conducting reconnaissance without a proper plan results in inconsistent intelligence and compromises the larger mission.

### Well-hidden

Intelligence operations are at the center of deception, and they are key in gaining an information advantage over the opponent. Carefully guarding intelligence gains, while simultaneously denying or deceiving the enemy's collection efforts, is considered an effort of the highest priority. The intelligence contest is viewed as zero-sum: allowing one's opponent to gain information is detrimental to one's own intelligence operations.

### Selective

Intelligence efforts must be prioritized; if everything is a focus, then nothing is a focus. Commanders must make clear what their most important pieces of information are, then efforts to match collection assets with targets must mirror this prioritization.

## Collection Process

The PLAA reconnaissance and intelligence collection cycle follows four steps. The process is considered continuous, beginning again immediately upon completion.

### Develop the Reconnaissance Plan

This step requires the intelligence officer to identify key terrain features, enemy formations, or other battlefield features that will contribute the most to the commander's decision-making cycle. Intelligence requirements are typically expressed as questions that have clear, unambiguous answers. Good requirements ask and answer questions that are of immediate use to commanders; questions considered of the most value are those which inform the commander about the morale, disposition, and potential weaknesses of the enemy force. PLAA commanders seek to get inside of their opponent's decision cycle, at which point they can disrupt, deceive, or otherwise manipulate the opponent to the commanders' advantage.

## Deploy Reconnaissance Forces

Reconnaissance forces are deployed in order to answer the questions asked by the reconnaissance plan. Commanders should employ all available reconnaissance capabilities and encourage their reconnaissance forces to employ as many different techniques and technologies as possible. If higher echelon reconnaissance forces are operating in the area, commanders should integrate or cooperate with them. Examples of critical information that reconnaissance forces may seek include enemy strength and disposition, enemy firepower, obstacles, locations of enemy command posts and communication hubs, and key battlefield terrain features.

## Screen and Integrate the Information

As information begins to flow into the command, the intelligence officer and staff must effectively filter and screen it for accuracy, consistency, and usefulness within the context of the reconnaissance plan. Conflicting reports must be resolved, and incomplete reports may be compared with other information in order to increase accuracy or thoroughness. This phase also emphasizes the importance of security and secrecy in the reconnaissance process: the intelligence officer must ensure that enemy forces do not gain access to recently gathered intelligence, nor be able to effectively collect on PLAA forces.

## Report

As intelligence is processed, it must be quickly and accurately reported both up and down the chain of command. Reports to higher echelons answer their intelligence requirements, while reports to lower echelons enable prosecution of the commander's intent. Traditionally, the PLAA was very slow to process intelligence, as it was deemed necessary to compute both the tactical and political elements of new information before acting on it. The reformed PLAA places great emphasis on rapid reaction to new data.

# IV. Reconnaissance Capabilities

*Ref: ATP 7-100.3, Chinese Tactics, (Aug '21), pp. 6-3 to 6-5.*

The PLAA considers reconnaissance an all-arms occupation, and basic training in reconnaissance, patrolling, reporting, and assessing is provided for all PLAA soldiers. Specialized reconnaissance units, along with SOF, reconnaissance systems, and national-level assets, provide a comprehensive suite of sensors intended to give the intelligence officer a wide-ranging menu of options with which to fulfill intelligence requirements.

**Soldiers and units** are the most basic reconnaissance assets. All soldiers are trained on the basics of observation, recording, patrolling, and assessing enemy formations. Collecting information of this type has not historically been a strength of the PLAA. Soldiers were often unable to report developments to their commander due to poor upward communication, a lack of literacy, and poor communications infrastructure. PLAA reforms seek to offset this historic weakness through enhanced training, better-quality recruits, a more decentralized command approach, and an emphasis on bottom-up messaging.

**Observer teams** are the most basic dedicated reconnaissance element. They may be trained scout troops or they may be soldiers tasked to perform observation. The observer team functions as a stationary element typically assigned to observe a particular target, most often a piece of key terrain. The observer team then informs its higher echelon command if and when an enemy unit occupies or traverses this terrain. Observer teams are usually not charged with conducting direct action, though they may be given a security mission in addition to their reconnaissance task. Placing and managing observer teams is a basic competency for lower-level PLAA leaders. Observer teams often form the first security line for a deployed unit, screening the unit's main body and giving the unit's commander decision time and space in the event that enemy contact is made. Observer teams may be augmented by unmanned aircraft systems (UAS) or other advanced surveillance systems; simple unmanned aircraft (UA) may be operated by the observer team without additional support.

The **scout team** is one of the oldest and most important reconnaissance capabilities in the PLAA. Previous generations relied heavily on the simple two-man dismounted team to provide most ground intelligence—probes to determine enemy disposition, weak points, and avenues of approach were virtually all performed by two-man scout teams. The basic mission of the scout team is the patrol: a movement designed to surveil one or more reconnaissance objectives over a specified geographic area. The scout team remains a key element of reconnaissance; trained scouts are present in all units down to the company level. Scouts today are often mounted, allowing them to conduct patrols over wider areas. They also often have direct communications back to their headquarters element, allowing for real-time updates to commanders. Scout teams may also be augmented by UAS or electronic surveillance capabilities, and many scout teams can operate small UASs without additional support.

The **CA-BDE's reconnaissance platoons**, housed within the reconnaissance battalion, provide the bulk of the brigade's long-range reconnaissance patrol capability. The reconnaissance platoon is capable of conducting independent patrols over several days, deploying numerous scout teams over a wide area to perform reconnaissance and security tasks. These platoons operate small organic UA to enhance ground surveillance operations. The reconnaissance platoon may have a direct communications link to the CA-BDE's artillery battalion in order to facilitate rapid artillery fire on time-sensitive targets or to facilitate reconnaissance-by-fire operations. Reconnaissance units also conduct raids, a form of hit-and-run direct action aimed at forcing a confrontation with enemy formations. Raids may be used to reveal enemy positions, to seize enemy prisoners, or to disrupt enemy maneuver. Reconnaissance platoons are likely fully motorized, with heavy CA-BDE reconnaissance platoons operating armored vehicles, such as eight-by-eight light armored vehicles or infantry fighting vehicles (IFVs), in a cavalry-type role. Specialized cavalry fighting vehicles equipped with radars and other sophisticated sensors are being fielded, but there are still of relatively few of them. Reconnaissance platoons work in concert with SOF units, particularly those SOF units assigned to the SOF brigade of the group army. In most cases, the reconnaissance battalion comprises the core of the CA-BDE's reconnaissance intelligence system.

Each CA-BDE houses numerous **UASs**, with some in the reconnaissance battalion and some in the artillery battalion. These units operate group two and possibly group three UA, enabling aerial surveillance with a sophisticated suite of sensors in a maximum combat radius of roughly 80 km to 100 km. Sensor options available include simple visual sensors, advanced electro-optical telescopic sensors, infrared sensors, radars, and passive SIGINT sensors. Many PLAA UA are reconfigurable based on mission requirements, allowing commanders to tailor sensor suites to support the collection plan. The reconnaissance battalion's UASs support the ground reconnaissance efforts of the battalion, and they may be task-organized to support reconnaissance platoon operations, particularly long-range or long-endurance patrols.

Each group army operates an **aviation brigade**, which in turn operates a significant number of scout helicopters. Basic aerial reconnaissance has been a key capability of the PLAA for decades, but dedicated scout helicopters are a relatively new addition. It is likely that aviation brigade assets will be task-organized to subordinate CA-BDEs based on mission priority. In many cases, a CA-BDE may be assigned a section of scout helicopters in order to support its collection plans. Scout helicopters are typically light utility helicopters equipped with advanced sensors and limited weaponry, and they can provide direct air-ground communications to CA-BDE leaders as they conduct reconnaissance patrols over wide geographic areas. Scout helicopters may also communicate directly to attack aviation or artillery units, providing elevated forward observation in support of firepower operations.

The CA-BDE employs limited **technical reconnaissance** capabilities. These include electronic intelligence (ELINT) and SIGINT systems that support collection efforts through surveillance of the electromagnetic spectrum. PLAA systems of this type were traditionally limited to simple radio direction-finding equipment; ELINT and SIGINT operations were clearly less important than ground patrols. PLAA reforms place much greater emphasis on ELINT and SIGINT, particularly in support of reconnaissance and counterfire operations. PLAA tactical-level ELINT and SIGINT capabilities include radio direction finding, network surveillance, radar detection, and fire finding. These capabilities are rare, and the doctrine and training surrounding their deployment is immature. The systems fielded may vary widely in capability. PLAA commanders also employ decoy systems, designed to spoof, jam, or deceive enemy collection systems, as part of their intelligence operations. CA-BDEs will likely receive significant ELINT and SIGINT intelligence support from group army collection efforts and regional military capabilities such as dedicated ELINT and SIGINT surveillance aircraft.

**SOF at the group army** provide an enhanced long-range ground patrol capability that can be employed either alongside or in addition to the operations of the reconnaissance battalion. Long-range patrolling is considered a key competence for PLA SOF. Small SOF units are capable of conducting days-long patrols both in direct support of CA-BDEs and in deep reconnaissance areas.

**Joint and national assets** may provide information to CA-BDE commanders and intelligence officers. These assets include rare and highly capable ELINT and SIGINT platforms, satellite surveillance, cyber intelligence, and political intelligence. The degree to which these collection assets are available to the CA-BDE vary significantly. A CA-BDE, for instance, may receive direct support from a space-based asset for a high-priority mission, but it may only passively access national-level intelligence data at other times. National-level assets also contribute heavily to deception campaigns, providing disinformation and decoy efforts that support tactical operations. Wide-area deception operations are considered a very high-priority mission, but it is unclear how effectively they are coordinated with activities at tactical echelons.

# V. Forms of Tactical Reconnaissance

The PLAA frames tactical reconnaissance in much the same way as the U.S. Army. Ground reconnaissance efforts are tailored to meet specific intelligence collection requirements using one or more forms of reconnaissance. Information superiority at tactical levels begins with a comprehensive and well-executed reconnaissance plan. Tactical reconnaissance efforts are integrated with operational and strategic intelligence efforts to provide the commander with a holistic picture of the battlefield, inform rapid and effective decision making, and deny or deceive enemy collection efforts.

The PLAA recognizes two primary forms of tactical reconnaissance: combat reconnaissance and search.

## A. Combat Reconnaissance

Combat reconnaissance consists of those reconnaissance activities that involve direct action against the enemy. These are implemented before or during a larger operation to inform the commander's decision making as the battle unfolds. Combat reconnaissance is the centerpiece of most PLAA reconnaissance activities, and great emphasis is placed on winning the information battle during aggressive reconnaissance activities. All combat reconnaissance techniques are designed to force the enemy to make a decision— usually related to whether or not to engage—and in doing so, revealing its disposition, strength, or other capabilities.

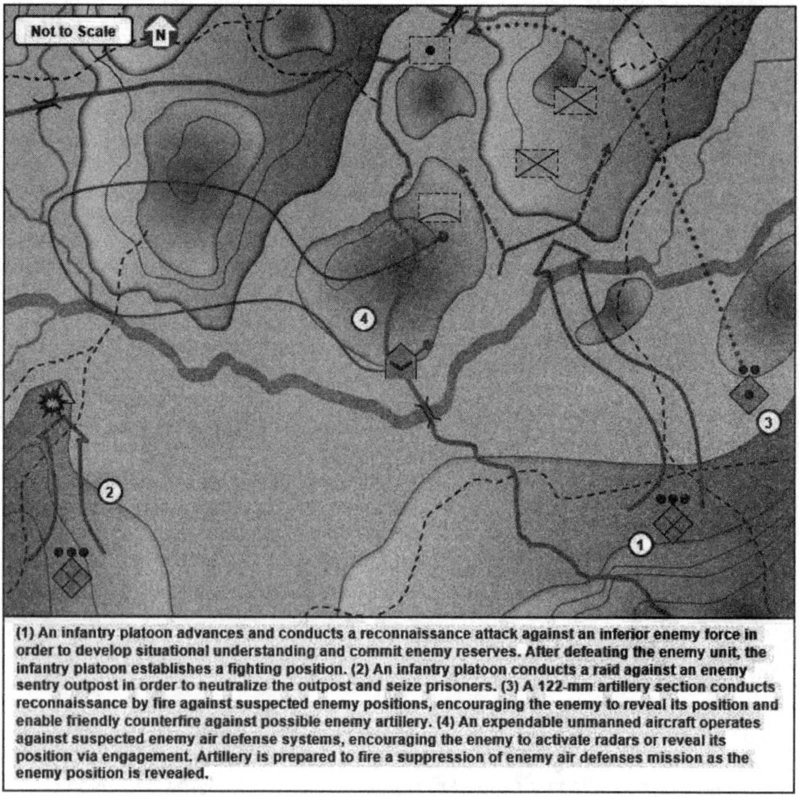


**(1) An infantry platoon advances and conducts a reconnaissance attack against an inferior enemy force in order to develop situational understanding and commit enemy reserves. After defeating the enemy unit, the infantry platoon establishes a fighting position. (2) An infantry platoon conducts a raid against an enemy sentry outpost in order to neutralize the outpost and seize prisoners. (3) A 122-mm artillery section conducts reconnaissance by fire against suspected enemy positions, encouraging the enemy to reveal its position and enable friendly counterfire against possible enemy artillery. (4) An expendable unmanned aircraft operates against suspected enemy air defense systems, encouraging the enemy to activate radars or reveal its position via engagement. Artillery is prepared to fire a suppression of enemy air defenses mission as the enemy position is revealed.**

*Ref: ATP 7-100.3 (Aug '21), fig. 6-1. Combat reconnaissance (example).*

## Reconnaissance Attack

A reconnaissance attack is a deliberate or hasty attack on an enemy position performed during a reconnaissance mission. The reconnaissance attack differs from a raid in that it is not hit and run; it is a reconnaissance-in-force that develops into a sustained attack. A commander's orders may grant subordinates the authority to decide, upon encountering an enemy force, whether to develop the situation to gain intelligence and then withdraw, or to press the attack if enemy forces seem weak or otherwise vulnerable.

## Reconnaissance by Fire

Reconnaissance by fire is the use of fire—typically indirect fire—to either reveal enemy defensive positions or to encourage the enemy to reveal its artillery assets by counterfire. The reconnaissance by fire usually involves delivering a large amount of indirect fire into a specific area believed to be occupied by enemy forces, then observing the area to confirm or deny enemy presence as it reacts to being fired upon.

Reconnaissance by fire may also be used to encourage enemy batteries to react with counterfire, allowing PLAA counterfire capabilities to effectively target them. It has the added effect of depleting enemy ammunition or forcing the enemy to displace due to the counterfire threat.

## Raid

A raid is a hit-and-run attack, usually performed as a probe to test or locate enemy defenses or dispositions. Raids may also attempt to seize prisoners, either as a primary mission objective or as a target of opportunity. Traditionally, the PLAA was heavily reliant on raids as part of the infiltration tactics it used against strong conventional forces. It is likely that the focus on raids has receded somewhat as more advanced reconnaissance platforms are fielded.

## Feint, Demonstration, Diversion, or Decoy Attack

Feints, demonstrations, diversions, and decoys employ either expendable assets or careful maneuvers designed to encourage the enemy to reveal dispositions or capabilities through engagement. Feints and demonstrations by ground forces involve intentionally engaging the enemy to fool, deceive, or trick it into making bad decisions. Feints seek enemy contact, while demonstrations seek to avoid contact. Diversions are activities designed to distract the enemy, causing it to focus on one part of the battlefield while secret operations are conducted elsewhere. Decoy operations employ expendable systems designed to fool an enemy into revealing key intelligence. A common example of decoy use is the employment of inexpensive, expendable UA flown over enemy positions, with the intent of sacrificing the vehicles in order to have the enemy's air defense sensors and shooters reveal their positions through engagement. These techniques may also employ cyber or other information warfare elements to manipulate enemy networks, sensors, or mission command nodes.



# B. Search

Search consists of those reconnaissance activities that are designed to assess the enemy, terrain, or other battlefield features without deliberate or direct action against enemy forces. The PLAA historically did not emphasize search missions—preferring direct action instead—but on the informationized battlefield, search is recognized as a critical component of overall reconnaissance efforts. The PLAA prescribes that search operations be energetic, active, and well-protected, as reconnaissance forces are high-value targets that are highly susceptible to attack or ambush by a vigilant enemy.

## Linear Search

A linear search is a form of tactical reconnaissance that focuses on a single route or lengthwise piece of terrain and various features related to that route. Linear search can be thought of as analogous to route reconnaissance in U.S. doctrine. Though any unit can conduct a linear search, for longer routes of sufficient size and condition for CA-BDE use, a reconnaissance platoon is likely required. Intelligence questions that a linear search can answer include condition of roadways, natural or manmade terrain features, presence of enemy forces, and potential hazards along the route. Linear search is considered a key task in support of the advance, the CA-BDE's movement from rear areas to an attack position. The PLAA emphasizes rapid movement and secrecy during approach marches, and high-quality reconnaissance of the route to be used is a major enabler to that end. Linear search may be enhanced by imagery or terrain analysis, map reconnaissance, and aerial surveillance, though scouts moving on the terrain in question are a necessary component to a thorough linear search. Figure 6-2 depicts a reconnaissance platoon conducting a linear search prior to the movement of a CA-BDE along the specified route.

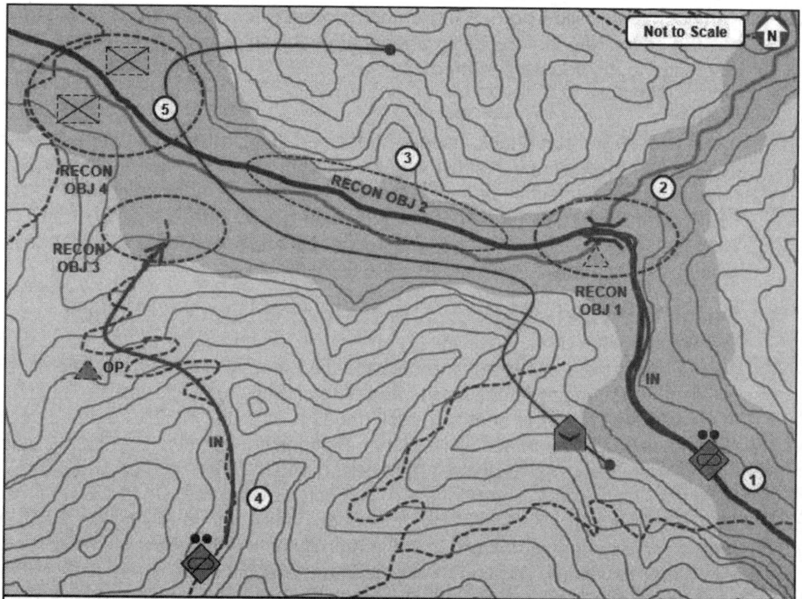


**(1) A mounted reconnaissance squad and attached unmanned aircraft system conduct a linear search in order to establish route conditions and assess enemy strength. One section moves up the main route, confirming road conditions and moving toward two main reconnaissance objectives. (2) RECON OBJ 1 is a bridge; the squad must assess bridge conditions, possible enemy presence, and determine if follow-on forces can use the bridge as a crossing enabler. (3) RECON OBJ 2 is a canalized portion of the route; the reconnaissance unit must assess the threat to follow-on forces as they move through this area. (4) A second section assesses a potential alternate route to RECON OBJ 3 while providing support to the main reconnaissance effort. This section reconnoiters and establishes an observation post to support the primary effort. (5) Both sections are to assess enemy strength at RECON OBJ 4, a suspected enemy position. Ideally, they conduct this assessment without being detected, avoiding contact with the enemy and withdrawing before any enemy response takes place.**

| | | | | | |
|---|---|---|---|---|---|
| EX | exfiltrate | IN | infiltrate | OP | observation post |
| RECON OBJ | reconnaissance objective | | | | |

*Ref: ATP 7-100.3 (Aug '21), fig. 6-2. Linear search (example).*

## Area Search

Area search is a form of tactical reconnaissance intended to obtain information about a discrete geographic area. This area is typically established by translating intelligence requirements into specific areas on the ground that must be surveilled to confirm or deny intelligence questions. Area search takes longer to plan and execute than other forms of tactical reconnaissance, but it provides detail over a wider area and can answer more questions than other forms. Due to the extended timeline required, area search is best used in support of deliberate operations, where adequate planning and time for execution can be devoted to reconnaissance operations. Area search for the CA-BDE is typically conducted by the reconnaissance battalion, with each reconnaissance troop given a specific area to reconnoiter. Active reconnaissance is conducted by mounted and dismounted scouts supported by UAS, scout helicopters, or electromagnetic sensors; passive reconnaissance is conducted by observer teams. Each subordinate reconnaissance unit is given one or more intelligence questions to answer: examples include confirming or denying enemy presence in a given area; assessing the condition of enemy defensive works, condition, or morale; assessing the condition of key terrain; or finding advantageous routes of attack. Figure 6-3 depicts a reconnaissance troop conducting an area search in support of future combined arms battalion (CA-BN) operations in the area.

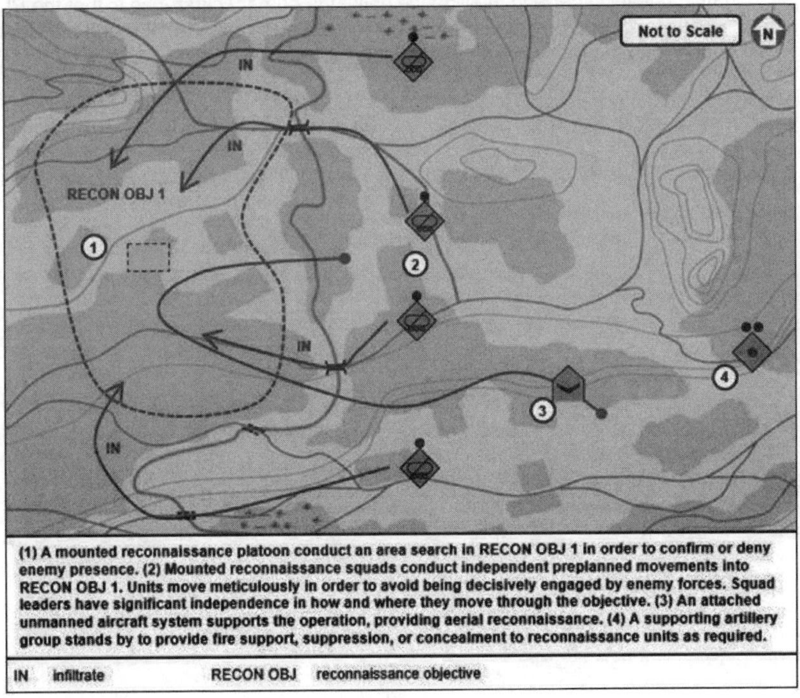


**(1) A mounted reconnaissance platoon conduct an area search in RECON OBJ 1 in order to confirm or deny enemy presence. (2) Mounted reconnaissance squads conduct independent preplanned movements into RECON OBJ 1. Units move meticulously in order to avoid being decisively engaged by enemy forces. Squad leaders have significant independence in how and where they move through the objective. (3) An attached unmanned aircraft system supports the operation, providing aerial reconnaissance. (4) A supporting artillery group stands by to provide fire support, suppression, or concealment to reconnaissance units as required.**

| IN | infiltrate | RECON OBJ | reconnaissance objective |
|---|---|---|---|

*Ref: ATP 7-100.3 (Aug '21), fig. 6-3. Area search (example).*

## Target Search

Target search is a tactical reconnaissance method by which a reconnaissance element—likely a scout team, forward observer, or scout vehicle—conducts surveillance on a specific target. This can be done to ascertain information about the target or to deliver location data about it to long-range shooters or other engagement elements. Target search is typically passive, as it is intended to feed information to other battlefield systems, rather than to develop the situation through action. As such, it is often conducted from concealment or from standoff ranges. UASs are a key contributor to target reconnaissance, particularly the UAS company that is part of each artillery battalion. These units act in concert with artillery shooters to create a reconnaissance-fires complex, a series of sensor-to-shooter linkages that facilitate rapid target engagement even against over-the-horizon or otherwise concealed target types.

## Electronic Search

Electronic search or digital search is a relatively new tactic that the PLAA recognizes as a distinct form of reconnaissance, and it has operationalized into long-range or deep electronic reconnaissance. Electronic search broadly refers to the use of collection assets to conduct reconnaissance on enemy signals, networks, and network content, such as social media. This approach is broadly similar to that of Western national-level cyber and intelligence units, the primary difference is that the PLAA intends to try and provide lower echelon units, such as CA-BDEs and artillery brigades, with the means to conduct electronic search.

# II. Security Actions

*Ref: ATP 7-100.3, Chinese Tactics, (Aug '21), pp. 6-12 to 6-24.*

## Security Fundamentals

PLAA security operations protect a supported force with a designated level of early warning and combat power. Security and reconnaissance operations complement one another, and together are the key to gaining an information advantage over the opponent. Security tasks are conducted by all PLAA units, but primary tactical security in support of CA-BDE operations is conducted largely by the reconnaissance intelligence system.

PLAA security operations are in a state of transition. Traditionally, nearly all domestic security functions were performed by the PAP, while the PLAA focused almost entirely on tactical security tasks associated with major combat operations. Chinese observation of modern regional conflicts—such as those in Iraq and Ukraine—revealed the importance of area security operations performed by conventional military forces as part of securing wider regional objectives. This prompted the PLAA to expand the scope of its security operations to include wide area security in support of stability tactical tasks and, in turn, regional stability. It is unclear how the PLAA and PAP might work together in a wide area security mission against a regional opponent.

The commander of the parent unit determines the priorities, engagement and disengagement criteria, withdrawal criteria, and information-gathering priorities for the security force. The security force commander develops a security plan informed by the intent of the higher-echelon commander. For tactical units, the priority is typically on protection of the main force, followed by information gathering, then by disruption of opponent activities. For units conducting wide area security operations not in direct support of a parent unit, these priorities may shift based on the needs of the operation.



## I. Security Operation Principles

PLAA security operations revolve heavily around support to maneuver formations. Five principles guide PLAA commanders as they develop and implement security plans: defend stubbornly, standoff, deceive, maintain contact, and protect.

### A. Defend Stubbornly

Like most security forces, PLAA security elements will likely face superior opponents as they execute their missions. Commander's guidance to security elements instructs them to fight fiercely even in the face of superior numbers or firepower. This principle is intended to ensure three things: enemy forces are adequately reconnoitered and assessed prior to their reaching their objective, enemy forces are disrupted or demoralized prior to reaching their objective, and friendly commanders are given decision space in the form of time and distance.

### B. Stand Off from the Opponent

This principle instructs security commanders to engage their opponents at or near the maximum ranges of their weapons systems. This ensures that enemy forces are engaged throughout their approach to the security line, while making it more difficult to effectively reconnoiter friendly forces. To help enable this principle, security forces are often equipped with longer-range weapons and sensors, such as antitank guided missiles, heavy direct fire guns, and airborne sensors. Effectively standing-off one's opponent is considered a critical step in the counterreconnaissance process, denying the enemy use of key positions or capabilities.

## C. Deceive the Opponent

Deception is a common theme in PLAA operations, including security operations. PLAA security commanders are encouraged to deploy their forces in such a way that the opponent overestimates or underestimates PLAA combat power. They are also encouraged to use techniques such as decoys, spoofing, false communications traffic, and feints in order to confuse or mislead enemy commanders. Advanced deception techniques that integrate higher-level capabilities, such as cyber assets, clandestine intelligence personnel, and space-based assets, are being integrated into operations, but their use at tactical levels has not been widely practiced.

## D. Maintain Contact with the Opponent

This principle is common throughout army security forces worldwide. Once contact with an opponent is made, security forces attempt to continue it in order to develop the situation through action and gain useful intelligence about the opponent's disposition and intent. Continued contact also ensures that the opponent expends personnel, fuel, and ammunition on the engagement, while further disrupting its plans and timelines. Traditional contact generally revolves around the use of direct fire systems and small arms; advanced techniques may involve the use of information warfare capabilities, deep SOF operatives, or targeting long- range fires by elevated or preplaced sensors.

## E. Protect the Main Body

This is ultimately the main objective of security operations, and it should be the task of greatest concern to the security force commander. Security forces conduct stubborn retrogrades in order to buy the main force commander decision space and ensure that the main force does not face a sudden or unexpected attack from an unobserved opponent. Security operations should be integrated with the larger mission of the protected force, enabling the commander to rapidly transition from defensive to offensive operations when the initiative is seized.



# II. Security Operation Tasks

The PLAA recognizes four forms of security tasks: combat security, security while marching, security while bivouacking, and garrison security.

## A. Combat Security

Combat security missions are those activities that actively defend larger units, assets, or key terrain from enemy activity. Missions inherent to combat security include counterreconnaissance and localized defense. Both offensive and defensive missions include a combat security element. There are two primary types of combat security missions: screen and cover.

### Screen

The screen is deployed to provide early warning to the main body, to inform the intelligence cycle through action, and to prevent or preclude enemy reconnaissance and intelligence operations. Screening forces may also disrupt and delay the enemy, but they are not intended to become decisively engaged. Counterreconnaissance is the most important mission of a screening unit. The primary targets are enemy reconnaissance, collection, and surveillance assets. Following the reconnaissance and counterreconnaissance operation, screening units typically conduct retrograde operations intended to give the main body commander decision space and inform the commander about enemy intentions. Screens are conducted during both offensive and defensive actions, and they may consist of either dedicated reconnaissance/

# III. The Security Operation Process

*Ref: ATP 7-100.3, Chinese Tactics, (Aug '21), pp. 6-13 to 6-14.*

PLAA security missions broadly follow four steps: assignment, position selection and occupation, security conduction, and retreat.

## Step 1 - Assign the Mission

This step is essentially identical to most PLAA operational planning. It involves assessing friendly and enemy forces, the higher echelon mission, coordination measures, and available time. Security missions place a high premium on assessing the enemy's courses of action—specifically the possible directions of attack.

## Step 2 - Choose and Occupy Positions

Security forces often have some latitude in choosing their positions, and orders may not include specific battle positions for subordinate units. As such, lower echelon commanders may be charged with choosing their own positions based on their best judgment. Positions should be selected based on the expected actions of the enemy, the strength of the position and its ability to blunt or disrupt enemy activity, and how favorable the terrain is to retrograde movement. Positions should be occupied secretly if at all possible, and subordinate units should have a clear concept of when to engage and when to retreat.

## Step 3 - Conduct Security Operations

There are three primary substeps to any security mission: defend against firepower assaults, defend against direct attacks, and guard against flanking maneuvers. Defending against firepower assaults involves making use of cover and concealment to blunt the enemy's firepower attack, defense against direct attack involves the use of direct and indirect fire to disrupt or destroy enemy offensive actions against the security position, and guarding against flanking maneuvers entails using maneuverable forces to counterattack enemy attempts to target vulnerable flanks of the security position.

## Step 4 - Retreat

After meeting their prescribed requirements for delaying the enemy, inflicting casualties, or disrupting enemy attacks, or when their position becomes untenable, security forces should retreat from their security positions. Retreat should be orderly and should include both rear guard actions and concealment actions to protect the main body. If possible, the retreating unit should sabotage any terrain features that may be of use to the enemy. After retreat, the security unit should consolidate and rejoin the combat action alongside the main body. If retreat is not possible, the security unit should conceal itself and then conduct sabotage operations in the enemy's rear areas.

scout units or nonspecialized units. Screens may operate independently at long ranges from their parent unit, or they may operate in close proximity. Screening units should be capable of conducting independent operations.

Screens may consist of observation posts; mounted, dismounted, or aerial patrols; or both. The distance from the screen to the main body varies based on unit size, terrain, and the commander's intent. A screen further away from the main body gives greater decision time and a longer period in which to assess the nature of the threat, but it is more difficult to communicate with and sustain. At the CA-BDE, the prescribed distance for a screen is between 5 km and 15 km, though this can vary significantly based on the aforementioned factors. Screening forces typically enjoy a high priority of indirect fire assets, and reconnaissance and surveillance sensors may have direct linkages to shooters to conduct rapid targeting. Indirect fire is considered a critical enabler of retrograde operations, disrupting enemy actions and preventing the screening force from being decisively engaged.

A screen is typically assigned to the flanks or rear of a main body. Screens can also be placed forward, but this is not employed when the main body is maneuvering; instead, security is provided by forward reconnaissance units as they conduct reconnaissance operations. Screen missions may be employed to close gaps between units, to influence or deceive enemy commanders, or to support wide area security operations. There are two primary types of screen mission: static and mobile. Figure 6-4 depicts a screen operation.

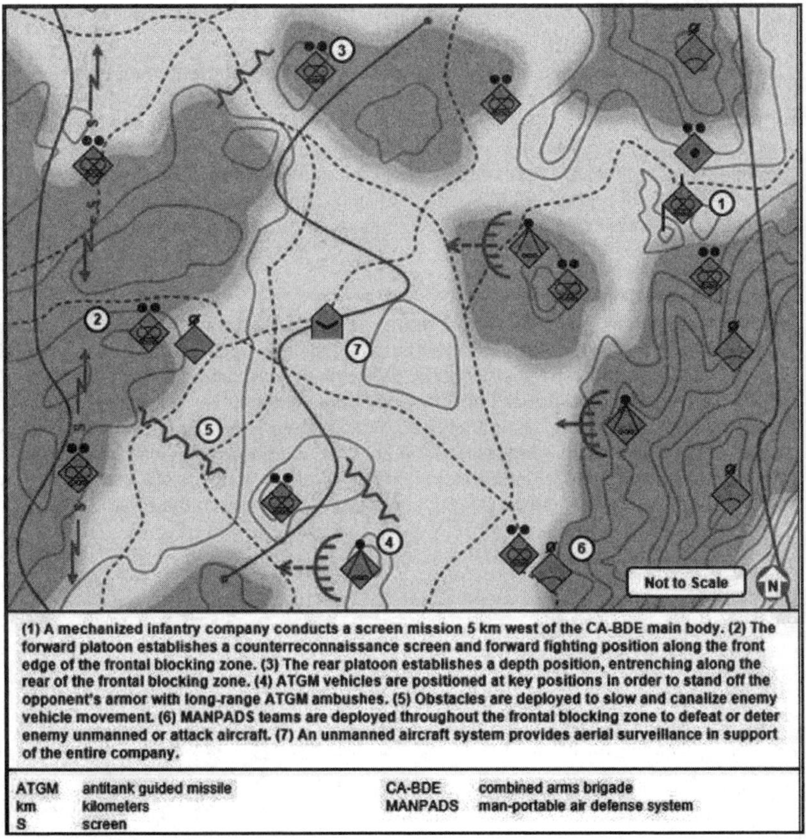


(1) A mechanized infantry company conducts a screen mission 5 km west of the CA-BDE main body. (2) The forward platoon establishes a counterreconnaissance screen and forward fighting position along the front edge of the frontal blocking zone. (3) The rear platoon establishes a depth position, entrenching along the rear of the frontal blocking zone. (4) ATGM vehicles are positioned at key positions in order to stand off the opponent's armor with long-range ATGM ambushes. (5) Obstacles are deployed to slow and canalize enemy vehicle movement. (6) MANPADS teams are deployed throughout the frontal blocking zone to defeat or deter enemy unmanned or attack aircraft. (7) An unmanned aircraft system provides aerial surveillance in support of the entire company.

| | | | |
|---|---|---|---|
| ATGM | antitank guided missile | CA-BDE | combined arms brigade |
| km | kilometers | MANPADS | man-portable air defense system |
| S | screen | | |

*Ref: ATP 7-100.3 (Aug '21), fig. 6-4. Screen (example).*

Static screen missions are most commonly employed during defensive operations. A screening unit is given a geographic area in which to operate, and it is oriented in a specific direction—usually facing away from the main body and toward a potential enemy approach. The screen is given a forward and rear boundary, engagement and disengagement criteria, supporting indirect fire, and reconnaissance objectives in support of intelligence requirements. The screening unit then assesses terrain, forces available, and commander's intent in order to develop a plan to screen the assigned area. The unit integrates multiple sensors, including organic sensors, assets from the force main body, and assets from higher echelon headquarters, to collect and monitor conditions throughout its assigned geographic area. Maneuver forces may also be task-organized with attack or reconnaissance aviation, UA or ground-based sensors, and SIGINT to create an in-depth sensor network. The sensor network facilitates cueing, mixing, and redundancy, providing enhanced sensor coverage throughout the security area. Aerial reconnaissance acts as long-range early warning. Direct and indirect fires and attack aviation are deployed to develop situational understanding through action, to disrupt enemy operations, and to attrit enemy forces. Engineers provide mobility, countermobility, and protection enhancements to the screening unit; obstacles that disrupt or delay an enemy in conjunction with indirect fire are a common approach. The screening unit is responsible for sustainment and support throughout its assigned security zone.

Mobile screen missions are broadly similar to static screens, but they are conducted in support of a main body as it maneuvers. Mobile screens are established using the same control measures as static screens, with the main body commander establishing rough geographic boundaries that support anticipated main-body maneuvers. Mobile screen operations are not typically conducted in front of a maneuvering main body, only to the flanks and rear. The screening force may remain in continuous motion alongside its supported main body, or it may conduct bounding movements with a sister unit.

Mobile screens are of critical importance in PLAA offensive doctrine. Secrecy in maneuvering is a top priority, a goal that can only be achieved if enemy reconnaissance efforts are effectively neutralized. Mobile screen operations are a key enabler of this goal. Mobile screens may be integrated with screening units from a higher echelon, particularly if a group army is conducting maneuvers over a wide area. Mobile screens may be used to deceive enemy commanders, serving as decoys or distractions to draw reconnaissance efforts and combat power away from the main body.



## Cover

A cover is a force that conducts independent reconnaissance and security operations in support of a large main body or geographic area. A cover force is self-sufficient and contains significant combat power. It must be capable of defeating enemy reconnaissance and security elements without augmentation or reinforcement from a higher echelon. Covering forces may be employed during offensive or defensive operations. A covering force is deployed to prevent reconnaissance and targeting of a main body through aggressive counterreconnaissance and security actions. It does not need to be tied to a specific main body; it may provide cover for an entire operation or be an operation in itself. The light CA-BDE is the most well-suited formation to conduct a cover operation—though other units, such as a heavy CA-BDE, a reconnaissance battalion, or an SOF brigade, can also effectively perform the cover mission. A cover force is likely organized as an operational system centered around a CA-BDE, with subordinate groups built around defense, fire support, antitank, and antiair missions.

The primary difference between a cover force and other reconnaissance and security forces is that a cover force is self-sufficient, able to conduct the full range of combat operations independent of a parent unit or main body. This gives a cover force far greater flexibility and combat power than other reconnaissance and security forces. The cover force is capable of extended defensive or delaying actions, and it can conduct its own transition and counterattack as required. The security zone established by a cover force may be dozens of km from its supported force and may cover a very wide geographic area. The cover force does not retreat or withdraw unless it is tactically necessary, and it does not allow enemy units to bypass its position without resistance.

A cover force represents a greater allocation of resources than a screen. It requires extended logistics and reconnaissance support over wide areas. Cover forces must usually be mobile, and they contain a high proportion of antitank and counterreconnaissance weapons and systems designed to increase combat power against enemy scouting and reconnaissance units. There are two forms of cover: offensive cover and defensive cover. Figure 6-5 depicts a cover mission.

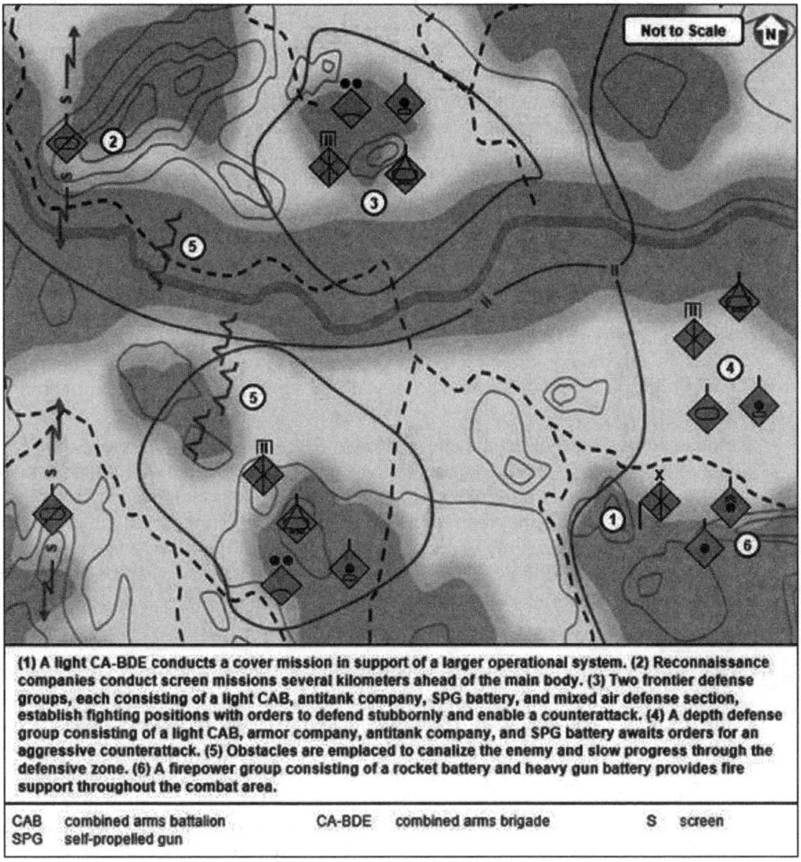


**(1) A light CA-BDE conducts a cover mission in support of a larger operational system. (2) Reconnaissance companies conduct screen missions several kilometers ahead of the main body. (3) Two frontier defense groups, each consisting of a light CAB, antitank company, SPG battery, and mixed air defense section, establish fighting positions with orders to defend stubbornly and enable a counterattack. (4) A depth defense group consisting of a light CAB, armor company, antitank company, and SPG battery awaits orders for an aggressive counterattack. (5) Obstacles are emplaced to canalize the enemy and slow progress through the defensive zone. (6) A firepower group consisting of a rocket battery and heavy gun battery provides fire support throughout the combat area.**

| | | | | | |
|---|---|---|---|---|---|
| CAB | combined arms battalion | CA-BDE | combined arms brigade | S | screen |
| SPG | self-propelled gun | | | | |

*Ref: ATP 7-100.3 (Aug '21), fig. 6-5. Cover (example).*

In support of offensive operations, a cover force may deploy on the flanks of an advance or to any potentially vulnerable area of terrain. The objective of the cover force is to ensure that the main effort of the attack does not suffer from enemy surveillance, direct attack, or indirect fire from the force's assigned security area. This is accomplished through the deployment of the force to impede enemy movement, defeat or degrade enemy reconnaissance capabilities, and defeat enemy units that enter the security area. A cover force may also deploy to the front of an advance, operating at an extended distance from the main body. When operating in this capacity, the cover force aggressively seeks weaknesses in the enemy position and exploits them as they appear, enabling follow-on operations into enemy rear areas.

Offensive cover units may have to rapidly displace to new security areas in order to properly cover a maneuvering main body. If operating at a significant distance from the supported force, integrating maneuver and reconnaissance operations with the main body is a challenge. Cover forces should generally be considered independent and be allowed to conduct operations within prescribed geographic limits without excessive input from higher echelons.

In support of defensive operations, a cover force protects a specific flank or geographic area from enemy attack and surveillance. The cover force's security zone extends across the full width of the vulnerable area, and the cover force is responsible for reconnaissance, counterreconnaissance, and security throughout. The cover force compels the enemy to reveal its disposition and intent, enabling an effective counterattack and transition to offensive operations. It first looks to transition to offensive operations by itself. If this is deemed impossible, then a counterattack force from the main body may be committed.

## B. Security While Marching

Security while marching encompasses those missions that protect a main body while it is on the move. These missions are collectively referred to as guard missions. Guard is a security mission to protect the main body by engaging and standing off the enemy, giving the main body decision space and reaction time. Guard missions differ from screens in that guard units are expected to actively protect the main force from a significant enemy attack, rather than simply delay or disrupt it. The primary mission of a guard force is to stand off enemy forces in order to prevent observation, direct fire, and indirect fire observation on the main body. The guard force also conducts reconnaissance and counterreconnaissance operations within its combat area, seeking to destroy or neutralize enemy reconnaissance assets while informing friendly reconnaissance objectives. The guard force must be prepared to conduct a decisive engagement, either to defend the main body from a superior force, or to counterattack and defeat a weaker opponent. There are three types of guard missions: advance, flank, and rear. In a noncontiguous situation, advance, rear, or flank guard forces may be deployed on cardinal directions or in a general orientation to an enemy.

Guard forces typically operate in closer proximity to the main body than screen forces. Guard forces also typically operate over narrower frontages, allowing for greater concentration of firepower and better command and communication. One of the key tasks of the guard mission is transition, the shift from defensive operations to offensive operations. Depending on the strength of the opponent, the guard force may transition to the offense and defeat the enemy on its own, or it may maintain contact with the enemy, develop situational understanding, and enable reinforcements from the main body to transition rapidly into an offensive posture. In either case, the key task of the guard is to seize and maintain tactical initiative, giving the main body commander the ability to dictate actions and timeline. The guard force may conduct a tactical withdrawal or lead, fix, or disrupt enemy formations to facilitate seizing the initiative.

Fire support to a guard force is similar to that of a screen. A guard force, however, typically has more fire support for a given unit size, and it will also integrate observation and aerial surveillance to create a reconnaissance-fire system that can rapidly engage targets with indirect fire. Engineers support the guard force with mobility, countermobility, and protection capabilities. Other combat support and combat service support units are task-organized to support guard operations throughout the security zone.

One of the primary emerging focuses of PLAA security operations is the development of air-ground integration. Modernized aviation brigades are to provide enhanced rotary-wing support to guard formations in security zones. Air support greatly expands the geographic area that the guard force can service, greatly increases available firepower, and reduces reaction time. Similarly, PLAA units conducting guard operations routinely employ air defense systems to neutralize or defeat enemy aviation assets operating in the security zone. Effectively integrating the twinned capabilities of aviation and air defense in the security zone is an area of tactical emphasis for the CA-BDE. Figure 6-6 depicts a guard mission.

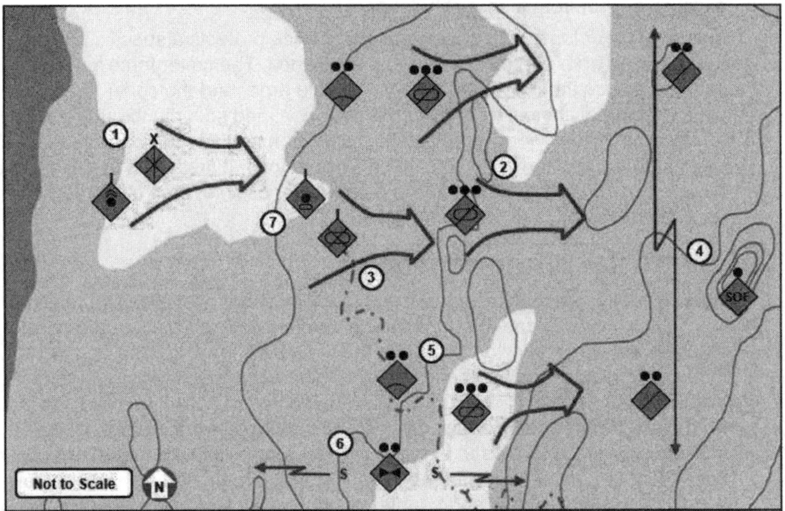


**(1) A light CA-BDE conducts an advance to an assault position. The reconnaissance intelligence group is given the task of security while marching. It employs an advance guard, flank guards, and rear guard. This graphic depicts the advance guard, which is built around a mechanized reconnaissance company, mechanized infantry company, 122-mm SPG battery, attack aviation section, and MANPADS squad. In addition, the group army provides dismounted scout patrols and SOF patrols in deep areas. (2) The three platoons of the reconnaissance company each advance on their own axis, with orders to scout the area and report any enemy presence without being decisively engaged. (3) The mechanized infantry company forms the advance guard's reserve and assists any reconnaissance platoons in breaking contact. (4) Group army dismounted patrols and SOF conduct deep reconnaissance in front of the advance. (5) MANPADS sections stand off enemy attack aviation and target any enemy unmanned aircraft that surveil the advance. (6) The attack helicopter section screens the southern flank, hunting unmanned aircraft and providing reconnaissance to the south. (7) The advance guard's indirect firepower consists of a 122-mm SPG battery, augmented by the CA-BDE's rocket battery.**

| | | | |
|---|---|---|---|
| CA-BDE | combined arms brigade | MANPADS | man-portable air defense system |
| S | screen | SOF | special operations forces |
| SPG | self-propelled gun | | |

*Ref: ATP 7-100.3 (Aug '21), fig. 6-6. Guard (example).*

## Advance Guard

An advance guard is a guard force that deploys ahead of a main body along the main axis of advance, establishing the advance security zone. The advance guard has three primary missions: streamline movement of the main body; conduct reconnaissance and counterreconnaissance in anticipation of the main body's primary mission; and engage the enemy if it appears along the main axis of advance. If the enemy is engaged by the advance guard, the latter should maintain contact with the former, develop the situation through action, and transition to offensive operations based on enemy strength and engagement and disengagement criteria. The advance guard likely represents the largest proportion of combat power for the CA-BDE's reconnaissance intelligence system, and it works closely with the frontline attack group or frontier defense group during periods of contact, engagement, and transition.

When functioning as part of an offensive group, the advance guard should maintain close contact with reconnaissance forces operating to the front and flanks of the main body. These units inform the advance guard about enemy dispositions and intent, enabling the advance guard commander to make rapid and effective decisions about positioning and the allocation of fires and reconnaissance assets. The advance guard commander remains in close contact with the main body commander in order to facilitate rapid decision making and transition to the offense. Typical advance guard force tasks while on the offense include deceiving the enemy about the main body's intent and disposition; forcing the enemy to commit reserves early; and targeting critical assets, such as command and communication nodes, fires systems, and lines of communications.

When functioning as part of a defensive group, the advance guard should deploy along the enemy's most likely avenue of approach. The primary mission of the advance guard in this situation is to create a disruption zone, with the objectives of slowing the opponent's progress, developing intelligence through reconnaissance by force, deceiving the enemy as to the main body's disposition and intent, and forcing the enemy to commit to one course of action. Once the guard force is in contact, it defends, delays, and disrupts in support of the main body force. Tactical offensive actions, such as ambushes, raids, or limited counterattacks, may be employed while part of a defensive group in order to facilitate the transition to offensive operations.

## Flank Guard

The flank guard protects the flank of the main body, establishing the flank security zone. Flanks are broadly defined as the areas parallel to the axis of the main body's direction of travel or, while stationary, areas not along the expected enemy main avenue of approach. A flank guard will generally have more geographic territory to cover than an advance guard, and it will do so with fewer forces. Thus, flank guards must have enhanced wide area reconnaissance capabilities and mobility, but they will likely have less overall combat power as compared to advance guards. The flank guard should stand off at sufficient distance to give the main body commander adequate decision space in the event of a flank threat, but it must maintain constant contact with the main body at all times.

Flank guards that are part of offensive groups move along with the main body, maintaining continuous surveillance of their assigned geographic areas while simultaneously conducting counterreconnaissance and security operations. If contact is made, the flank guard maintains contact while defending the main body from enemy offensive operations; this defense buys decision time for the main body commander to react. Maneuvering the flank guard as part of offensive operations is challenging. Bounding or continuous march are the two primary methods of maneuvering the flank guard. Bounding involves one unit passing another in sequence to maintain contiguous security across the flank, while continuous march involves moving constantly along with the main body. Movement of the flank guard should be carefully integrated with both the main body and outlying reconnaissance elements to maintain a constant flow of intelligence information and to prevent fratricide.

As part of a defensive group, flank guards establish a flank security zone and conduct reconnaissance throughout their assigned geographic areas. A flank guard may employ battle positions, observation posts, and ambushes enabled by a comprehensive reconnaissance and surveillance effort. The flank guard ensures that its assigned flank is protected against enemy intrusion, and it is responsible for conducting counterreconnaissance in its assigned area. Deployments maximize weapon ranges. Extended-range engagements are designed to delay the enemy and prolong the decision cycle for the main body commander. If engaged, the flank guard conducts retrograde operations, coupled with limited spoiling attacks, in order to disrupt the enemy attack and enable transition to offensive operations.

## Rear Guard

The rear guard protects the rear of the main force. This includes traditional guard tasks, such as reconnaissance, counterreconnaissance, and screening, and it may also include supporting rear area operations, such as sustainment and engineer efforts. The traditional rear guard is most likely to be used during offensive operations. However, in the event that the main body must withdraw or retire, the rear guard may be called upon to conduct a delaying or screening action to enable the maneuver. The rear guard has responsibility for maintaining the main body's lines of communications to higher echelon headquarters and supporting security units in rear areas.

# C. Security While Bivouacking

Security while bivouacking, also called security at the halt, includes those security activities that protect a main body when it has stopped moving, but it is not located in a garrison or rear area. These missions are designed to protect the main body from enemy reconnaissance and attack. It is up to commanders to balance the security needs of the unit with available forces by carefully assessing the nature of the threat, the vulnerability of the main body, and the number of troops available to conduct security missions. There are two types of security missions conducted while bivouacking: patrol and sentry.

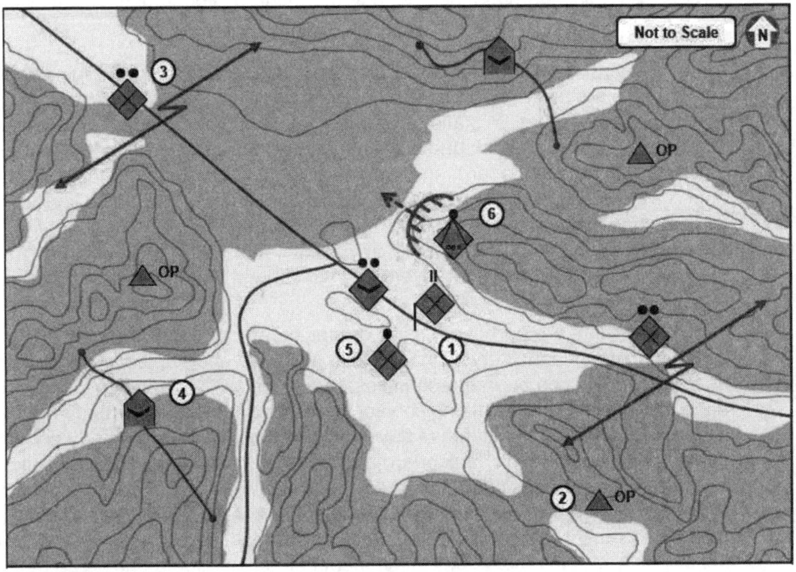


*Ref: ATP 7-100.3 (Aug '21), fig. 6-7. Security While Bivouacking (example).*

## Patrol

Patrol missions are movements designed to facilitate surveillance of a given geographic area through the use of small, mobile scouting teams. For a main body at the halt, patrols should conduct wide-ranging assessments of the surrounding terrain, then maintain a reconnaissance presence ahead of the security zone. These patrols should locate and fix potential enemy activity targeting the main body, allowing the commander decision space to commit security or defensive group forces.

In general, patrols are not to take direct action against enemy forces and should instead rely on cover, concealment, and deception to break contact. However, if a patrol encounters an enemy patrol of similar size and strength, the patrol should either destroy or force the withdrawal of the enemy unit in support of the wider counterreconnaissance mission.

## Sentry

Sentry missions are security missions that employ small, static units carefully positioned in and around the security zone in order to provide advanced warning of enemy activity; reconnaissance and counterreconnaissance support; and, if necessary, early engagement of enemy forces. Sentry missions are similar to patrols, except that they remain in place or move comparatively little. Sentry missions are conducted by observer teams, which should be carefully concealed while occupying positions of tactical advantage. Observer teams should have direct communications links to their supported headquarters.

# D. Garrison Security

Garrison security operations are those security operations that take place behind frontline units. These missions differ from security while bivouacking missions in that they take place in more permanent areas: in and around supply depots, assembly areas, air and sea transport facilities, and in civilian areas. This security mission should not contact enemy frontline units, and so it focuses heavily on defeating small-scale enemy penetrations, civilian disruption in rear areas, and air intrusions. The PLAA recognizes the importance of wide area security as part of its transition to an informationized force, and it is developing capabilities that enhance this mission. In many cases, the PAP will play an important role in garrison security operations, freeing up PLAA resources for more-demanding missions.

## Area Security

Area security is conducted to protect friendly forces and critical assets across a specified geographic area. Area security can be assigned to a unit of any size or composition, but it is best performed by mobile units that can effectively operate while dispersed. Though security is critical to operations of all types, generally speaking, area security missions are tied closely to stability operations.



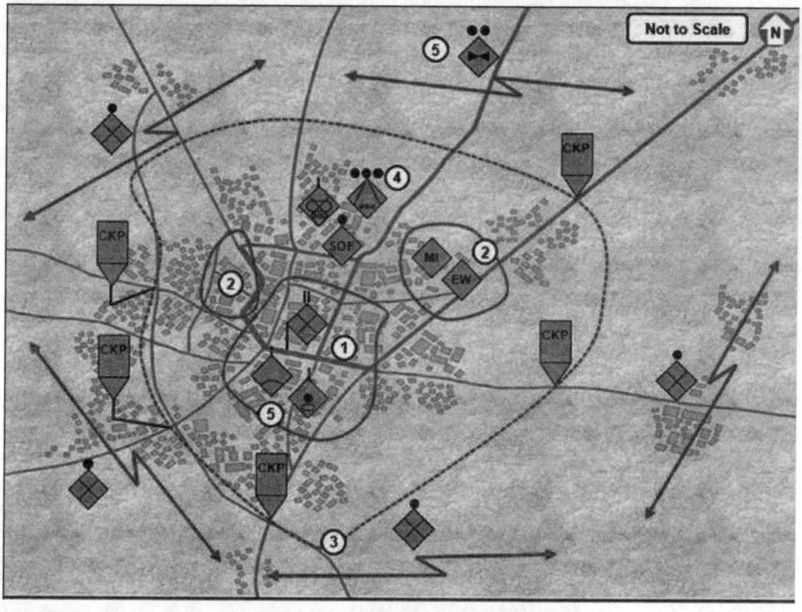


*Ref: ATP 7-100.3 (Aug '21), fig. 6-8. Area security (example).*

Area security differs from other security missions in that it is oriented around security objectives, rather than around a unit, route, or geographic region. Security objectives are those high-value assets in the security zone that the commander deems most critical. Area security operations orient around these objectives. When conducting area security, a unit performs many of the tasks associated with screen, guard, and cover missions. These include—but are not limited to—area reconnaissance, fires integration, air-ground integration, route reconnaissance, observation posts, and security checkpoints. Observation posts and security checkpoints are established to control traffic and canalize potential threats. A mobile reaction force, likely a depth attack group or depth defense group, remains on call to rapidly deploy and defeat enemy forces. The same basic security principles for all PLAA forces are applied to checkpoints in an area security operation.

Area security nearly always involves a perimeter, a boundary around the security zone that represents the limit of the area security operation. The unit assigned to the area security mission is responsible for creating and maintaining the perimeter. The perimeter may focus either on keeping enemy forces out of the security area or on keeping persons of interest within, for example, a cordon around a separate operation. Unlike other security operations, area security may require a strong visible presence. Instead of seeking to conceal friendly forces and deceive an enemy, a unit conducting area security may choose instead to prominently display its combat power to dissuade or deter enemy actions within the perimeter. Area security is likely to be a peacetime or transition task, and it is actively conducted within Chinese borders. This requires integration with PAP and local police forces, as both of these organizations may have roles in maintaining security in a given geographic area. PLAA involvement in domestic security matters is typically limited, but the potential for an expanded role following ongoing reforms seems probable.

## Route Security

Route security is a security mission performed along a specific route. The main objective of route security is to ensure that units using the route can move rapidly and safely from point to point along its length. Route security operations are terrain oriented and defensive in nature. They typically involve clearing of enemy presence, maintaining a constant patrol presence, ensuring that the route remains trafficable, and conducting limited offensive operations to stand off enemy actions.

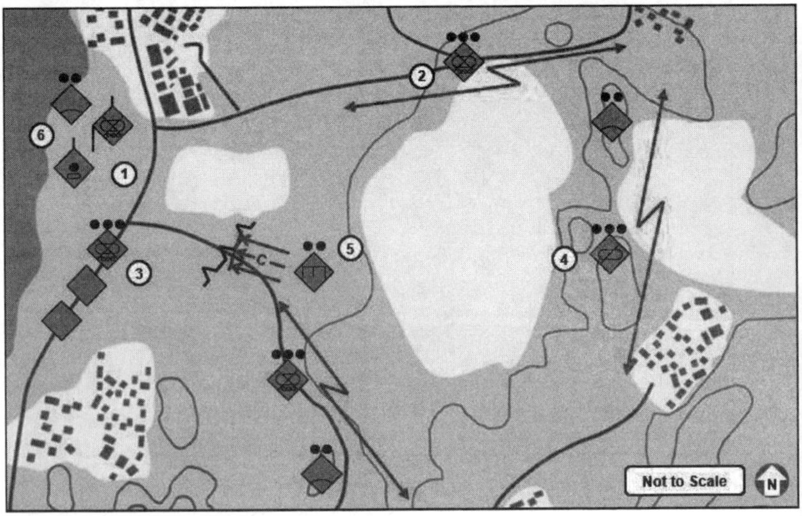


*Ref: ATP 7-100.3 (Aug '21), fig. 6-9. Route security (example).*

Chap 6

# I. PLAA Offensive Actions (Overview)

*Ref: ATP 7-100.3, Chinese Tactics, (Aug '21), pp. 7-1 to 7-4.*

Chinese forces consider tactical offensive actions to be the decisive form of land operations. The active defense strategy relies on effective and credible tactical offensive actions as the basic contribution of ground forces. Tactical offensive actions destroy an opponent's will to fight through a combination of firepower, maneuver, deception, and information warfare. This chapter outlines the People's Liberation Army Army's (PLAA's) methodology for planning, preparing, and executing tactical offensive actions.

## I. Overview of PLAA Offensive Operations

Throughout its history, the People's Liberation Army (PLA) has emphasized offensive operations. The idea that war can only be won by attacking is fundamental to People's War theory, and it has been tested time and again throughout the Chinese Civil War and the Korean War. PLA leaders historically emphasized the spirit of the attack—what Westerners would call élan or esprit de corps—as the only way their military forces could overcome the technological, firepower, and training superiority of their opponents. Developing this spirit in their formations was a fundamental skill of PLA leaders. Indeed, the role of political officers and commissars was in large part to help develop a unity of purpose underpinning the spirit of the offensive.

The PLAA today takes a similar view toward offensive operations, though with a greater focus on firepower, joint integration, and maneuver instead of the more traditional approach of infiltration followed by close combat. In keeping with Mao's principles, PLAA operations focus on destroying enemy formations rather than taking ground. As such, objectives are often described as enemy formations. PLAA forces seek to use a mix of maneuver, deception, and firepower to preclude enemy actions, isolate enemy units, and then fight the isolated enemy to annihilation. PLAA units integrate advanced deception and information warfare capabilities to fix enemy forces and then conduct decisive attacks on enemy weak points. They employ firepower not only as an enabler of maneuver, but also as an offensive tactic in itself: employing massed fires to destroy, neutralize, or fix opponents.

PLAA offensive operations are performed to accomplish one or more of the following objectives:

- Destroy, defeat, or neutralize enemy formations, personnel, or equipment.
- Enable friendly freedom of maneuver.
- Restrict enemy freedom of maneuver.
- Gain information.
- Gain control of key terrain.
- Disrupt enemy operations.

PLAA leaders traditionally preferred to operate using a grand battle plan—an extensively planned and prepared operation that demanded adherence to a rigid hierarchy and a complex, often inflexible plan. This approach was well-suited for armies consisting largely of undertrained and underequipped conscripts, and leaders who emphasized party loyalty and political enthusiasm over tactical competence. PLA reforms are attempting to significantly change this approach, employing a decentralized approach to leadership, greater tactical flexibility in planning, and more empowerment and better resourcing of leaders at lower echelons.

# II. Principles of the Offense

*Ref: ATP 7-100.3, Chinese Tactics, (Aug '21), pp. 7-3 to 7-4.*

PLAA offensive operations are informed by seven basic principles: concentration, perspective, depth, coordination, adaptation, bravery, and focus. PLAA commanders make maximum use of these principles to multiply the effectiveness of their force, keep the enemy off balance, seize and retain the initiative, and preempt enemy reactions.

**Principles of the Offense**

- **Concentration**
- **Perspective**
- **Depth**
- **Coordination**
- **Adaptation**
- **Bravery**
- **Focus**

## Concentration

All PLAA offensive operations seek one essential characteristic: concentration of overwhelming combat power against the enemy at key times and locations. This is a basic and timeless principle, descended from Sun Tzu and Mao, and modified to fit the informationized battlefield. Concentration historically involved maneuver forces and possibly artillery conducting a coordinated action with vast local numerical superiority. Today, however, concentration requires the integration of numerous capabilities across all domains, carefully synchronized and decisively employed. The PLAA now characterizes concentration as an exercise in quality rather than quantity. Instead of focusing on gaining numerical superiority, the command should focus on getting the correct capabilities to achieve the desired effects at the right time and place. Similarly, concentrated forces are seen as potentially vulnerable to enemy counteractions, and they should be carefully guarded by security efforts and counterreconnaissance until their mission is complete. Once complete, forces should either rapidly disperse or swiftly move on to the next target.

## Perspective

The principle of perspective demands two different but interrelated characteristics from PLAA leaders. First, they must be aware of their higher-echelon mission and shape their own decisions based on the objective of their higher echelon. Second, they should seek to know their own strengths, along with enemy weaknesses, to maximize the effectiveness of offensive operations. Offensive objectives should be carefully chosen and offensive capabilities carefully synchronized to create the best possible environment for paralyzing, dismembering, and collapsing the enemy's systems. Once these systems have been defeated, enemy units can be defeated in detail.

## Depth

PLAA doctrine at all echelons places great emphasis on attacks in depth. Depth can refer to the physical—for instance, operating deep behind enemy defenses; the virtual—utilizing offensive cyber or EW capabilities; or the psychological—getting into the enemy's decision cycle, preempting its actions, or breaking its morale. The primary objective of attacks in depth is to isolate enemy formations physically, psychologically, or temporally. Conceptually, the principle of depth is closely related to the ancient axiom of divide and conquer. Enemy units are isolated through envelopment or penetration, weakened, and then annihilated in detail. Attacks in depth keep the enemy off balance, forcing it to fight in numerous locations and preventing it from effectively concentrating forces for an effective counterattack. Depth attacks should be conducted jointly, using all services to create a combined arms effect across the breadth and depth of the enemy formation. The end state is a series of isolated and besieged enemy units and strongholds that can then be annihilated or reduced in detail.

## Coordination

The PLAA considers the informationized battlefield to be incredibly complex, fast moving, and intense. Proactive cooperation between different organizations is a key enabler of offensive actions on the modern battlefield. The PLAA identifies three different components of effective coordination: First, plans should include both a primary plan and numerous backup plans, covering a variety of possible contingencies, and outlining cooperation efforts between different units in each contingency. Second, any changes that take place as situations evolve should be carefully coordinated with adjacent and other concerned units. Third, efforts should be coordinated across the force as a whole, with every subordinate unit conducting operations that support the main effort and commander's intent.

## Adaptation

Adaptability is not a traditional strength of the PLA, but ongoing reforms are trying to change this dynamic in order to better fit the informationized battlefield. Interestingly, while there is an emphasis on growing independence in commanders and a decentralized command approach, the principle of adaptation calls for extensive contingency planning and a scientific approach to unexpected occurrences. Adaptation also emphasizes mobility as a key capability; reaction speed of troops is a major enabler of an adaptive force. Adaptability calls for flexible use of tactics and techniques, and it demands that commanders quickly respond to changing situations on the battlefield with new and creative approaches to the employment of tactics.

## Bravery

The PLAA places a high premium on bravery and aggression in ground actions of all types. Only through audacious and aggressive action will the initiative be gained, and only by gaining and maintaining the initiative can the enemy be annihilated. Though lethality of weapons systems and the complexity of the modern battlefield have changed, the fundamental principle of rapidly closing with the enemy, then aggressively attacking weak points of its formation, remains the basic approach to PLAA offensive tactical actions. The PLAA believes that bravery from its leaders and soldiers leads to decisive engagements, which in turn shorten the length of battles. Bravery in the attack—particularly in the initial assault—helps to take full advantage of the very small windows of opportunity available on the informationized battlefield. One of the most important roles of the political officer is building and maintaining a culture of bravery down to the lowest echelons of the PLAA.

## Focus

The principle of focus has two subordinate concepts. First, it instructs leaders to carefully prioritize their efforts, concentrating their operations against only the most important targets. The PLAA believes that one who prioritizes everything, prioritizes nothing. Second, it demands that commanders properly resource their subordinate units in accordance with the tasks asked of them. Commanders must make their priorities clear to both subordinates and superiors in order to ensure effective prioritization. Subordinates, if given adequate focus by their leaders, are better able to act properly in the absence of direct orders or in other ambiguous situations.

# III. Informationized Battlefield (Offense)

*Ref: ATP 7-100.3, Chinese Tactics, (Aug '21), pp. 7-2 to 7-3*

The PLAA recognizes six important trends that significantly impact offensive operations on the informationized battlefield:

## The Transparency of the Informationized Battlefield

Advanced multispectrum surveillance, networked intelligence, and the connected world have combined in such a way that true deception is much harder to achieve than it was in the past. Units will find it far more difficult to assemble, move, and prepare for offensive operations without the enemy uncovering their intent. Commanders are instructed to integrate deception plans across the entirety of the information spectrum..

## The Proliferation of Precision Munitions

Precision munitions enable rapid, deep, and precise targeting of critical assets all across the informationized battlefield. Firepower no longer requires mass; effective synchronization of precision munitions can achieve the same effect that formerly required dozens of attack sorties or hundreds of guns. Precision platforms usually overlap one another to achieve a combined arms effect, and they are enabled by advanced multispectrum surveillance in support of targeting.

## The Importance of Electronic Warfare

The PLAA not only puts a high priority on its own EW capabilities; it also anticipates enemy capabilities will contest its network and communications capabilities. EW has the potential to offset precision munitions, sensors, and joint communications networks, creating an environment where aggression and short-range firepower can prove decisive. As this is equally true for both Chinese forces and their opponents, electromagnetic protection is a high-priority mission for the PLAA. Chinese leaders and units are instructed to train in communications blackout conditions, relying on ingenuity and tactical competency to overcome the effects of communications isolation, and wherever possible, use communications means that are not susceptible to enemy EW efforts.

## The Rapid Tempo of Operations

The informationized battlefield moves quickly, and changes in the environment can be unpredictable and sudden. Motorized and mechanized ground units, along with aerial units, can traverse significant distances in short periods of time. Firepower systems can range distant targets, and can they react quickly to targeting information. All of these developments likely mean that windows of opportunity for commanders to seize the initiative are getting smaller and shorter.

## The Multidimensional Battlefield

The informationized battlefield involves all domains and all dimensions. Commanders must think three dimensionally, and they must effectively integrate capabilities throughout all domains. The PLAA fears enemy airpower, multispectrum surveillance, and precision firepower, and it seeks to offset these capabilities through integrating its different operational systems. At the same time, commanders and planners must ensure that PLA forces can strike through all domains in a synchronized, intelligent manner, to effectively isolate enemy units and defeat advanced enemy systems.

## The Financial Cost Of Modern War

The weapons that populate the informationized battlefield are high technology and often very expensive. Outfitting an entire force is an exercise in national economics as much as military strategy. Precision munitions are lethal and effective, but expensive, and stockpiles of them are minimal. The logistics train, which begins with manufacture and ends on the battlefield, must be up to the task of supplying the Chinese force with sufficient modern systems and munitions. Once initial stockpiles are depleted, whichever nation or military that can best sustain its forces will enjoy a significant advantage in combat.

# II. Planning the Offense

*Ref: ATP 7-100.3, Chinese Tactics, (Aug '21), pp. 7-4 to 7-15.*

The PLA historically stressed the importance of careful planning and preparation before an offensive operation. This focus continues in the post-reform PLAA, where offensive actions are developed, rehearsed, revised, and carefully executed whenever possible. Meticulous planning allows for integrating capabilities, developing tight security measures, and building an effective deception plan. While devoting more time for planning may reduce speed in the short term, it is the PLAA's belief that a well-constructed and well- rehearsed plan ultimately saves time and increases operational speed.

PLAA offensive operations are objective-based: the higher echelon commander designates an objective and specifies a force charged with accomplishing it. People's War principles dictate that objectives should typically be enemy formations, though key terrain or other valuable assets may also be considered. The attacking force identifies key features in the defense and terrain including disposition of enemy forces, weak points and points of strength, locations of reserves and fire support, and key terrain features. The offensive action is then designed to rapidly confuse and isolate the enemy, swiftly reducing its morale and enabling freedom of action for PLAA commanders. Through a combination of careful planning, effective deception, rapid movement, and application of decisive combat power at key times and locations, the enemy comes to believe its position to be untenable and either withdraws or faces annihilation. Once the enemy force is in retreat, reserve forces maintain contact and pursue the enemy while additional forces secure the area and prepare for follow-on operations.

## I. Build the Command System

An offensive operation makes use of at least two, and ideally four, command posts. The base command post is the commander's primary location, and it should be located to best facilitate coordination between the frontline, depth, and thrust maneuver groups. The advance command post, if established, is led by the deputy commander, and it is typically located near the main defensive line. The rear command post's primary role is to organize logistics and reinforcements and to create the backup defensive line supporting the offensive action. If possible, a reserve command post is established along a possible route of egress or in a well-defended rear location, ready to take over for the base or advance command post should either of them come under threat. At the combined arms battalion (CA-BN) echelon, command post operations may be informal. The command and limited staff at the battalion may fully decentralize and not physically co-locate.

The offensive zone is further subdivided into two or more secondary zones, each with a specific set of objectives and different set of tactics. While the PLAA used to be highly prescriptive about the physical sizes of these zones, it has gradually moved to a more flexible approach. The various zones should account for terrain, friendly and enemy capabilities, and higher echelon missions, and they should enable careful integration of various units and capabilities.

*Fig. 7-1 (following pages) depicts a combined arms brigade's (CA-BDE's) genericized offensive zone.*

# Offensive Zone (Notional)

*Ref: ATP 7-100.3, Chinese Tactics, (Aug '21), pp. 7-5 to 7-8.*

An offensive operation makes use of at least two, and ideally four, command posts. The base command post is the commander's primary location, and it should be located to best facilitate coordination between the frontline, depth, and thrust maneuver groups. The advance command post, if established, is led by the deputy commander, and it is typically located near the main defensive line. The rear command post's primary role is to organize logistics and reinforcements and to create the backup defensive line supporting the offensive action. If possible, a reserve command post is established along a possible route of egress or in a well-defended rear location, ready to take over for the base or advance command post should either of them come under threat.

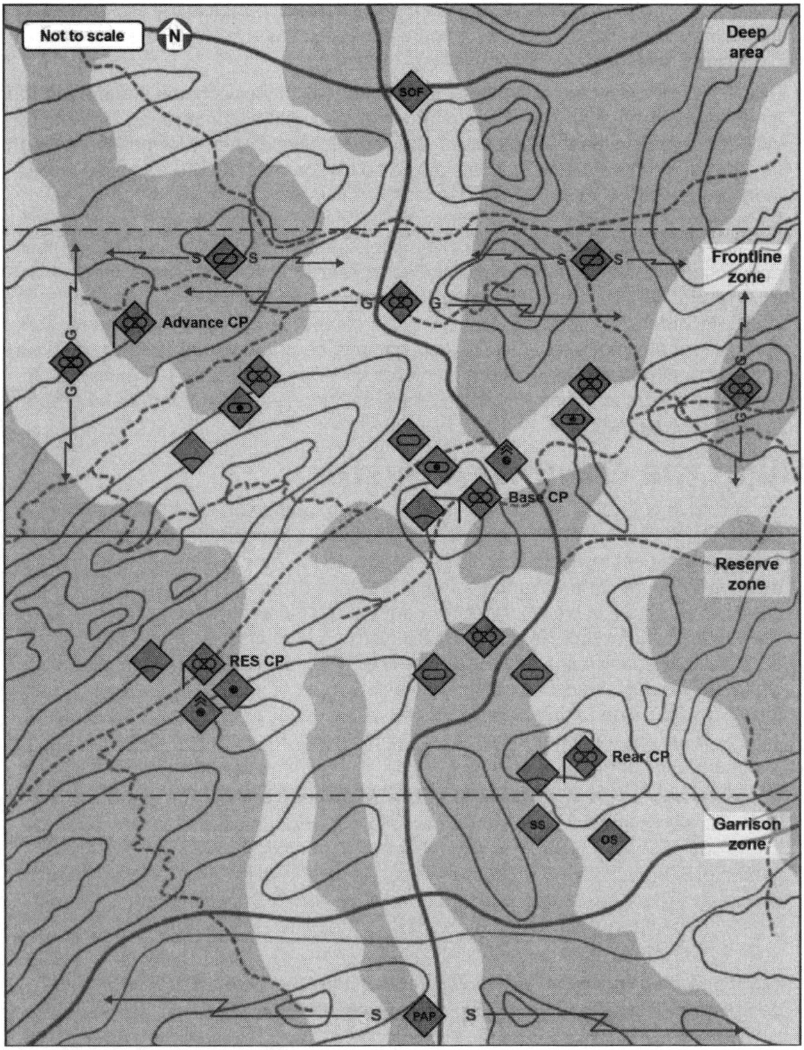


*Ref: ATP 7-100.3 (Aug '21), fig. 7-1. Offensive zone (notional).*

Offensive Actions

At the combined arms battalion (CA-BN) echelon, command post operations may be informal. The command and limited staff at the battalion may fully decentralize and not physically co-locate.

The offensive zone is further subdivided into two or more secondary zones, each with a specific set of objectives and different set of tactics. While the PLAA used to be highly prescriptive about the physical sizes of these zones, it has gradually moved to a more flexible approach. The various zones should account for terrain, friendly and enemy capabilities, and higher echelon missions, and they should enable careful integration of various units and capabilities.

## Deep Area

The deep area is the territory past which a unit's organic sensors and weapons can operate. For a CA-BDE, this typically means the area past which its rocket artillery and targeting support can operate. The fight in the deep area usually consists of independent special operations forces (SOF) or scout units supported by manned or unmanned aircraft (UA), possibly augmented by supporting fires from long-range shooters assigned to support the offensive action. Reconnaissance, counterreconnaissance, fire, counterfire, screening, and blocking all take place in deep areas. The purpose of deep-area operations is to provide early reconnaissance, target long-range preparatory fires, and carefully assess enemy strength and disposition in preparation for an offensive action.

## Frontline Zone

The frontline zone contains the territory in which the main offensive action is to occur. Early objectives, along with the enemy's main defensive line, are typically located in the frontline zone. The frontline attack group is the primary occupant of this zone, and the depth group may also occupy the area, depending on terrain and enemy disposition. The advance command post, if present, is usually forward in the zone, and the base command post is typically located either rearward in this zone or in the zone immediately behind it. The entire frontline zone should be within the range of the offensive group's organic fire support. The frontline zone typically contains a security zone on its forward edge, where security, reconnaissance, and counterreconnaissance activities take place. The primary battle takes place in the frontline zone, with the intent of breaching the enemy's main defensive line and enabling the depth and thrust maneuver groups to move into enemy rear areas.

## Reserve Zone

The reserve zone lies just to the rear of the frontline zone and typically houses the depth attack group, thrust maneuvering group, reserve group, command groups, firepower groups, and forward logistics bases. The reserve zone also usually contains a defensive line intended to resist enemy counterattacks into rear areas, and serves as the anchor for the offensive action. The reserve command post, if present, is typically located in this zone, as is the rear command post.

## Garrison Zone

Rear areas not actively occupied by the offensive group make up the garrison zone. Augmentations and reinforcements may reside in this zone, or it may serve as an assembly area for another offensive group preparing to conduct follow-on operations. Supporting capabilities such as logistics, EW, and long-range artillery reside here. Garrison zones typically contain one or more security zones that surround key positions such as bases, supply routes, or command posts. The People's Armed Police (PAP) may take on much of the security load in garrison zones in order to free up PLAA forces for more-intense duties.

# II. Phases of the Offense

PLAA offensive operations are divided into phases in much the same way that U.S. Army operations are organized. While an operation may have many phases depending on the breadth and complexity of the mission, in most cases it will involve five primary phases: advance, unfold, initiate, annihilate, and continuing operations.

**Phases of the Offense**

- **A** Advance Phase
- **B** Unfold Phase
- **C** Initiate Phase
- **D** Annihilate Phase
- **E** Continuing Operations

## A. Advance Phase

The advance, also called the moving-in, is the initial phase of an offensive operation. This traditionally referred to the movement of a main body from a staging or assembly area to the initial point of attack. For a modern force, this understanding has expanded to include the full breadth of activities that occur between the time the mission is received until initial contact is made. These activities include, but are not limited to—

- Security throughout the combat area, including assembly areas, staging areas, axes of advance, and flank areas.
- Reconnaissance operations focusing primarily on the objective: assessing the enemy force and probing for possible weak points or vulnerabilities.
- Counterreconnaissance operations attempting to deny the enemy key information about PLAA forces' dispositions and objectives.
- Deception operations designed to mislead and confuse the enemy, conceal friendly movements, fix enemy formations, and manipulate the enemy's mindset.
- Artillery groups delivering preparatory fires, reconnaissance by fire, and counterfire.
- Air defense groups seeking to deter air attack against the main body and to deny aerial reconnaissance to the enemy.
- Engineer groups conducting both mobility and countermobility operations along enemy and friendly axes of advance.
- Protection operations seeking to maintain the force, especially during periods of vulnerability to air or artillery attack.
- EW activities seeking to degrade or neutralize enemy sensors and communications, while protecting friendly network systems and emitters.

The advance begins upon receipt of an order. Reconnaissance groups are rapidly deployed in order to determine possible routes of advance, enemy strength and disposition, and key terrain features. The commander develops an initial scheme of maneuver that outlines objectives, establishes a basic concept, and enables subordinate units to establish contact and rapidly close with enemy defenses. Concealment of movement and deception operations are of paramount importance during the advance phase. As the main body moves to the point of contact, supporting capabilities conduct concurrent missions, including preparatory fire support, counterfire, information warfare activities, and mobility and countermobility activities. Commanders continuously assess enemy positions to decide upon a final course of action while attempting to manipulate the enemy's mindset and conceal their own intentions.

Security during the advance is also of the utmost importance. Deception operations can only be successful if enemy reconnaissance operations are neutralized, spoofed, or defeated by friendly counterreconnaissance. Security elements ensure that main-body movement is unhindered by enemy attack, countermobility efforts, or deception activities. Air defense and protection operations ensure that the main body is not attritted by enemy air attack and artillery during staging or movement.

The advance phase ends when the main body makes contact with the enemy and the commander initiates actions on contact.

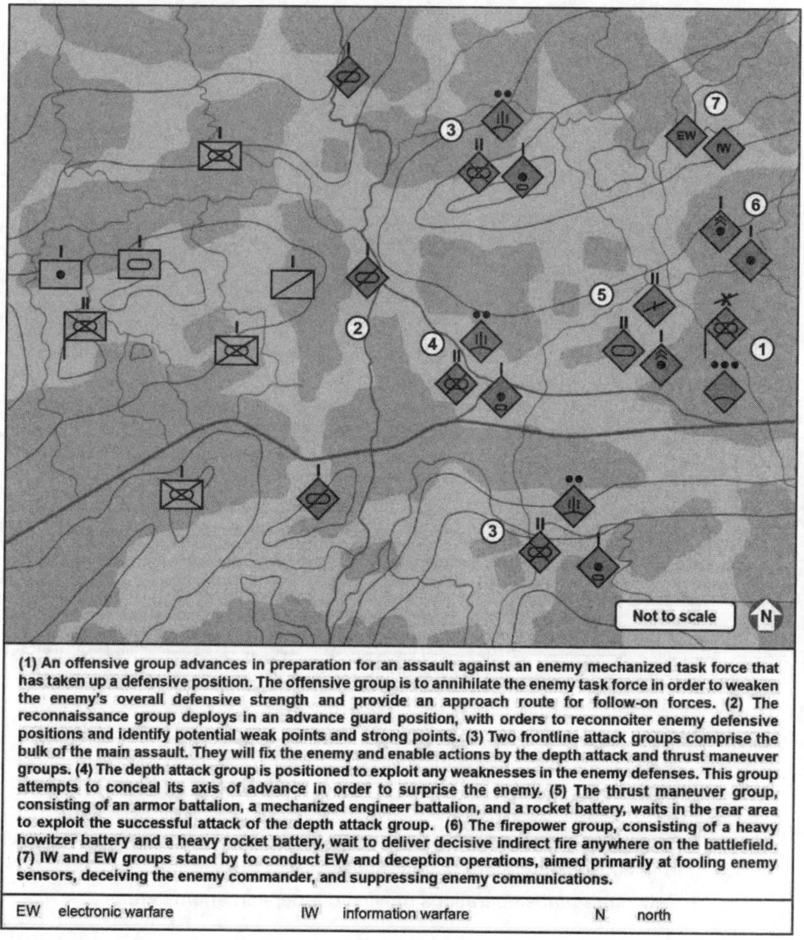


**(1) An offensive group advances in preparation for an assault against an enemy mechanized task force that has taken up a defensive position. The offensive group is to annihilate the enemy task force in order to weaken the enemy's overall defensive strength and provide an approach route for follow-on forces. (2) The reconnaissance group deploys in an advance guard position, with orders to reconnoiter enemy defensive positions and identify potential weak points and strong points. (3) Two frontline attack groups comprise the bulk of the main assault. They will fix the enemy and enable actions by the depth attack and thrust maneuver groups. (4) The depth attack group is positioned to exploit any weaknesses in the enemy defenses. This group attempts to conceal its axis of advance in order to surprise the enemy. (5) The thrust maneuver group, consisting of an armor battalion, a mechanized engineer battalion, and a rocket battery, waits in the rear area to exploit the successful attack of the depth attack group. (6) The firepower group, consisting of a heavy howitzer battery and a heavy rocket battery, wait to deliver decisive indirect fire anywhere on the battlefield. (7) IW and EW groups stand by to conduct EW and deception operations, aimed primarily at fooling enemy sensors, deceiving the enemy commander, and suppressing enemy communications.**

EW electronic warfare | IW information warfare | N north

*Ref: ATP 7-100.3 (Aug '21), fig. 7-2. Advance (example).*

# B. Unfold Phase

The unfold phase consists of actions designed to set conditions for annihilation that occur upon initial contact by the main body with enemy defenses. Subordinate units conduct rapid movement in accordance with the commander's scheme of maneuver, seeking to position themselves appropriately for the decisive phases of the operation. Following initial contact, units maintain contact with the enemy and continue to develop situational understanding. Key intelligence requirements not yet answered by reconnaissance groups are resolved by main-body actions. During initial contact, the commander attempts to conceal the direction, strength, and objective of the main effort through deception, information warfare activities, feints, and demonstrations.

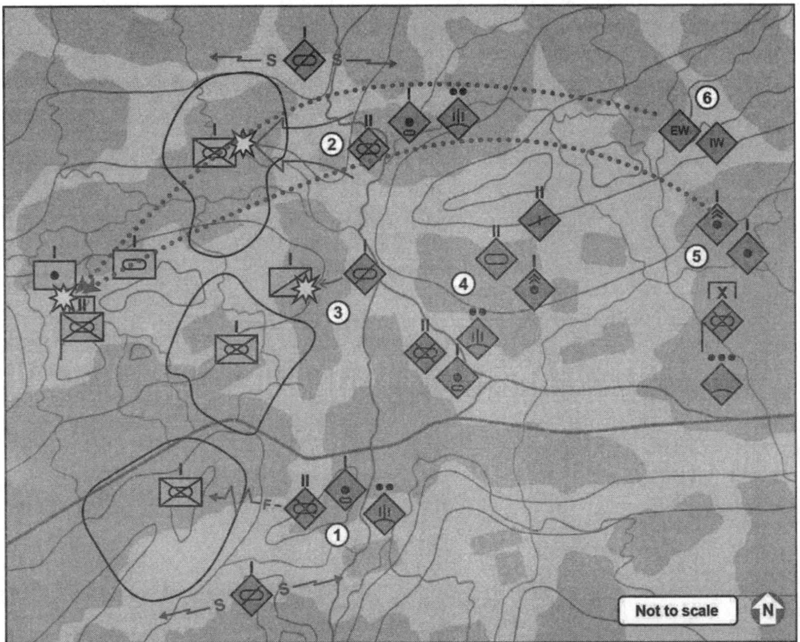


**(1) One of the frontline attack groups positions in a support-by-fire position, engaging an enemy unit and fixing it in position. (2) At the same time, the other frontline attack group conducts a limited attack against the enemy's left flank, testing the strength of the enemy defenses. These two actions, conducted in concert with one another, are intended to confuse the enemy commander about the location and axis of the main assault, forcing him to commit reserves early. (3) Reconnaissance units engage enemy scouts, preventing them from effectively reconnoitering friendly units and ensuring the enemy commander remains ignorant of the location and direction of the main assault. (4) The depth and thrust groups move under concealment to their initial attack positions. As the frontline attack groups conduct their attacks, the offensive group commander develops his understanding of the situation and finalizes the axis of the main assault. Combat power is concentrated along this axis. (5) The firepower group conducts a fire assault against the enemy's rear area, disrupting enemy command and communication and causing casualties. (6) EW and IW units conduct electronic and information warfare against the enemy command, with their main effort focused on deceiving the enemy about the location and axis of the main assault.**

| | | | | | |
|---|---|---|---|---|---|
| EW | electronic warfare | IW | information warfare | S | screen |
| F | fix | N | north | | |

*Ref: ATP 7-100.3 (Aug '21) fig. 7-3. Unfold (example).*

Other activities during the unfold phase include, but are not limited to—

- Security operations, particularly to the flanks and rear, to protect the main body from spoiling attacks or counterattacks.
- Reconnaissance operations moving deeper into enemy areas.

- Counterreconnaissance operations that defeat enemy attempts to identify the main effort and objectives while reinforcing deception efforts through feints and demonstrations.
- Deception operations attempting to fool the enemy commander about the time and place of the main effort; the size, strength, and disposition of friendly forces; and the size and proximity of reinforcements for both sides.
- Artillery groups shifting fire support focus to counterfire, obstacle reduction, and disruption efforts in deep areas.
- Air defense groups deterring or defeating aerial attack and surveillance throughout the area, with an emphasis on the main body and main effort.
- Engineer groups shifting focus to obstacle reduction and maintaining friendly mobility.
- EW focusing on jamming or spoofing enemy communications to impede rapid response to the attack.

The primary objective of the unfold phase is to ensure that the objective is isolated and contending with numerous dangers, preferably from multiple directions. The enemy commander should be confused and enemy forces disrupted. The enemy, however, should not yet know either the direction or strength of the main effort. The entire unfold phase should be orderly and well planned in order to ensure concealment and correct positioning of friendly units. Movement should be rapid, but not so fast that confusion sets in or concealment is broken. The best practice of the unfold phase is to encircle the enemy. Encirclement refers to the complete isolation of the enemy force through a combination of friendly ground unit actions, firepower, deception, psychological warfare, and EW. Encirclement does not necessarily imply that an enemy is actually physically surrounded. A combination of lethal and nonlethal capabilities may be sufficient to cause the enemy to believe itself encircled. For example, a well-timed artillery raid may cause the enemy to believe that a route is cut off or unavailable, without requiring the commitment of additional PLAA ground forces. Similarly, disinformation passed over communications channels may deceive the enemy into believing it does not have access to a particular route or terrain, without requiring friendly forces to physically occupy the area in question.

The unfold phase ends when the initiate phase begins with the commencement of the main effort. Figure 7-3 depicts a PLAA CA-BDE conducting the unfold phase in preparation for an attack on an enemy position.

## C. Initiate Phase

The initiate phase commences when the main effort begins its assault against the objective. This phase often begins with a massive assault by fire, wherein fire support groups target enemy command nodes, critical systems, mobile reinforcements, and other high-value targets. Supporting efforts continue to develop the situation. Units charged with fixing enemy units continue to prevent them from reinforcing the main objective, while counterreinforcement groups preclude the commitment of enemy reserves. The main effort is typically the primary assault upon the enemy's center of gravity: that thing—be it a headquarters, a unit, or a piece of terrain—that is most essential to the enemy's morale and mindset. Ideally, the enemy will either withdraw, collapse, or be routed once the center of gravity is seized or destroyed. Other activities during the initiate phase include, but are not limited to—

- Security operations continuing with the primary objective of protecting the main body from spoiling attacks or counterattacks.
- Reconnaissance operations focusing on intelligence objectives that support the main effort.
- Counterreconnaissance operations continuing to spoil enemy efforts to discover the composition, axis, and objectives of the main effort.

- Deception operations continuing to fool the enemy commander about the time and place of the main effort; the size, strength, and disposition of friendly forces; and the size and proximity of reinforcements for both sides.
- Artillery groups conducting a firepower assault, delivering massed fire support to the main effort, targeting defending enemy units, interdicting movement of reinforcements, and destroying or suppressing command and communication nodes.

Attacks are launched from multiple directions simultaneously, forcing the enemy commander to make decisions about force deployment. If possible, the enemy commander's decision cycle is interrupted—or, better yet, influenced—so that reinforcements or other maneuvers do not concentrate in opposition to the main effort. The main effort employs substantial force concentration to achieve local firepower and numeral superiority over the enemy, allowing the former to defeat the latter's forces in detail and maintain the initiative. Fire support is weighted heavily in support of the main effort's assault.

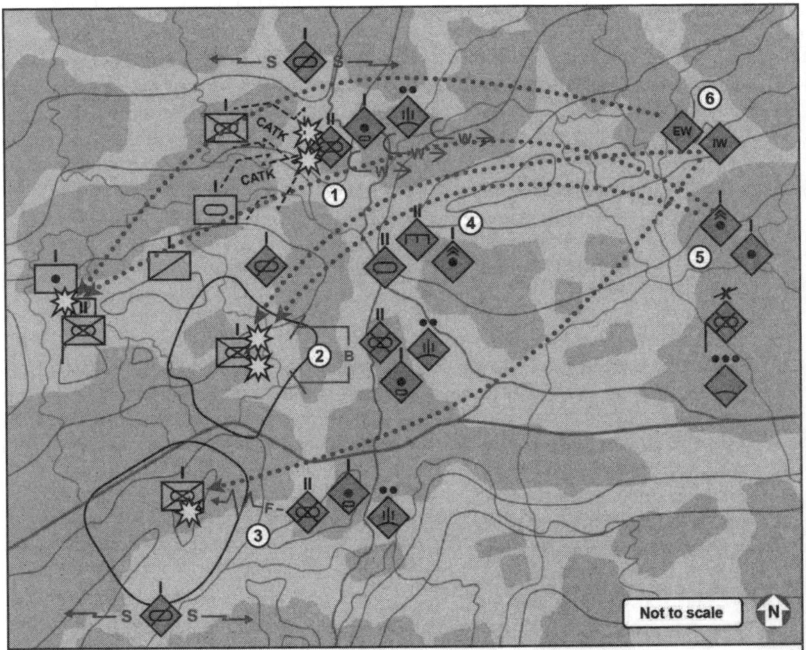


**(1) The enemy commander, deceived into thinking that the main effort is targeting his left flank, commits his reserve in a counterattack against the northern frontline attack group. The latter rapidly transitions to a defensive posture and begins a deliberate retrograde operation, intended to overextend the enemy's counterattack force and eventually isolate it from its command. (2) The friendly commander, having identified the enemy's center as vulnerable, commits the depth attack group in a decisive assault. Its objective is to breach the enemy's main defensive line and isolate the two enemy flanks. (3) The other frontline attack group conducts an assault against the enemy's right flank, fixing the enemy unit and preventing it from reinforcing the group under assault in the center. (4) The thrust group positions itself to exploit the breach created by the depth attack group. (5) The firepower group conducts a fire assault against the enemy center in support of the depth group's assault. (6) The EW group commences its decisive effort, suppressing enemy communications in order to electronically isolate enemy units and confuse response to the main assault. The IW group transitions to information attack operations, attempting to increase enemy units' perception of isolation.**

*Ref: ATP 7-100.3 (Aug '21), fig. 7-4. Initiate (example).*

The main effort targets one or more primary objectives as a part of the assault on the enemy's center of gravity. These may include enemy command posts, assembly areas, reserve forces, artillery units, or network nodes. The main effort should seek weak points in the enemy's defense, then move rapidly to penetrate them. Follow-on forces help to secure the main effort, fix or deceive remaining enemy forces, and preclude enemy reinforcement. Groups—possibly airborne, air assault, SOF, or militia—located to the flanks and rear of the enemy secure key terrain to prevent outside reinforcement of the enemy formation while encircling the objective. Supporting groups ensure that enemy counterattacks—particularly counterencirclement—are spoiled, disrupted, neutralized, or defeated before they can affect the main effort.

The initiate phase is complete once the enemy center of gravity has been seized by the main effort.

## D. Annihilate Phase

The annihilate phase commences after the enemy is confused and demoralized following the seizure or fall of its center of gravity. Enemy units should be isolated, with communications to their higher echelons disrupted and routes for reinforcement either cut off or unusable. The enemy is encircled, either physically or psychologically, and mutual support between enemy units is no longer possible. Decisive attacks are conducted throughout the combat area, seeking to destroy isolated enemy units before retreat and a prepared defense are possible. Fire support is used aggressively in brief but violent volleys in order to suppress and demoralize enemy units prior to assaults. Special focus is given to ensuring that the enemy cannot break out of the annihilation zone. Security and countermobility efforts slow or stop movement before enemy units can escape.

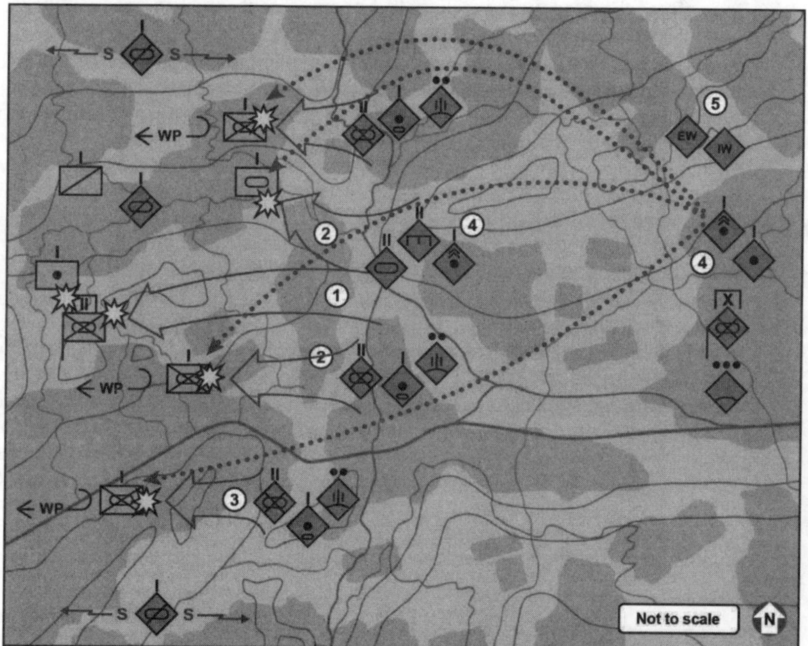


**(1) The thrust maneuver group conducts the decisive deep assault into the enemy's rear area, targeting command posts, supply areas, artillery units, and potential escape routes. This completes the isolation of enemy units and compromises the enemy's overall defensive position. (2) The thrust maneuver group also conducts a hasty attack against the exposed flank of the enemy's counterattack force, ensuring it cannot be redeployed and maintaining its isolation. (3) The frontline attack and depth groups conduct storming attacks against the isolated and depleted enemy units. Local fire support is integrated with maneuver to either destroy or force the withdrawal of enemy units. Assaults are coordinated as much as possible to ensure the enemy cannot reinforce units under attack. (4) Firepower assaults target retreating enemy units, disrupting their movement and ensuring that retrograde actions cannot be mounted. (5) EW and IW groups shift their focus to disrupting adjacent enemy units from reinforcing the now-defeated enemy unit and preventing the enemy's higher echelon from communicating with the defeated unit.**

*Ref: ATP 7-100.3 (Aug '21), fig. 7-5. Annihilate (example).*

Once the enemy recognizes that one of its units is facing annihilation, a strong response is likely inevitable. Enemy air and artillery units will attempt to target friendly formations to suppress or disrupt friendly attacks. Antiair and artillery groups must be prepared to conduct counterair and counterfire operations in response. Other enemy ground formations will attempt to reinforce their besieged units Friendly security forces must defend stubbornly and disrupt, slow, or defeat these attempts until the objective force has been annihilated.

The annihilate phase is concluded once the objective force has been destroyed, routed, or has surrendered.

## E. Continuing Operations

Continuing operations commence immediately when it is clear that the objective force has been destroyed, routed, or has surrendered. The remainder of the enemy force loses cohesion in the face of decisive attacks throughout the combat area, and widespread withdrawal or retreat follows. The main focus during this phase of the operation is pursuit of these fleeing enemies. High-tempo operations continue to maintain contact with these enemy forces during their retrograde, defeating rear-guard security actions and preventing the enemy from consolidating to form a coherent defense. Counterreinforcement groups continue to defeat enemy attempts to reinforce the area by either defeating enemy forces, spoiling enemy attacks, or conducting countermobility operations.

When continuing operations commence, the commander begins assessing follow-on opportunities. For example, if defeating the enemy unit created a breach in the enemy's main defensive line, an opportunity for envelopment or penetration may be present. Similarly, the enemy unit's defeat may create an opportunity for a Public Opinion Warfare offensive, highlighting the success of the operation. Units may begin transition from combat postures to movement postures in accordance with the scheme of maneuver. The decision to do so is a key decision point. If a unit redeploys into march order too soon, it may be vulnerable to enemy counterattack. If it waits too long, opportunities for penetration or envelopment may be lost.

The PLAA unit must also consolidate. Consolidation consists of three primary activities: perform security activities, reorganize and reconstitute friendly units, and conduct passage of lines. Security activities consist of securing the combat area against enemy reconnaissance and counterattack and ensuring that friendly forces entering the area are not subject to threat from bypassed enemy elements or irregular forces. Reorganization and reconstitution are those activities that enable the unit to conduct follow-on operations: resupply, replacement of lost or damaged equipment, casualty processing, and receipt of reinforcements. Passage of lines takes place when one unit passes through another's combat area, typically with the intent to resume offensive operations against the enemy. Passage of lines can be a complex task, and it is enabled by effective control measures, communications, and planning.

Chap 6

# III. Types of Offensive Operations

*Ref: ATP 7-100.3, Chinese Tactics, (Aug '21), pp. 7-15 to 7-22.*

The PLAA recognizes at least five different types of offensive operations: attacking a fortified enemy, attacking an unprepared enemy, amphibious operations, urban offensive operations, and airborne operations. Though they are broadly similar to deliberate and hasty attacks, the primary difference between attacking a fortified enemy and attacking an unprepared enemy is the enemy's level of preparation in contrast to the friendly forces' level of preparation. Other offensive operations focus on specific situations—amphibious, airborne, or urban. In addition to the five types of offensive operations, particular attention is given to operations taking place in special conditions, such as jungles or extreme cold.

**Offensive Operations (Types)**

- I Attacking a Fortified Enemy
- II Attacking an Unprepared Enemy
- III Amphibious Landing Operations
- IV Urban Offensive Operations
- V Airborne Operations

## I. Attacking a Fortified Enemy

Attacking a fortified enemy is offensive action in which the unit assaults an enemy force that occupies a prepared position. The attacking unit likely has ample time to conduct all five phases of the offensive operation. Reconnaissance, firepower, and maneuver are all carefully coordinated to achieve maximum surprise, deception, and concentration. Enemy positions are carefully reconnoitered, and enemy reconnaissance is effectively suppressed. The advance phase is done deliberately, with great care used to conceal friendly movement, strength, and disposition from the enemy. The unfold phase is also performed carefully, with strong emphasis on deception and concealment of friendly forces. Once the initiate phase commences, tempo is increased rapidly. Movements are rapid and aggressive, and the unit seeks decisive contact with the enemy. Attacking a fortified enemy may allow time for complete envelopment of the enemy, extensive preparatory fires, positioning of forces in deep areas, and extensive information warfare operations upon enemy leaders and soldiers.

Attacks against a fortified enemy must make the most effective use of stratagems. Commanders can carefully manipulate the enemy's morale and situational understanding through the use of deception, information warfare, counterreconnaissance, and concealment. Effectively employed stratagems can convince the enemy to cease resistance, to conduct an early retreat or retrograde, or to counterattack at the wrong time or place. The development of stratagems begins early in the planning process, and it is a central objective in any offensive action.

If a commander has the time and decision space, the ideal objective in attacking a fortified enemy is to achieve its annihilation. Such opportunities are relatively rare and valuable, especially during high-intensity operations. The commander must ensure that the unit's operational tempo, planning, rehearsals, and execution are all carefully optimized, but still fit within the available timeframe. Synchronization between units should be well executed, with careful timelines designed to keep the enemy off balance and confused throughout the engagement.

# II. Attacking an Unprepared Enemy

Attacking an unprepared enemy is an offensive operation in which the unit must rapidly conduct an offensive operation against an unprepared or surprised opponent, possibly without the benefit of careful positioning, preparatory fires, information warfare support, or reconnaissance. Such an operation may be one of opportunity—such as a friendly unit discovering a previously unseen weakness in an enemy's position. It may also be in response to enemy actions, such as a spoiling action against an enemy counterattack. Rapid tempo is of great importance during an attack against an unprepared enemy. The window for decisive action may be very limited, requiring rapid decision making and action on behalf of the attacking unit.

Unprepared enemies are typically vulnerable, or they hold disadvantageous positions. Careful reconnaissance may not be possible on a limited timeline, so attacking an enemy in a strong defensive position without proper preparation should be discouraged. The rapidity and aggressiveness of the attack is the key enabler to the mission. It keeps the enemy off balance and grants the initiative to the friendly unit commander.

Attacks against an unprepared enemy still include all five phases of an offensive operation, but one or more phases may be very short in duration or limited in scope. The advance phase is short, with units forgoing careful movement, deception, and concealment for rapidity—they seek to close with the enemy as quickly as possible, ensuring it cannot construct a strong defensive position. The unfold phase is conducted in a similar manner, with rapidity of action being prioritized over deception and concealment. While in an attack against a fortified enemy, units may be positioned on a number of different approach axes. An attack against an unprepared enemy may allow for as few as two. From the initiate phase onward, operations are broadly similar to those in an attack against a fortified enemy. The enemy is struck with speed and aggression, and the attacking units seek to isolate, overwhelm, and then annihilate enemy units in detail.

*Note. Chinese military theory suggests no attack be made with fewer than two axes of advance against a single point.*

A unit conducting this type of attack is particularly vulnerable to enemy counterattack. The unit likely did not have the time or resources to deploy a comprehensive security element, and it may be deployed in a vulnerable or disorganized fashion due to the rapid push through the unfold phase of the attack. Friendly commanders must be prepared to react to a well-timed enemy counterattack, either through fire support, information warfare, or direct action by ground or air forces. Developing and maintaining security following the attack is a critical component of the continuing operations phase.

Successful attacks against unprepared opponents require a significant degree of autonomy by the attacking units. The small time window for success and the lack of careful preparation often preclude careful coordination with higher echelons. Enabling this level of autonomy is a significant focus area for the PLAA. Historically, it was rare for units above the battalion echelon to conduct any such actions. Modern PLAA CA-BDE and CA-BNs are expected to be able to conduct quick attacks with only limited coordination and support from their group army.

# III. Amphibious Landing Operations

The PLA defines amphibious landing operations as those that make use of watercraft to cross a body of water, enabling the attacking unit to conduct an assault and exploitation against an enemy defending a littoral area or island. The PLAA recognizes two distinct kinds of amphibious landing operations: coast to coast and vessel to coast. Coast-to-coast operations are those in which staging and embarkation occur on shore, while vessel-to-coast operations stage and embark from a larger seagoing vessel.

Amphibious landing operations use a similar approach to operations across land. The largest differences are that the advance phase occurs across water and reconnaissance efforts are likely to be less comprehensive. An amphibious landing operation consists of an advance phase, an assault and penetration phase, and an exploitation and consolidation phase. Secrecy and deception during embarkation and advancing are considered paramount, as embarked troops are highly vulnerable to counterattack and firepower attack. Ideally, the initial wave of an amphibious assault strikes an undefended or lightly defended beach, allowing the first wave to establish a beachhead security area while under minimal enemy fire. Once the beachhead is established, follow-on waves land within the secure area, then conduct penetrations of deeper enemy positions. If at all possible, multiple beachheads should be established, enabling multiple follow-on penetrations and dispersing the effects of enemy counterattack and enabling.

Amphibious landing operations are to be supported by massive firepower and information attack. Firepower may be generated from missiles, aircraft, surface vessels, or long-range artillery. Information attacks maximize the effect of deception operations and seek to extend enemy response times. Once ashore, the initial wave establishes substantial firepower superiority over local enemy defenses, forcing enemy retreat and enabling the establishment of a beach security area. Reconnaissance likely occurs concurrently with the assault. Reconnaissance elements should deploy and conduct reconnaissance in force to ascertain enemy intent and dispositions and inform the command about how and where to commit penetration efforts.

The decision of where and how to commit the follow-on efforts is the most important decision in an amphibious landing operation. Commanders in the initial wave must communicate their situational understanding—enabled by on-the-spot reconnaissance efforts—to their higher echelon, allowing the higher echelon to decide how and where the second wave should attack. Commanders in the first wave must also coordinate firepower operations to support follow-on waves and clear any remaining obstacles or other enemy countermobility assets to enable the penetration effort.

*See following pages (pp. 6-18 to 6-20) for an example amphibious operation (figures 7-6 through 7-8).*

# Amphibious Operation (Example)

*Ref: ATP 7-100.3, Chinese Tactics, (Aug '21), pp. 7-5 to 7-8*

## Amphibious Landing Operation (example, part 1 of 3)

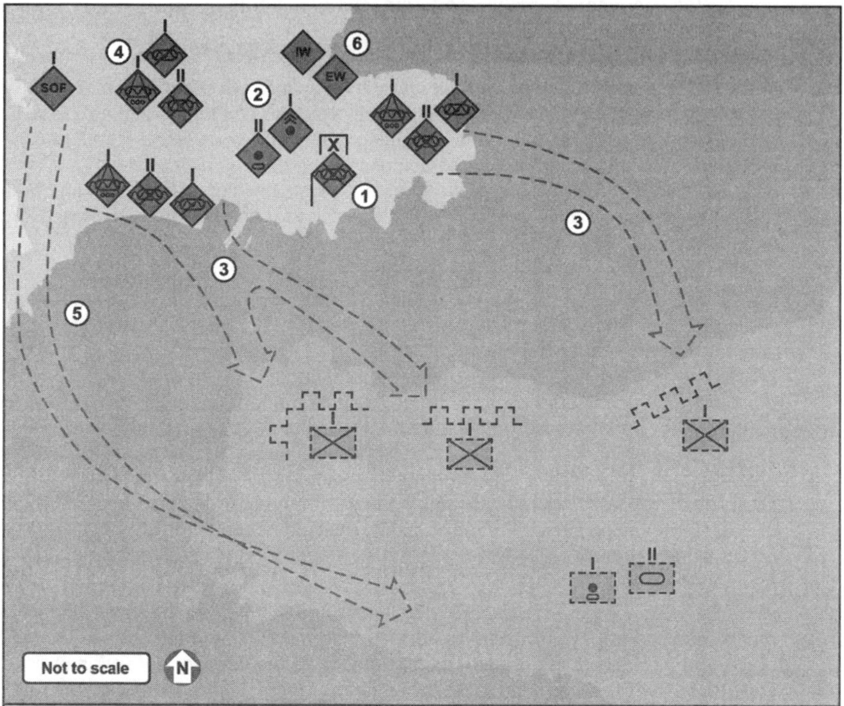


**(1) An amphibious group built around a PLANMC brigade conducts an amphibious assault on a contested enemy shoreline. The group is to establish a beachhead, drive enemy forces from the area, and secure the area to enable follow-on operations. (2) Long-range fire support for the operation is provided by a heavy MRL battery, while a SPG battalion prepares to move across the strait once the beachhead is established. (3) Two frontline attack groups, each consisting of an amphibious reconnaissance company, amphibious infantry battalion, and amphibious tank/assault gun company, are to transit the strait and engage enemy shore defenses. (4) A depth attack group stands by to exploit the weakest part of the enemy's defenses and strike inland. (5) A SOF company is to conduct an air insertion into the enemy's depth area. (6) EW and IW groups begin EW and deception operations, attempting to suppress enemy sensors and communications and fool the enemy about the time and place of the main effort's assault.**

| | | | |
|---|---|---|---|
| EW | electronic warfare | PLANMC | People's Liberation Army Navy Marine Corps |
| IW | information warfare | S | screen |
| MRL | multiple rocket launcher | SOF | special operations forces |
| N | north | SPG | self-propelled gun |

*Ref: ATP 7-100.3 (Aug '21), fig. 7-6. Amphibious landing operation (example, part 1 of 3).*

## Amphibious Landing Operation (example, part 2 of 3)

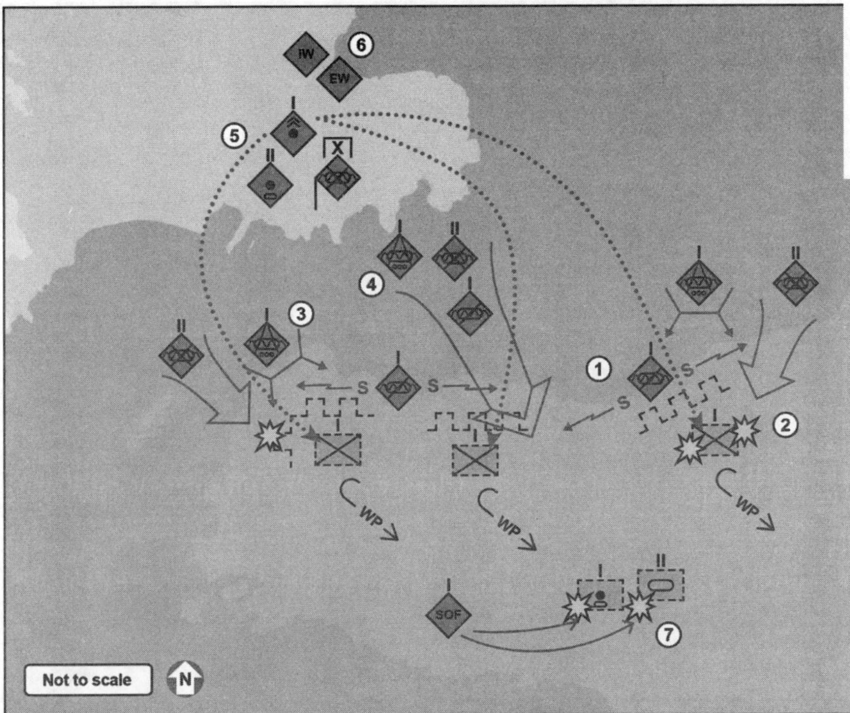


**(1) Reconnaissance companies establish a foothold on the beach and begin screening and reconnaissance operations, fighting as skirmishers, and establishing enemy positions. (2) Frontline attack group main bodies conduct assaults against enemy weak points, seeking to penetrate enemy shore defenses and fix enemy units. (3) Frontline attack group assault guns provide support by fire both from shore and while embarked. (4) The depth attack group conducts a decisive assault through the enemy center, forcing an enemy retreat and isolating remaining enemy defenders. (5) The firepower group delivers long-range fire support at key points on the battlefield. (6) EW and IW groups continue to dominate the electromagnetic spectrum and deceive the enemy about the time and place of the main assault until the last possible moment. (7) The SOF company conducts raids in depth areas, disrupting enemy reinforcement and fire-support missions.**

| | | | | | |
|---|---|---|---|---|---|
| EW | electronic warfare | N | north | SOF | special operations forces |
| IW | information warfare | S | screen | WP | withdrawal under pressure |

*Ref: ATP 7-100.3 (Aug '21), fig. 7-7. Amphibious landing operation (example, part 2 of 3).*

Continued on next page

# Amphibious Operation (Example) Cont.

*Ref: ATP 7-100.3, Chinese Tactics, (Aug '21), pp. 7-5 to 7-8*

Continued from previous page

## Amphibious Landing Operation (example, part 3 of 3)

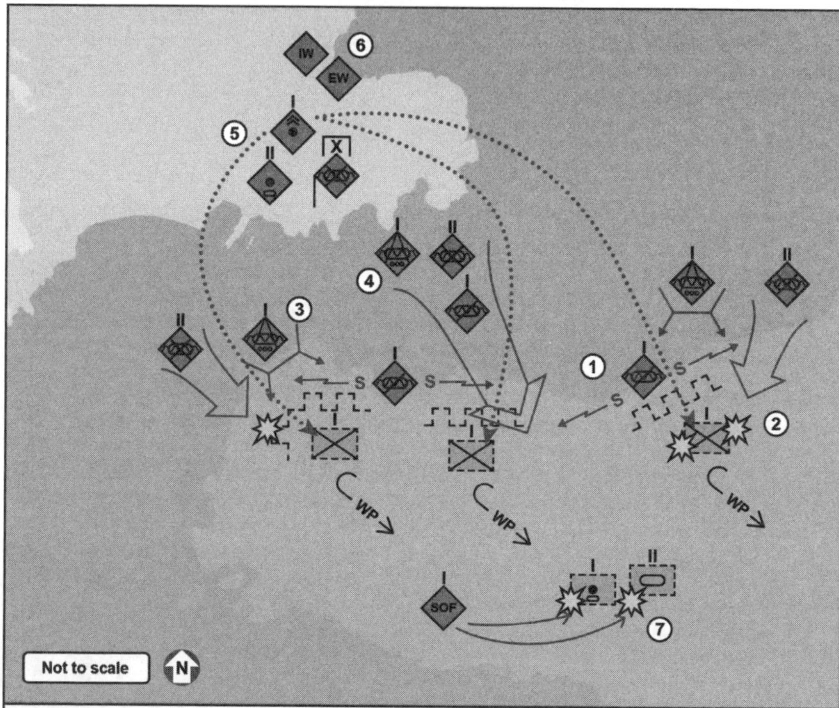


**(1) Reconnaissance companies establish a foothold on the beach and begin screening and reconnaissance operations, fighting as skirmishers, and establishing enemy positions. (2) Frontline attack group main bodies conduct assaults against enemy weak points, seeking to penetrate enemy shore defenses and fix enemy units. (3) Frontline attack group assault guns provide support by fire both from shore and while embarked. (4) The depth attack group conducts a decisive assault through the enemy center, forcing an enemy retreat and isolating remaining enemy defenders. (5) The firepower group delivers long-range fire support at key points on the battlefield. (6) EW and IW groups continue to dominate the electromagnetic spectrum and deceive the enemy about the time and place of the main assault until the last possible moment. (7) The SOF company conducts raids in depth areas, disrupting enemy reinforcement and fire-support missions.**

| | | | | | |
|---|---|---|---|---|---|
| EW | electronic warfare | N | north | SOF | special operations forces |
| IW | information warfare | S | screen | WP | withdrawal under pressure |

*Ref: ATP 7-100.3 (Aug '21), fig. 7-8. Amphibious landing operation (example, part 3 of 3).*

Continued from previous page

# IV. Urban Offensive Operations

Urban offensive operations are those offensive operations that take place in complex urban terrain. In contrast to maneuver warfare in more open terrain, the PLAA stresses careful planning and execution, slow and deliberate actions, and meticulous intelligence assessments when conducting offensive operations in urban areas.

Urban offensive operations are characterized by bounding overwatch techniques, using streets or other thoroughfares as avenues of approach. Following an extended period of reconnaissance across the area, a forward element attacks and secures a position of overwatch. Ideally, this element occupies an elevated position in order to provide fire and reconnaissance support to assault elements. Assault elements then move down avenues of approach while targeting enemy weak areas. Urban operations are viewed as fundamentally slower and more deliberate than other operations for both sides, Surprise and deception are even more important, as actions take longer and attacking units are more vulnerable.

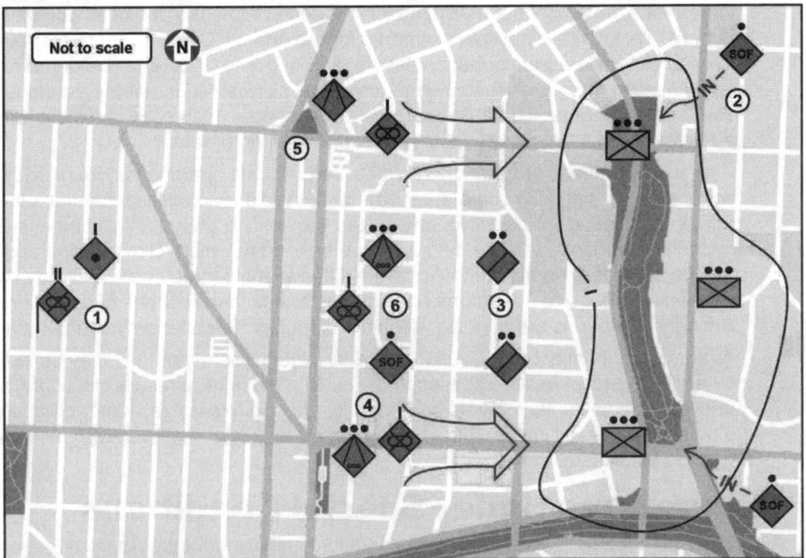


**(1) An offensive group built around a medium CA-BN conducts an assault on an entrenched enemy infantry company in dense urban terrain. Firepower in the form of short-range mortars is the primary indirect fire capability. (2) SOF units infiltrate enemy positions through a mix of deception and concealment. They provide a mixture of reconnaissance and sniper support to the main operation. SOF units seek the high ground by occupying upper levels of tall buildings, if possible. (3) Dismounted reconnaissance units occupy key terrain with good visibility on the objective, seeking high ground wherever possible. (4) The primary attack consists of a mechanized company supported by a platoon of assault guns. This force moves slowly and meticulously down a main road, using a bounding overwatch technique and the assault guns as support by fire. (5) The second assault group times its attack to commence once the enemy is committed to repulsing the first attack. It must move more quickly but meticulously down an adjacent avenue of approach. (6) The depth group provides support by fire to both assaults, and will conduct its own assault if required.**

| | | | |
|---|---|---|---|
| CA-BN | combined arms battalion | N | north |
| IN | infiltrate | SOF | special operations forces |

*Ref: ATP 7-100.3 (Aug '21), fig. 7-9. Urban offensive operation (example).*

Fighting in urban areas is also characterized by increased complexity, particularly of terrain. High-rise buildings are areas of special focus. If a large building must be taken, it should be carefully cordoned, isolated, and then aggressively stormed by specially trained troops. Underground facilities are defeated in a similar manner. The employment of specialized troops to deal with these difficult targets is a critical enabler of urban offensive operations.

Complex urban terrain also changes the nature of the enemy's counterattack. Buildings and other hardened pieces of terrain make concealed movement simpler for the enemy's counterattack force, but they also enable and enhance ambush operations. PLAA commanders should seek to defeat or demoralize enemy counterattack forces through ambushes that use the urban terrain. Security units should defend stubbornly and protect the main body as it deliberately and thoroughly defeats any lingering enemy units within the combat area. In addition, urban terrain enhances the use of countermobility capabilities, particularly in tight, canalized areas. Commanders should liberally employ countermobility systems in careful concert with ambush and security operations.

# V. Airborne Operations

Airborne operations are offensive operations using fixed-or rotary-wing aircraft to move friendly troops into enemy territory. Following airborne insertion, units conduct otherwise typical maneuver warfare. Airborne operations are one of the most important elements of the multidimensional penetration, the use of all domains and dimensions to create multiple dilemmas for the enemy. Currently there are relatively few PLAA airborne transportation assets, and airborne units are considered high-value assets. They are prescribed to target the highest-value enemy assets within their operating area: the flanks and rear of the enemy main body, airfields, artillery, air defense systems, or key routes for enemy reinforcements.

Airborne operations are broadly similar to amphibious operations, though they are conducted on a smaller scale. After careful but rapid reconnaissance and a firepower attack, an assault team attacks and seizes the landing area, annihilating enemy forces in the vicinity and creating a secure base from which follow-on operations can be launched. Additional forces are airlifted to the secure base, and they conduct aggressive offensive operations against key targets in their areas of operations. Security operations are ongoing, seeking to spoil or defeat enemy counterattacks.

Airborne forces may be relieved in place by ground forces, or they may be exfiltrated by air. Both processes are challenging. Relief in place requires careful coordination between friendly units to avoid both friendly fire and excessive vulnerability to enemy attack, while air exfiltration must be conducted rapidly and efficiently to minimize the risk to friendly aircraft.

# * Operations Under Special Conditions

The PLAA recognizes eight special conditions: hills, mountains, jungles, extreme cold, deserts, rivers, swamps, and nighttime. Each of these conditions requires modifications and adaptations to baseline operations. Primary considerations include ensuring that troops are properly prepared for the special condition-with cold weather gear or bridging equipment, for example. Units should also train for potential special conditions that may occur in their areas of operations.

Chap 6

# IV. Offensive Tactics

*Ref: ATP 7-100.3, Chinese Tactics, (Aug '21), pp. 7-22 to 7-38.*

The PLAA employs several different offensive tactics: envelopment, penetration, pursuit, firepower, ambush, and raid. Use of these tactics is informed by the principles discussed earlier in this chapter. Each tactic may be employed by virtually any echelon, but the figures in this chapter show employment at the CA-BDE level against a notional enemy. Tactics may be used independently, sequentially, or simultaneously. Figures in this section depict genericized units conducting simplified variants of the tactics described.

**Offensive Tactics**

- I Envelopment
- II Penetration
- III Pursuit
- IV Firepower
- V Ambush
- VI Raid

## I. Envelopment Tactics

Two primary envelopment methods are used by the PLAA: simple envelopment and complex envelopment—both of which occur on enemy flanks.

### A. Complex Envelopment

A complex envelopment is an offensive tactic wherein an assault occurs against multiple flanks of the enemy. The complex envelopment has long been a mainstay of Chinese maneuver tactics, employed regularly by Chinese generals throughout the various wars of Imperial China. More recently, it was used regularly—and with good effect—by Communist Chinese forces in both the Chinese Civil War and the Korean War, most famously at the 1946 Battle of Guanzhong and the 1951 Battle of the Soyang River. A successful complex envelopment creates multiple dilemmas for the enemy commander, enables effective isolation of enemy units, and enables concentrated forces to defeat enemy strongpoints in detail. For a contemporary maneuver unit, a complex envelopment requires advanced coordination of multiple maneuvers along with effective integration of fires, information warfare, and mobility and countermobility capabilities.

The definition of flank traditionally refers to the right or left side of a ground formation. The PLAA recognizes two distinct types of complex envelopment: the flank envelopment, wherein attacks target the enemy flanks, and the separating envelopment, wherein attacks target enemy positions of weakness in depth. The PLAA's multi-domain approach to fighting recognizes that "flanks" can exist not only on the sides of an enemy formation, but in the air, below ground, or in the virtual world. A modern complex envelopment may make use of any or all of these in a coordinated, synchronized manner in order to get the enemy off balance, preclude effective enemy response, gain and maintain the initiative, and get inside the enemy's decision cycle. In addition, a penetration attack may create flanks where they did not previously exist; a network attack may threaten or disrupt the enemy's digital or virtual flank; and air or missile forces may threaten an enemy's flank in the third dimension. A flank, in other words, is defined by the PLA as any area in which the enemy has not concentrated defenses, does not expect an attack, or is vulnerable. In general, the more flanks are threatened or engaged simultaneously, the less coherent and effective the enemy's response will be. This in turn enables isolation of enemy units and eventually leads to their destruction.

The complex envelopment employs one or more depth attack groups enabled by one or more frontline attack groups. A frontline attack group attacks, feints, or demonstrates against one portion of the enemy formation to fix or deceive enemy forces, engage enemy artillery, and drive commitment of enemy reserves. Once the enemy has been fixed, the depth attack groups conduct aggressive, penetrating assaults against weak points of the enemy's flank, while reserve and artillery groups continue to fix other enemy forces, preventing effective reaction to the penetration. Depth attack groups seek to disrupt enemy rear areas and isolate enemy units, enabling follow-on annihilation attacks. Depth attack groups must achieve substantial overmatch against their initial opposition, integrating numerical superiority, firepower superiority, and information superiority to overwhelm the enemy and enable rapid maneuver into and through enemy rear areas.

Success of a complex envelopment depends heavily on deceiving the enemy about the strength and axis of the depth attack groups. Deception operations and careful concealment enable the assault forces to strike from an unanticipated direction against areas of weakness. The operations of the frontline attack groups prevent the enemy from counterattacking or reinforcing the units opposing the depth assault until it is too late to adequately respond.



## B. Simple Envelopment

The simple envelopment is an offensive tactic wherein an assault occurs on a single flank of the enemy. A simple envelopment involves two primary groups: a depth attack group and a frontline attack group. The frontline attack group feints or demonstrates against one portion of the enemy formation, with the intent of fixing enemy maneuver forces, engaging enemy artillery, and driving enemy commitment of reserves. Frontline attack group actions can be thought of as a deception operation, intending to prevent the enemy from concentrating its forces against the depth attack group.

Once the enemy is engaged by the frontline attack group, the depth attack group advances to its attack position and commences an assault. The depth attack group seeks to achieve vastly superior concentration of combat power against the specified flank of the enemy. Maneuver forces, enabled by integrated fires and information warfare support, overwhelm any prepared defenses, and penetrate enemy rear areas. Once the enemy's flank is defeated and its rear areas are threatened, isolated enemy units can be annihilated by follow-on operations.

Much like with a complex envelopment, success of a simple envelopment depends heavily on deceiving the enemy about the strength and axis of the depth attack group. Deception operations and careful concealment enable the assault force to strike from an unanticipated direction against areas of weakness. The operations of the frontline attack group prevent the enemy from counterattacking or reinforcing the units opposing the depth attack group until it is too late to adequately respond.

The simple envelopment is employed in situations where a complex envelopment is not possible due to terrain, strength, or disposition. Use of only a single penetration makes it easier for the enemy to respond and effectively counter the envelopment attempt. If possible, the enemy should face multiple penetrations to complicate and confuse responses. Simple envelopments are very common at the squad and platoon levels, where one element creates a base of fire to fix and suppress the enemy, while another element assaults the objective. Figure 7-11 depicts a simple envelopment.

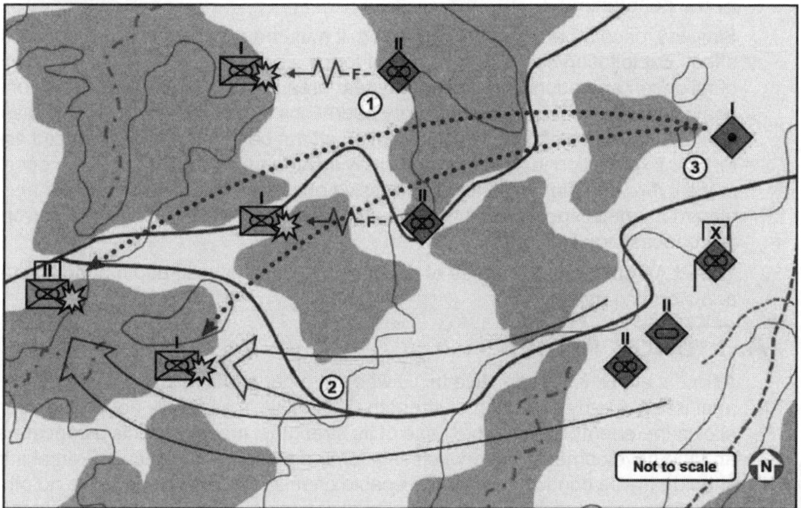


**(1) Frontline assault groups conduct demonstration or feint attacks against enemy positions in order to fix enemy defenses and deceive the enemy commander about the axis and strength of the main assault. (2) The depth and thrust groups envelop the enemy's vulnerable flank, penetrating the enemy's defensive line and moving rapidly into the enemy's deep area, targeting headquarters, key mobility routes, and retreating enemy units. (3) The firepower group provides fire support to the main assault, disrupting enemy responses and demoralizing enemy units at critical times and places.**

| F | fix | N | north |
|---|---|---|---|

*Ref: ATP 7-100.3 (Aug '21), fig. 7-11. Simple envelopment (example).*

# II. Penetration

A penetration is an offensive tactic wherein a mobile or fast-moving element defeats an enemy line of defense—preferably through an area of weakness—and continues to press forward into the enemy's rear area, bypassing enemy areas of strength. A penetration is rarely employed by itself. Instead, it is typically employed as one component of a wider operation, such as an envelopment. Ideally, an offensive campaign successfully conducts numerous multidimensional penetrations, employing a variety of capabilities to penetrate enemy defenses in a synchronized and mutually supporting manner. Units best suited to conduct a penetration as part of a ground action are armored or mechanized units due to their combination of mobility, protection, and firepower. Penetrations may also be employed in the air or cyber domains or in the electromagnetic spectrum using the same basic principles: target enemy weak points and bypass strong points in order to threaten vulnerable areas behind the main lines of defense.

Most contemporary PLAA tactics involve one or more penetrations as the basis for offensive operations. This is true for all echelons and unit types. Penetrations are the means by which enemy units are isolated, enemy actions and counterattacks are disrupted, and additional enemy flanks are exposed. Aggression and mobility are the key enablers of a successful penetration, along with effective integration of enablers that preempt enemy response to a penetration, deceive the enemy as to the axis or strength of a penetrating force, or create confusion or the perception of isolation among enemy units during a penetration.

As it requires a unit to bypass active enemy positions, a penetration by itself leaves the conducting unit highly vulnerable to counterattack on its flanks and rear. In order to maximize the effect of a penetration, the main effort must be well supported by security operations and follow-on operations designed to secure the areas bypassed by the penetration.

Similarly, once a penetration is achieved, it must be exploited in order to maximize effect. Exploitation can take a variety of forms. For a ground action, it may consist of follow-on operations against enemy rear area assets, annihilation operations against isolated enemy units, security operations that offset enemy reconnaissance and counterattack efforts, or consolidation efforts centered on newly seized areas or terrain. Exploitation in the air domain may include ground attack or strike operations; in the cyber domain, anything from passive observation to network attack; and in the electromagnetic spectrum, jamming or meaconing (the interception and surreptitious rebroadcasting of signals).

The PLAA practices four types of penetration: frontal attack, depth attack, infiltration, and storming attack.

## A. Frontal Attack

A frontal attack is an offensive tactic wherein a penetration is attempted directly against an enemy's position of strength. This tactic is generally not preferred, as it allows the enemy to make best use of its strengths, making effective concentration and the development of overwhelming combat power more difficult. Frontal attacks should only be conducted either to enable offensive actions or because no other alternative is available, but an offensive action is necessary. Frontal attacks need not involve maneuver forces; any offensive action that attacks an enemy's area of strength can be considered a frontal attack.

The frontal attack is opposed to both Sun Tzu and Maoist theories about warfare, which prescribe attacking the enemy at weak points. In a frontal attack, friendly forces should be highly concentrated and should close as rapidly as possible with the enemy in order to minimize exposure to defensive measures and stand-off weapons. If possible, a frontal attack should be carefully planned and executed to maximize the use of enabling capabilities. Any frontal attack should be enabled by massed fire, substantial information warfare and deception support, and feints or demonstrations designed to preempt enemy actions, confuse and demoralize the enemy, or suppress the enemy. Enabling capabilities should be carefully synchronized with the attack to maximize their effectiveness at key times and places. Even with these enablers effectively used, a frontal attack will likely face significant resistance and will suffer higher casualties than an attack targeted at an enemy weak point. Figure 7-12 depicts a frontal attack.

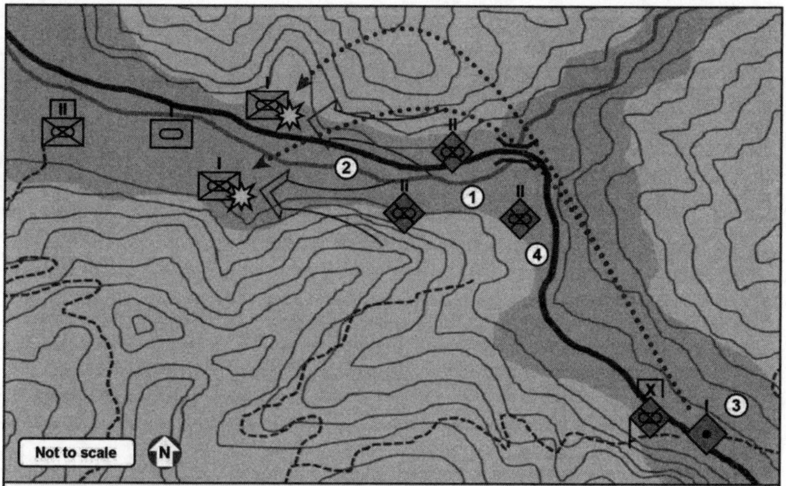


**(1) An offensive group is canalized by terrain but must make an assault against a strong enemy position. The commander orders a frontal attack against the entrenched enemy. Two frontline attack groups are massed against an outnumbered enemy. (2) The primary assault concentrates a full battalion in an attempt to penetrate and compromise the enemy's main defensive line. A supporting attack is synchronized in order to prevent the enemy from massing greater combat power against the main effort. (3) Firepower is massed in support of the frontal attacks, attempting to disrupt enemy defenses as much as possible. (4) A depth group awaits successful penetration of the enemy's main defensive line, prepared to exploit the advantage and assault the enemy's rear area.**

N north

*Ref: ATP 7-100.3 (Aug '21), fig. 7-12. Frontal attack (example).*

## B. Depth Attack

A depth attack is an offensive tactic wherein a force focuses on attacking targets deep behind the enemy's forward positions through a series of deep penetrations. A depth attack targets enemy high-value assets in enemy rear areas, such as command nodes, supply bases, reserve troop concentrations, and artillery units. The depth attack also seeks to further isolate enemy units throughout the battlefield, allowing their encirclement and destruction in detail.

The depth attack is typically the main effort of a given battle plan, and it often involves the best-equipped and -trained troops available to a commander. The depth attack group is typically employed as the second line of troops, after the initial line engages the enemy and creates one or more breaches in the enemy's defensive line. The timing and direction of the depth attack are key decision points in the battle, and commanders are encouraged to position themselves such that these decisions are not delegated. The depth attack should achieve maximum concentration of combat power against its key targets. If at all possible, security functions should be performed by other units in support of the depth attack group.

Depth attacks may employ one point of entry or multiple points, they may be staged and synchronized, and they may employ third-dimension mobility. The depth attack should be arranged to meet few obstacles and entrenched enemy forces in order to maximize the rapidity of movement into enemy rear areas. It may be supported by demonstrations, feints, assaults, or other penetrations designed to occupy the enemy's attention and divert its combat power. Figure 7-13 (following page) depicts a depth attack.

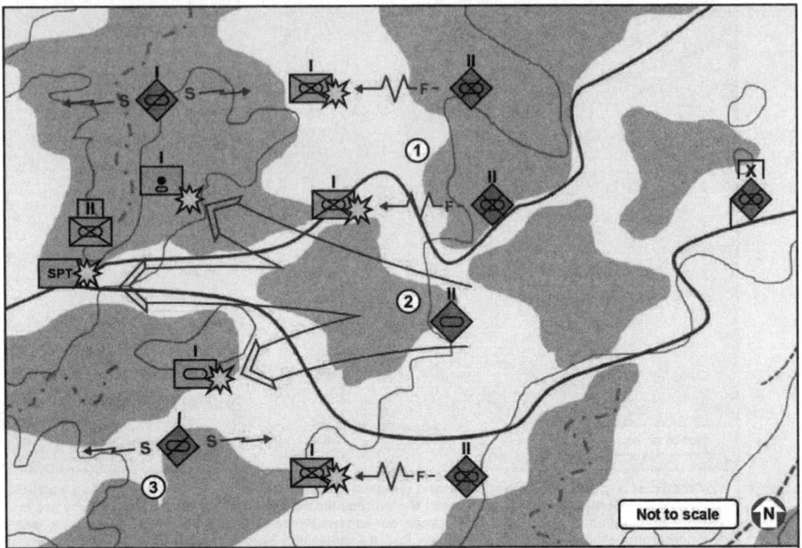


(1) Frontline attack groups engage and fix the enemy main defensive line. Previous reconnaissance efforts located weak point in the enemy defensive line, enabling the commander to deploy the depth group to best effect. (2) The depth group bypasses the enemy main defensive line and penetrates into the enemy rear area, targeting enemy reserves, support, headquarters, and artillery units. (3) Security units carefully screen the flanks of the depth attack group, ensuring that the enemy cannot counterattack during the assault.

| | | | |
|---|---|---|---|
| F | fix | S | screen |
| N | north | SPT | support |

*Ref: ATP 7-100.3 (Aug '21), fig. 7-13. Depth attack (example).*

# C. Infiltration

Infiltration is an offensive tactic wherein a force closes with and conducts an undetected penetration of enemy defenses by using stealth, rapid movement, and deception. Infiltration tactics were the backbone of PLA tactics for much of the Chinese Civil War and Korean War, and they remain an important element of light infantry operations today. The primary benefit of infiltration tactics is that they offset enemy advantages in firepower. By closing distances and conducting penetrating maneuvers covertly, enemy firepower cannot be brought to bear prior to close contact. This enables friendly units to surprise the enemy, creating shock and the perception of isolation. The main limitation of infiltration is that it is very difficult to do with larger formations or with any unit that is not light infantry—large formations and vehicles are far easier to detect, and thus have great difficulty infiltrating enemy positions. Infiltration tactics are also very time consuming and limited in both scope and range, due to the slow movement rate of dismounted infantry and the need to maintain secrecy and concealment.



The PLAA today employs infiltration tactics alongside penetrations or other assaults to isolate and annihilate enemy units. For example, a light infantry depth attack group may infiltrate one flank of an enemy formation while an armor-heavy depth attack group conducts an aggressive penetration of another flank.

Successfully executing an infiltration attack requires considerable skill and initiative by small-unit leaders. These leaders are charged with identifying enemy weak points, then choosing their own avenues of approach and methods of attack. By its nature, an infiltration attack requires less concentration of forces than other forms of offensive action, but it requires greater skill in the advance and unfold phases of the operation, as detection by the enemy essentially changes the action from an infiltration to a frontal assault. Artillery and other enablers are also largely unavailable to an infiltration due to the need for secrecy and concealment, though information warfare and deception operations may provide vital support by fooling the enemy about the strength or direction of the attack or by directing enemy reconnaissance focuses elsewhere. Figure 7-14 depicts an infiltration.

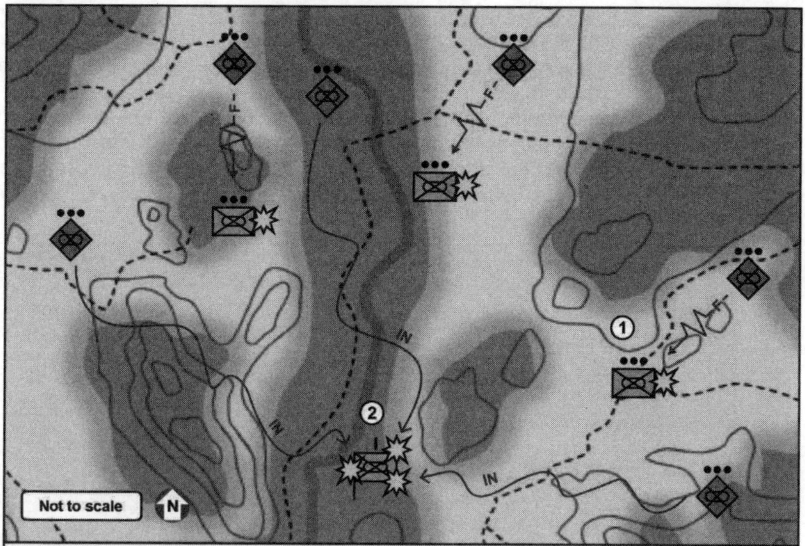


**(1) Infantry platoons conduct probing attacks on the enemy perimeter in order to fix enemy units and disrupt enemy defenses. These attacks are intended to reveal enemy positions and open opportunities for infiltrating units to penetrate the enemy's perimeter. (2) Infiltrating units carefully assess the situation, and commanders exercise initiative as to when and where their attacks will take place. Infiltration efforts should be synchronized, but will necessarily happen on their own specific timelines. Infiltrating units move rapidly but carefully into rear areas, targeting key command nodes and attempting to disrupt enemy command and communications. If effective, this isolates the forward defensive units, enabling their destruction or forcing their withdrawal.**

| F | fix | IN | infiltrate | N | north |
|---|---|---|---|---|---|

*Ref: ATP 7-100.3 (Aug '21), fig. 7-14. Infiltration (example).*

# D. Storming Attack



A storming attack is a thoroughly prepared, high-intensity assault against an entrenched, isolated opponent. Storming occurs after penetrations defeat the enemy's main defensive line, most enemy units have been forced to retreat or been annihilated, and the PLAA unit must complete the annihilation phase by reducing the remaining enemy strongpoints. Storming attacks place great emphasis on obstacle defeat, firepower and maneuver integration, and aggressive action at the point of contact.

Storming attacks are preceded by a meticulous and integrated effort to clear enemy obstacles through a combination of firepower and engineering capabilities. These breaches provide the points of entry for the final assault. At the same time, careful reconnaissance of the enemy's position informs the commander of the enemy's strength and disposition. Firepower and information attacks against enemy positions are conducted throughout the operation, attempting to reduce enemy strength and cohesion, reduce morale, and suppress enemy systems. The enemy should be surrounded if possible, and it should be isolated not only physically, but also electronically and psychologically.

The main decision the commander must make in the assault phase is the deployment of armor. Three methods for armor deployment are prescribed: in front, in close support, and in rear support. The commander's decision on how to employ armor is dictated by terrain and enemy antitank capabilities. In general, tanks in front is preferred in more open terrain, while tighter terrain demands tanks in close support. If the enemy is in highly restrictive terrain—or if it has substantial antitank capabilities available— tanks are deployed in rear support.

The assault itself is carried out bravely and aggressively. Penetrations are carried through breaches and seek to further isolate smaller enemy units. As units are suppressed and surrounded, they face the choice of surrender or annihilation. If possible, multiple penetrations are created, and assaults are coordinated and synchronized. As enemy will begins to break, the breaches are exploited by second-echelon troops, who complete the annihilation phase and secure the objective. Figure 7-15 depicts a storming attack.

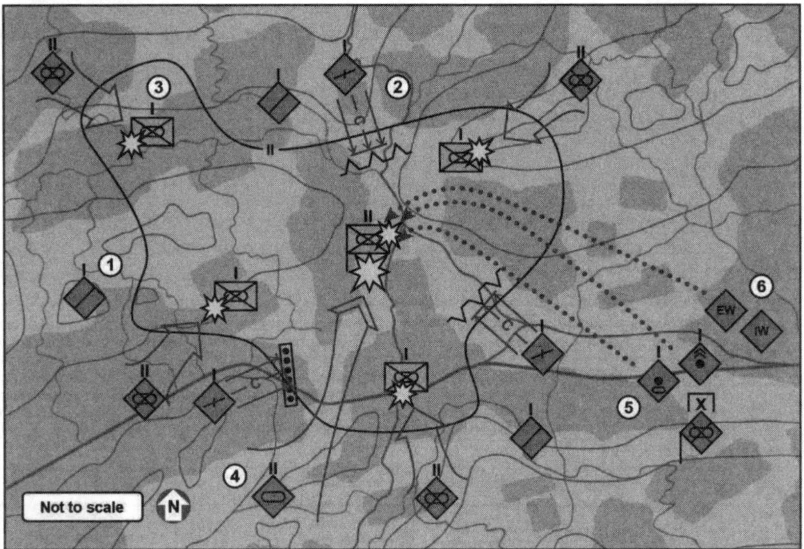


**(1) Reconnaissance units carefully assess the surrounded enemy's position in order to discover potential weak points and establish the enemy's disposition. Once the attack commences, these units transition to screening the assaulting force. (2) Engineer units rapidly clear obstacles and minefields that lay along key avenues of approach. These efforts are carefully integrated with reconnaissance in order to develop the best possible course of action for the assault. (3) Frontline attack groups assault isolated and vulnerable forward enemy units. These attacks gradually increase in intensity as the enemy command is destroyed and isolation increases. (4) The depth attack group conducts a penetration into the enemy's center, destroying the enemy's command and completing the isolation of remaining enemy units. (5) The firepower group conducts mass fire assaults on the enemy's main body in order to disrupt enemy operations and reduce enemy morale. (6) EW and IW groups conduct electronic and information attacks on the enemy formation in order to amplify the effects of isolation and prevent communications with possible relief forces.**

| | | | |
|---|---|---|---|
| C | clear | IW | information warfare |
| EW | electronic warfare | N | north |

*Ref: ATP 7-100.3 (Aug '21), fig. 7-15. Storming attack (example).*

# III. Pursuit Attack

Pursuit attack is an offensive tactic wherein a friendly formation maintains constant contact with an enemy formation in retrograde. This most often occurs when an enemy force is defeated but not surrounded, and it is thus able to retreat. Pursuit entails maintaining contact with the retreating unit and conducting quick attacks as required or where possible, with the objective of preventing effective enemy rear security operations—thus preventing the enemy unit from consolidating or reforming a strong defensive position.

Aggressively pursuing a defeated enemy formation is a basic component of mechanized warfare. The PLAA historically employed both mechanized and armored forces and fast-moving infantry forces in this role; on a modern battlefield, a mechanized force is likely required. Following an initial penetration, envelopment, or isolation, the enemy force is dislodged from its primary defensive position and begins retrograde operations to attempt to regain the initiative. Aggressive pursuit of the enemy allows friendly forces to retain the initiative while keeping the enemy force isolated and off balance.

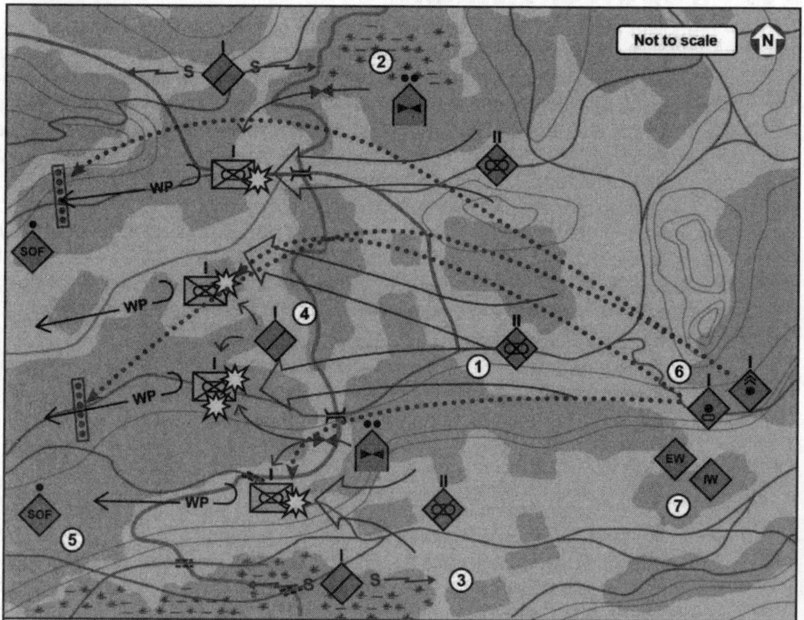


**(1) An offensive group conducts a pursuit attack against a retreating enemy. CA-BNs maintain continuous contact with retreating enemy formations, not allowing then to conduct rearguard or delaying actions, and not allowing them to establish a new entrenched defensive position. (2) Attack helicopters are well-suited to support pursuit attacks; they continually harass and disrupt enemy units on the move. (3) Reconnaissance elements screen the flanks of the offensive group, ensuring that the enemy cannot mount a surprise attack on the extended friendly line. (4) Other reconnaissance elements maintain contact with the retreating enemy forces. (5) SOF teams in deep areas sabotage roads and bridges, disrupt enemy reinforcement efforts, and conduct deep reconnaissance. (6) The firepower group employs howitzers to attack retreating enemy units and rocket artillery to deploy hasty minefields across probably routes of retreat. (7) The EW group ensures that enemy forces cannot contact possible reinforcements, maintaining their isolation. The IW group concentrates on maximizing the propaganda value of the victory, transmitting media of destroyed and retreating enemy units.**

| | | | | | |
|---|---|---|---|---|---|
| CA-BN | combined arms battalion | N | north | SOF | special operations forces |
| EW | electronic warfare | S | screen | WP | withdrawal |
| IW | information warfare | | | | |

*Ref: ATP 7-100.3 (Aug '21), fig. 7-16. Pursuit attack (example).*

Pursuit attack does not necessarily need to take the form of ground maneuver. Other capabilities can achieve the pursuit effects of maintaining contact with the enemy, preventing enemy consolidation, and retaining friendly initiative. Artillery fire, air attack, and countermobility activities can disrupt enemy formations or canalize or direct enemy movements. Network or electromagnetic attack can suppress enemy communications or disseminate disinformation to influence enemy actions. SOF or guerrilla forces can conduct raids, feints, or demonstrations to degrade enemy cohesion or influence enemy decision making. Any pursuit operation is best conducted using a variety of different capabilities to create a combined arms effect and maximize the impact of friendly operations on the strength and mindset of the enemy.

A unit conducting pursuit operations is vulnerable to counterattacks or other spoiling operations, especially if it is a maneuver unit strung out over a wide area. Security operations or supporting attacks should work to ensure that a unit conducting a pursuit operation is secure against enemy counterattack, reconnaissance, and countermobility activities.

# IV. Firepower Attack

A firepower attack is a tactic wherein a number of firepower capabilities are massed on a specific target to achieve an offensive effect. Firepower capabilities include, but are not limited to field artillery, ballistic or cruise missiles, air attack, antiair attack, network attack, electromagnetic attack, and direct fire. The defining characteristic of the firepower attack is that it does not require any physical occupation of key terrain or close contact with enemy forces. Instead, massed firepower is used to isolate or annihilate enemy units, canalize enemy movements, penetrate enemy defenses, reduce enemy morale, or affect enemy decision making. The PLAA uses firepower attacks in a variety of ways, including to reduce cohesion or strength of enemy forces, to support assaults or advances, to fix the enemy, to attack enemy flanks or rear areas, or to spoil or defeat enemy counterattacks. Firepower attacks can be employed either alone or in an integrated and synchronized manner with other actions.

The PLAA recognizes at least seven different forms of firepower attack: advance fire, fire support, information and firepower assault, depth firepower, counter-counterattack firepower, annihilation firepower, and air defense firepower.

*See facing page for an overview and further discussion.*

# V. Ambush



An ambush is an operation in which a unit deploys to a location in advance, then strikes the enemy from a concealed position in a surprise attack. The PLAA historically emphasized ambush tactics as a highly effective way to offset enemy advantages in firepower or protection, and it used ambushes extensively during the Second Sino-Japanese War and the Chinese Civil War. The ambush is still a popular tactic, but the PLAA advises that the informationized battlefield makes conducting an effective ambush more challenging.

### Ambush Types

- **Waiting Ambush**
- **Decoy Ambush**
- **Forced Ambush**

The PLAA recognizes three distinct types of ambush: the waiting ambush, the decoy ambush, and the forced ambush. The waiting ambush is employed when an enemy travels along a predictable route, and friendly forces can deploy and lie in wait in a concealed position. A decoy ambush is similar, but instead of making use of a known route, it employs a decoy that tricks the enemy into moving into the ambush zone. A

# Firepower Attack (Types)

*Ref: ATP 7-100.3, Chinese Tactics, (Aug '21), pp. 7-32 to 7-34.*

**Advance Fire.** Advance fire is fire that targets enemy positions or other key assets prior to follow-on operations. In Western operations, it is typically referred to as preparatory fire. Advance fire targets enemy reconnaissance platforms, command and communication nodes, security units, and artillery units. The purpose of advance fire is to neutralize, destroy, or degrade key enemy capabilities in support of future operations.

**Fire Support (Advance or Assault).** Fire support consists of those fires that are integrated closely with the advance phase or initiate phase of an operation. During the advance phase, the key mission for firepower is to ensure that enemy reconnaissance platforms are suppressed or destroyed—thus preserving the secrecy of the attack—along with enemy artillery units that may target the vulnerable advancing force during its movement operation. In support of assaults, firepower targets key enemy positions and units, creating weak points or breaches in the enemy's defensive line that enable effective penetration. Key targets of this mission type include enemy armor and antitank units, obstacles, fortified positions, and exposed units.

**Information and Firepower Assault.** The information and firepower assault is also referred to as the blitz—a specialized tactic wherein physical fires and information attack are carefully synchronized to maximize the combined arms effect against the enemy. Physical fires destroy, degrade, or neutralize key targets through their destructive power, while information attack targets enemy morale, cohesion, and communications. The blitz is employed at the key point in an overall offensive operation, targeting the enemy's center of gravity. When and where to employ the blitz is a key consideration when planning offensive operations.

**Depth Firepower.** Depth firepower targets enemy forces and key assets in deep areas, supporting follow-on operations that are exploiting successful initial assaults. Key targets for depth firepower include enemy artillery reserves attempting to spoil the exploitation effort, mobile enemy reserves conducting retrograde operations, enemy command posts, and enemy reconnaissance platforms. Depth firepower's primary mission is to create or enhance the isolation effect on enemy forces and enable the annihilation phase to commence.

**Counter-counterattack Firepower.** Counter-counterattack firepower specifically targets enemy counterattacks. This includes not only targeting ground forces, but also air power through antiair attack and information attack against enemy communications and command structures. The counter-counterattack firepower mission is reserved for the point in time when the enemy's counterattack has been clearly detected and identified. It can be thought of as a miniature blitz, specifically targeting the enemy's commitment of reserves to the counterattack.

**Annihilation Firepower.** Annihilation firepower attempts to destroy isolated or encircled troops. This may be in support of storming attacks, or it may be to annihilate the enemy independently. Annihilation firepower targets not only the enemy's isolated formations, but also any possible routes for retreat, any enemy forces attempting to relieve the besieged unit, and any obstacles that might inhibit a storming attack.

**Air Defense Firepower.** Air defense firepower uses ground-based defensive counterair systems—surface-to-air missile systems, self-propelled antiaircraft guns, and EW systems—in an aggressive, forward-positioned manner. There are two possible missions for air defense firepower: restrict the enemy's use of a specific volume of airspace through deterrence, or destroy or attrit enemy aircraft. Though it employs systems that Western armies view as defensive, the PLAA characterizes air defense operations as fundamentally offensive.

forced ambush is somewhat different: it employs direct actions such as feints, demonstrations, or countermobility activities to force the enemy into the ambush zone.

Choosing the ambush site is the most important decision when planning an ambush. Terrain should canalize the enemy formation while helping to conceal the friendly unit. If possible, enemy forces should be carefully reconnoitered to ensure friendly forces will enjoy local combat superiority. Engineers should set up obstacles to maneuver the enemy into the ambush zone and complicate its retreat. Artillery is deployed in concealed positions and registered across the ambush zone.

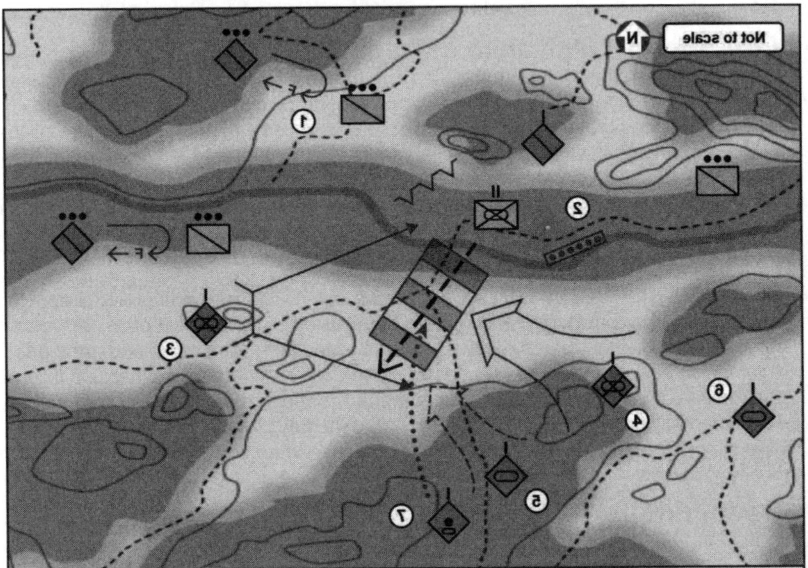

(1) A combined arms battalion sets a waiting ambush for an enemy battalion. Reconnaissance units carefully engage enemy scouts with feint attacks, drawing them away from the ambush zone and ensuring the main ambush force remains undetected. (2) Mines and obstacles laid along the enemy's potential routes are sued to canalize enemy movement into the ambush zone. (3) The concealed group occupies a well-protected support-by-fire position overlooking the ambush zone. This group initiates the ambush when the enemy enters the ambush zone. (4) The flank group occupies an adjacent position and engages the enemy's vulnerable flank once the enemy is heavily engaged by the concealed group. (5) The pursuit group waits just beyond the ambush zone and prepares to assault and pursue the enemy as it retreats from the ambush zone. (6) The interception group waits in a concealed position and prepares to aid in pursuit or intercept any reinforcements being sent to aid the ambushed unit. (7) The firepower group delivers preregistered artillery fire into the ambush zone.

F feint N north

*Ref: ATP 7-100.3 (Aug '21), fig. 7-17. Ambush (example).*

Maneuver units on ambush missions are divided into four primary groups: concealed, flank, pursuit, and interception. The concealed group initiates contact, engages, and fixes enemy forces from at least one flank. The flank unit assaults an alternate flank once the enemy force is fixed by the concealed group. The pursuit group is a reserve that engages and maintains contact with the enemy force as it retreats from the ambush zone. The interception group is deployed along potential reinforcement avenues of approach in order to prevent support or reinforcement to the ambushed enemy formation.

Concealment and deception are key enablers of the ambush. Reconnaissance and surveillance on the informationized battlefield make moving to ambush positions in secret a difficult task. Commanders must carefully move forces into their ambush positions, and units must remain highly disciplined in order to avoid detection. Once the ambush is complete and the enemy unit annihilated, secure retrograde operations should be conducted quickly.

# VI. Raid

A raid is an offensive operation designed to strike and surprise an unsuspecting enemy. The primary defining characteristic of the raid is the employment of a hit-and-run approach. The enemy is struck and injured, then, in most cases, the attacking force breaks contact and rapidly retreats. Though raids may evolve into full-scale assaults in certain circumstances—encountering a weaker-than-expected enemy, for instance—the typical approach to a raid is a fast-moving, intentionally abbreviated attack.

Raids are aggressive actions that have clearly defined parameters. They may seek to harass the enemy, to probe it, or to destroy isolated enemy units. A raid requires skillful use of concealment and deception, rapid movement, and sudden, aggressive action within the commander's parameters. Subordinate units must be aware of the commander's intent and must not over-engage the enemy unless a clear opportunity presents itself.

**Raid Types**

- **Rapid Raid**
- **Long-range Raid**
- **Sabotage Raid**
- **Sneak Raid**
- **Harassing Raid**

The PLAA recognizes five different types of raids: rapid, long-range, sabotage, sneak, and harassing.

## Rapid Raid

The rapid raid is analogous to the Western idea of a meeting engagement. A unit unexpectedly comes upon an enemy, transitions to an attack posture, then conducts an aggressive attack that is pressed until the commander deems it complete. Rapid raids are specifically prescribed as spoiling actions against counterattacking forces or newly arrived airborne or amphibious forces.

## Long-range Raid

The long-range raid is not necessarily defined by the distance the unit travels to conduct the raid. Rather, it refers to how deep behind the enemy's main defensive line the raid takes place. Long-range raids are conducted by light forces or SOF who infiltrate the enemy's territory to strike at predetermined targets. This may be a separate action, or it may be integrated as part of a larger operation.

## Sabotage Raid

A sabotage raid is one in which a specific high-value asset is targeted for destruction. The unit infiltrates enemy territory with the intent of locating and destroying one or more high-value assets through covert, aggressive action. Potential targets include command nodes, key intersections or roadways, lines of communications, ammunition stockpiles, or high-value weapons systems such as combat aircraft or surface- to-air missile systems.

## Sneak Raid

The sneak raid is a small-scale, short-distance raid designed primarily to make use of local conditions such as nighttime or bad weather to conceal the raiding force. Sneak raids may attempt to seize enemy prisoners, probe enemy defenses, or reconnoiter. The sneak raid does not attempt to strongly engage the enemy, and the sneak raiders retrograde quickly back to concealed positions. Figure 7-18 depicts a sneak raid against an enemy position.

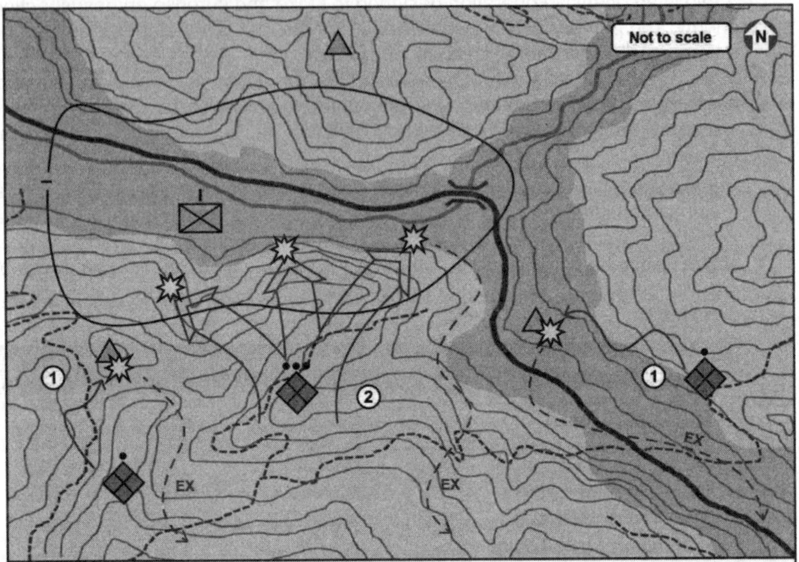


**(1) An infantry company conducts a sneak raid on an enemy position, attempting to seize prisoners, reconnoiter the enemy's position, and disrupt enemy defenses. Infantry squads conduct rapid hit-and-run actions against enemy outposts in order to seize prisoners, disrupt enemy counterreconnaissance, and gauge the enemy's strength. These units rapidly attack and withdraw quickly before the enemy can organize a response. (2) The main body of the raid strikes the enemy's main defensive line in multiple areas, complicating response and confusing the enemy as to the purpose and direction of the attack. The attack only continues for as long as the enemy cannot organize a strong response; once several prisoners are taken and the enemy's position carefully assessed, the main body rapidly retrogrades back to the friendly security area.**

| EX | exfiltrate | N | north |
|---|---|---|---|

*Ref: ATP 7-100.3 (Aug '21), fig. 7-18. Sneak raid (example).*

## Harassing Raid

The harassing raid is one intended to annoy, disrupt, or confuse the enemy. Harassing raids are intended to provoke an enemy response, wear out enemy soldiers, or spoil enemy maneuvers. Most importantly, they are used to disperse the enemy, forcing it to defend a wider geographic area and reducing its presence at potential key places on the battlefield.

Chap 7

# I. PLAA Defensive Actions (Overview)

*Ref: ATP 7-100.3, Chinese Tactics, (Aug '21), pp. 9-1 to 8-4.*

A well-planned and -coordinated defense is a critical component of every combat action. The defense is carried out in order to attrit enemy forces, retain key positions or terrain, buy the commander time and decision space, seize the initiative from the enemy, and transition to offensive operations. The People's Liberation Army Army (PLAA) believes that defense is an inherently stronger form of war, as the defender enjoys advantages of terrain and time unavailable to the attacker. The informationized battlefield, however, has reduced many of these traditional advantages. While it is not possible to destroy an opponent through defensive actions alone, a tenacious and well-executed defense enables decisive follow-on offensive operations.

## I. Overview of PLAA Defensive Operations

The PLAA takes the position that defensive actions are ultimately to preserve one or more assets, such as friendly forces, key terrain, or the initiative. In addition, defensive operations can play a key role in a wider operational or strategic sense by attritting the enemy's strength, forcing it to commit greater forces in an attempt to achieve an objective, and reducing or restricting the options available to enemy commanders. Though the PLAA considers the offense to be the decisive form of warfare, centuries of invasion and occupation have led to considerable emphasis on the importance of defensive operations: the People's Liberation Army's (PLA's) most sacred mission is defending Chinese territory from outside aggression.

## II. The Informationized Battlefield and Defensive Operations

The PLAA still maintains that defense is a fundamentally stronger form of warfare than offense, but it acknowledges that many elements of the informationized battlefield have changed the traditional dynamics between attack and defense. Most—though not all—of these changes benefit the attacker, making defensive operations more difficult than they have been historically. The PLAA has identified four major trends on the informationized battlefield that influence defensive operations: increasing arduousness, fewer traditional advantages, more dynamic, and increasing importance of offensive actions.

### Increasing Arduousness

Multiple factors have combined to make combat for the defender more difficult than in the past. While ground commanders once only had to concern themselves with enemy land forces, enemies can now strike simultaneously or consecutively from multiple domains. This requires, in turn, a comprehensive multi-domain defense that can effectively blunt or check enemy actions, even when coming from unexpected directions. Commanders must also defend their forces not only from physical attack, but also from information, electromagnetic, and psychological attack: a capable opponent will target enemy troops' morale and cohesion through a variety of channels. Finally, the depth and variety of enemy firepower systems have increased substantially in recent years. Long-range artillery and missile strikes, air strikes, attack helicopter operations, direct action by special operations forces (SOF), and electromagnetic and network attack capabilities allow an enemy to target the full

Defensive Actions

# III. Principles of the Defense

*Ref: ATP 7-100.3, Chinese Tactics, (Aug '21), pp. 8-2 to 8-4.*

The PLAA has identified four key principles that should guide commanders' planning and operations when conducting a defensive operation on the informationized battlefield: depth, consolidation, integration, and flexibility. When applied skillfully, these principles enable the commander conducting a defensive action to effectively blunt the opponent's attack and then rapidly transition to decisive offensive operations. .

## Principles of the Offense

- **Depth**
- **Consolidation**
- **Integration**
- **Flexibility**

## Depth

While defense-in-depth is an ancient concept, its modern application requires a somewhat different approach than the traditional ground-only concept. Defense-in-depth now requires PLAA commanders to integrate defensive measures throughout all domains while placing a strong emphasis on the counterattack. It is impossible to defend all areas of the modern battlefield from all potential avenues of attack, so commanders must effectively prioritize key areas for defense and make best use of available space— trading space for time is now more important than ever before. Commanders should encourage enemies to ingress through unimportant or noncritical areas, extending their lines of communications (LOCs) and exposing their flanks. Decisive counterattacks should target these formations at key times and places, seeking to disrupt or destroy enemy formations as they move through deep areas. These efforts are simultaneously aided by raids, firepower operations, and information warfare operations that target enemy reinforcement movement into the combat area. A properly executed defensive operation makes use of depth to achieve isolation of enemy forces while effectively integrating all available assets into a cooperative defensive effort throughout the combat area.

## Consolidation

The defensive principle of consolidation mirrors the offensive principle of concentration: it prescribes massing combat power at key times and places to make best use of available forces. Effective consolidation is underpinned by the thoughtful deployment of friendly forces to key points around the defensive zone before wider defensive operations commence. This deployment, in turn, requires the commander and staff to conduct extensive analysis of the enemy's probable approach, then prescribe the primary defensive line and axis, key defense points (KDPs), and key targets for the counterattack. Defensive operations must take advantage of favorable terrain or information conditions as combat multipliers, as counterattack groups are expected to be outnumbered by the forces they engage. The main defensive combat capability should take the form of a powerful mobile force able to conduct independent operations anywhere in the defensive zone. The commitment of this force typically represents the key point in a defensive battle. In addition, powerful mobile and armored forces should seek to conceal their movements as much as possible to deceive the enemy about the time and place of their commitment, and they should remain as dispersed as possible to protect against enemy firepower and to further deceive the enemy.

## Integration

A defensive operation must be carefully integrated throughout the defensive zone. This includes not only careful planning and unification of purpose, but also preservation of combat power, effective use of terrain and firepower as means of resistance, and use of small-unit actions across the defensive zone to harass and distract the enemy. Protection of friendly forces is a key element of integration. This involves not only traditional hardening of defensive positions, but winning the information battle as well—keeping friendly forces from being affected by enemy psychological attack and propaganda. Physical resistance should be enhanced by the use of obstacles and other countermobility capabilities, which are used to slow, stop, or canalize enemy formations in accordance with the overall scheme of defense. The appropriate use of firepower and terrain can allow vastly overpowered units the ability to resist enemy attacks for long periods, forcing the enemy to commit reserves and overextend its offensive to achieve its attack goals. Antitank and air defense weaponry must be fully integrated into the defense in order to offset enemy advantages in armor, firepower, and air power. Coordinated firepower assaults mixing direct fire and artillery can attrit or disrupt enemy formations at stand-off distances. Depth operations target enemy LOCs and reinforcements, attempting to isolate powerful enemy offensive groups as they penetrate into the defensive zone.

## Flexibility

The defensive principle of flexibility seeks to achieve a state of localized initiative where defensive forces, despite fighting on the defensive, successfully maneuver, consolidate, and achieve localized combat power at key times and places. It requires commanders to take prudent risk in their defensive plans, to use stratagems effectively in order to gain the advantage of surprise, and to attack and defend resolutely as the situation develops. A flexible defense is able to absorb enemy offensive actions by trading time and space, then rapidly adjusting to enemy tactics and strategy to formulate a response. A flexible defense is, in many ways, a series of small offensive actions, seeking to target enemy weak points with friendly strength and seizing localized initiative at key times and places. The sort of flexibility that the PLAA seeks requires superior lower-echelon leadership, with lower echelons capable of independent actions and lower echelon commanders able to make good decisions even in the absence of complete information.

depth of friendly formations and defended areas. There is no safe space on the informationized battlefield.

## Fewer Traditional Advantages

Historically, a defender enjoyed several fundamental advantages that translated to nearly every combat action: greater ability to conceal one's forces and deceive the enemy; better use of terrain to cover and harden one's position; improved communications and coordination due to interior lines and known terrain; and more time to make the best use of the deployment area and battlefield depth. Conditions on the informationized battlefield have eroded all of these traditional advantages. Because of advanced wide-area multispectrum reconnaissance, intelligence, and surveillance capabilities, concealing one's own forces and deceiving the enemy about deployments and dispositions is far more difficult. In the past, one only needed to win the ground reconnaissance and counterreconnaissance battle, but the informationized battlefield leverages air, space, and cyber intelligence collection in addition to more advanced ground-based capabilities. Precision munitions and long-range strike capabilities have eroded several basic defensive advantages offered by terrain, making it far more difficult to effectively harden defensive positions. The enemy can also effectively target the command and communication systems of a defensive position in ways never before available. Electronic warfare methods, such as communications jamming and meaconing, and network attack are capable of destroying, disabling, degrading, or manipulating the command and communication backbone of a defensive position. Finally, the defender's traditional advantage of depth has been eroded or eliminated by capabilities that enable the enemy to strike into deep areas. Depth areas that were once generally considered secure and untouchable can now be targeted with both lethal and nonlethal attack methods, even when many miles from forward areas.

## More Dynamic

As modern offensive operations demand rapid thrusts that target weak points from unexpected directions, the traditional approach of defending strong points with a relatively static approach is now obsolete. Static formations, even in a strong defensive position, will quickly be bypassed, isolated, and then annihilated by a competent attacker. Instead, defense must be considered a dynamic action characterized by rapid movement and decisive concentration of combat power at key times and places. Information superiority coupled with mobility enables the defender to rapidly detect and appropriately respond to enemy offensive actions, regardless of where they might occur. This approach to defense is in keeping with the PLAA's new emphasis on decentralization. A mobile, rapidly reacting defensive strategy requires lower-echelon commanders able to react appropriately to unexpected events without input from a higher echelon. While the hasty defense was once considered a last-ditch option, it is today the cornerstone of a PLAA defensive operation versus a capable opponent.

## Increasing Importance of Offensive Actions

While seeming somewhat paradoxical, offensive actions within a broader defensive operation are considered to be of heightened importance. Due to the conditions of the informationized battlefield, offensive actions have increased in their lethality and unpredictability; this trend extends to counterattacks. A well-timed and -executed counterattack can disrupt, defeat, or spoil an enemy's offensive action just as effectively as a well-planned defense, and the counterattack enjoys all the advantages of other offensive actions: enhanced firepower, enhanced reconnaissance, multi-domain options, and the elements of surprise and initiative. The PLAA approach to defense-by-offense prescribes the use of depth attacks and aggressive maneuver to put the counterattack group in the best possible position, concentrating combat power against enemy weak points and enabling the isolation and destruction of enemy offensive groups.

Chap 7

# II. Planning the Defense

*Ref: ATP 7-100.3, Chinese Tactics, (Aug '21), pp. 8-4 to 8-17.*

PLAA doctrine prescribes a large number of steps for planning and executing a defensive operation, which can be consolidated into seven major phases:

**Planning the Defense**

- I Build the Command System
- II Organize Reconnaissance
- III Organize the Defensive Group & Deploy
- IV Spoil the Enemy's Preparations
- V Resist the Enemy's Assaults
- VI Counterattack
- VII Consolidate or Withdraw

Each phase is part of every defensive operation to some degree, but the phases are not necessarily sequential. Because they are largely in response to enemy actions, they may be conducted in varied order or concurrently. This planning includes organizing the defensive battlefield and building the defensive groups that comprise the operational system.

## I. Build the Command System

As with offensive operations, the PLAA prefers to carefully plan and organize major defensive operations. Meticulous planning enables greater redundancy and depth in the defense, improved overall security measures, a more effective deception plan, and a greater chance of a decisive counterattack. The centerpiece of the defensive plan is the deployment of the command posts. A defensive operation makes use of at least two, and ideally four, command posts. The base command post is the commander's primary location, and it will typically be located to best coordinate between frontier defense and depth defense units, oriented toward the enemy's anticipated main axis of attack. The commander may also establish an advance command post

# Defensive Zone (Notional)

*Ref: ATP 7-100.3, Chinese Tactics, (Aug '21), pp. 8-4 to 8-7.*

The defensive zone is further subdivided into two or more secondary zones, each with a specific set of objectives and different set of tactics. While the PLAA used to be highly prescriptive about the physical sizes of these zones, it has gradually moved to a more flexible approach. The various zones should account for terrain, friendly and enemy capabilities, and higher echelon missions. They should also enable careful integration of various units and capabilities.

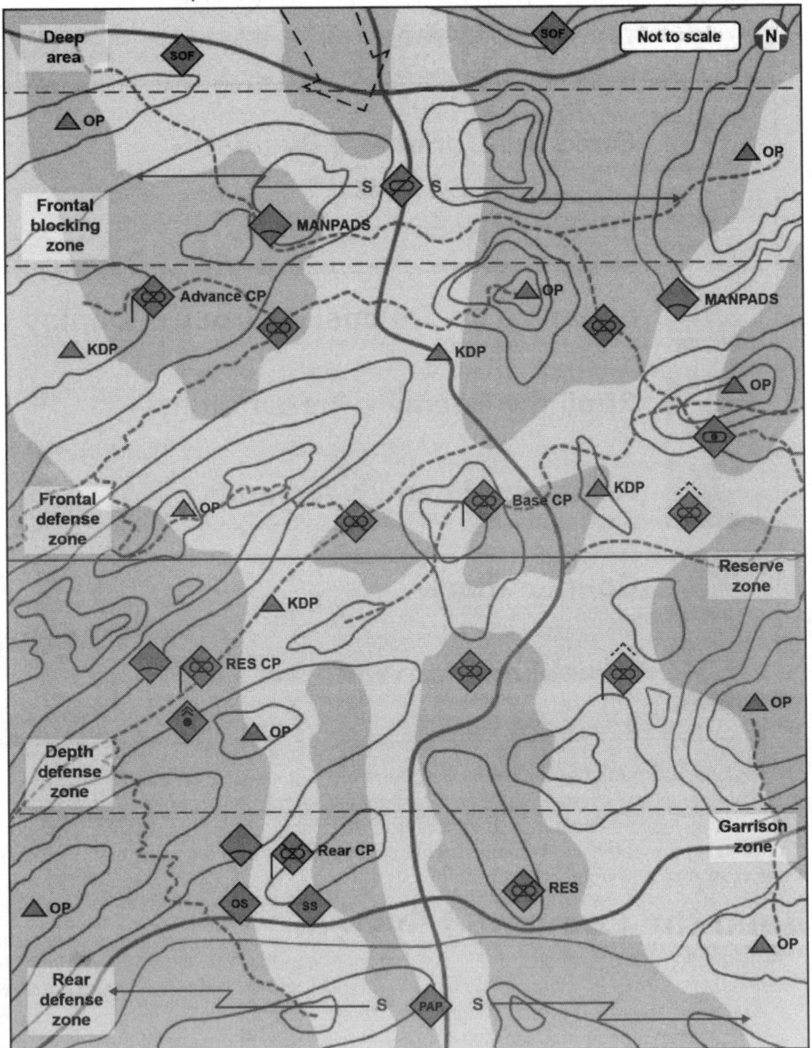


| CP | command post | OP | observation post | S | screen |
|---|---|---|---|---|---|
| KDP | key defense point | OS | operational support | SOF | special operations forces |
| MANPADS | man-portable air defense system | PAP | People's Armed Police | SS | service support |
| N | north | RES | reserve | | |

*Ref: ATP 7-100.3 (Aug '21), fig. 8-1. Defensive zone (notional).*

## Deep Area

Deep areas are those areas that are not targetable by a defensive group's organic weapons systems. They may still be occupied by elements of the defensive group conducting reconnaissance, counterreconnaissance, and screening missions, but these elements operate independently. SOF elements also occupy deep areas, and both air and missile support may be employed in support of deep-area operations. The primary purpose of deep area operations is to disrupt and slow the enemy advance while providing the defensive group commander with critical intelligence about the enemy's strength, disposition, and possible objectives.

## Frontal Blocking Zone

The frontal blocking zone is the forward-most area of the defensive zone. It is analogous to the Western security zone, and it is designed in much the same way. The frontal blocking zone is occupied by units performing screen or cover missions, and it serves as the primary early warning, disruption, reconnaissance, and counterreconnaissance zone for the main body. The frontal blocking zone should orient toward the enemy's primary axis of advance—particularly its primary reconnaissance axis. This zone should be positioned sufficiently in front of the main body to give the commander time and decision space, usually between 3 kilometers (km) and 5 km for a CA-BDE. The frontal blocking zone is likely occupied primarily by the cover group.

## Frontier Defense Zone

The frontier defense zone is typically the combat group's primary defensive area. It should contain most KDPs and the preponderance of combat power. Units should seek to conduct a strong defense within the frontier defense zone, attempting to force the enemy to commit most of its combat power and leave its forces vulnerable to counterattack. The frontier defense zone is primarily occupied by one or more frontier defense groups, and possibly a depth group. Operations in this zone are centered largely on active resistance to enemy assaults. Fortifications and entrenchments should be as extensive as time and resources allow, and units should be prepared to conduct stubborn and brave resistance, possibly in the face of significant odds. The base command post is typically located in this zone, while the advance command post will be either forward in the zone or rearward in the frontal blocking zone.

## Depth Defense Zone

The depth defense zone extends behind the frontier defense zone, and it serves as the deep area of the defensive zone. Depth defense groups—the heart of the counterattack force—are deployed here, to react quickly and appropriately to the enemy's assaults in the forward defensive areas. In addition, combat reserve groups may be stationed in this zone to reinforce or assist forward units. A single depth defense zone may serve as the deep area for multiple frontier defense zones. Operations in the depth defense zone should ensure protection of the depth defense group from air or artillery assault, ensure mobility, and ensure the concealment and secrecy of counterattack operations. The reserve command post, if established, will likely be located in this zone.

## Rear Defense Zone

The rear defense zone is the deepest area of the defensive zone, and it houses logistics, equipment support, and other rear area units and capabilities, including the rear command post. This zone is occupied primarily by rear area security units—possibly People's Armed Police (PAP) units in addition to PLAA security forces—and it may also house depth or combat reserve groups. The rear defense zone ensures protection against enemy deep artillery, air strikes, and enemy SOF or guerrilla actions in rear areas, and it ensures mobility to enable retrograde or reinforcement movements through the zone.

in a more forward position. The rear command post is led by the deputy commander, and it is deployed in a well-defended location. Its primary role is to organize logistics and rear area defense. If possible, a reserve command post is established along a possible route of egress or in a well-defended rear location, ready to take over for the base or advance command post should either of them come under threat. Additional information on command post operations can be found in chapter 4.

The defensive battle makes careful use of terrain and geographic control. There are two primary features that a commander must identify to make best use of the terrain. First, the main direction of defense informs the combat group as to the orientation of the enemy's anticipated primary effort. Commanders placing the main direction of defense should take into account higher echelon and adjacent-unit missions, specified defensive tasks, the unique characteristics and tendencies of the opponent, and terrain throughout the defensive zone. Second, the commander must establish KDPs. These are the specific features within the defensive zone that are of greatest importance to the integrity of the overall defense. Commanders should keep the numbers of KDPs small, choosing only those points absolutely critical to successfully prosecuting the defensive battle. The defensive plan should be centered on maintaining these KDPs, which assists commanders in prioritizing their available resources. KDPs should also be phased, allowing for them to change during the course of the battle. Finally, these points are not limited to key terrain. They may also include a critical command or network node, a key unit or leader, or even a piece of information on a computer network.

The defensive zone is further subdivided into two or more secondary zones, each with a specific set of objectives and different set of tactics. While the PLAA used to be highly prescriptive about the physical sizes of these zones, it has gradually moved to a more flexible approach. The various zones should account for terrain, friendly and enemy capabilities, and higher echelon missions. They should also enable careful integration of various units and capabilities.

*See previous pages (pp. 7-6 to 7-7). Fig. 8-1 depicts a combined arms brigade's (CA-BDE's) genericized defensive zone.*

# II. Organize Reconnaissance

The reconnaissance effort is a critical enabler of a defensive operation. Reconnaissance must be comprehensive and well planned, extending through all domains and the depth of the battlefield. These efforts concentrate on three primary intelligence objectives: disposition and intent of enemy forces; terrain; and conditions of the battlefield—civilians, weather, the electromagnetic environment, and so on. Reconnaissance commences immediately upon receipt of warning orders: commanders rapidly build and deploy the reconnaissance and intelligence group to facilitate rapid, accurate, and continuous scouting for the main body. Like all other aspects of a defensive battle, reconnaissance must be conducted in depth. Deep and forward reconnaissance units surveil and disrupt enemy forces in deep areas. As the situation develops, additional reconnaissance forces, having been held in reserve, are deployed to key areas of the battlefield to enhance situational understanding.

Reconnaissance groups are often charged with disrupting enemy operations, typically as part of a screen or cover force. This requires a mixture of stubborn resistance and prudent tactical offensive actions, usually in the form of raids coupled with entrenched units. These activities are intended to force the enemy to deploy earlier than it would like, buying the commander decision space and revealing the enemy's plan of action. At the same time, counterreconnaissance actions attempt to spoil the enemy's efforts to ascertain friendly force dispositions, leaving it ignorant as to the strong and weak points of the defense. Reconnaissance screens may attempt to canalize or steer enemy forces toward areas of strength.

Reconnaissance information should be carefully analyzed and processed to give the commander the clearest possible picture of the battlefield. The commander must identify the most important elements of information, and the reconnaissance and intelligence group must strive to answer these questions as clearly and accurately as possible. Intelligence filtering enables the commander to judge friendly and enemy situations; make best use of terrain, weather, and the conditions of the electromagnetic spectrum; then distill the ground truth from the "fog of war." In contrast to its historical approach of top-down intelligence, the post-reform PLAA places greatest value on the lower-echelon commander's situational understanding, above that which is handed down by upper-level leadership. Lower-level commanders are encouraged to act decisively based on their own judgment, though they must consider the higher-echelon mission and intent. Additional information on reconnaissance and security operations and the PLAA's intelligence process can be found in chapter 6.

# III. Organize the Defensive Group and Deploy

The combat power that comprises the defense is organized into a defensive group, the combat group assigned to the defensive mission. This operational system is built in the same way as any other tactical operational system, using the CA-BDE as the primary force provider and the combined arms battalion (CA-BN) as the primary building block. Capabilities are organized into one of four primary groups, each with a distinct mission: the frontier defense group, the depth defense group, the combat reserve group, and the cover group. A combat group may contain some or all of these, and it may contain more than one of each. Additional groups that may be employed include a firepower strike or artillery group, an air defense group, an electronic and network warfare group, and a combat support group. A specific description of each of these groups can be found in chapter 4. In addition, the commander may designate teams for specific missions, such as antitank, mobile artillery, obstacle construction or reduction, or rear defense.

Once the defensive group is assembled, groups deploy to the defensive zone. This action should be undertaken quickly, allowing time for adjustments and amendments to orders after the groups reach their initial positions. The cover group typically has the mission of scouting, occupying, and securing the defensive area, enabling the other groups to move into place quickly and safely. Any enemy reconnaissance or scouting elements in the defensive area must be aggressively attacked and driven off, ensuring secrecy of movement and disposition of friendly forces. These operations are likely performed in tandem with ongoing friendly reconnaissance efforts.

Secrecy of movement is a critical consideration as follow-on friendly forces occupy their assigned defensive positions. Commanders must always assume that the enemy is watching, and they should employ concealment and deception as much as possible to confuse or deceive the enemy. Commanders must also be aware of the threats posed by artillery bombardments, airborne or ground-based raids, and information operations efforts targeting the combat group throughout this phase.

Engineering construction begins immediately as the groups enter the defensive position. Engineer priorities are determined primarily by terrain, enemy capabilities, and time available. A hasty defense may only allow simple structures, such as improvised obstacles and firing positions. With more time available, a more elaborate defense may allow for extensive tunnels, entrenchments, field structures, permanent shelters, and fallback positions. The PLAA emphasizes the use of irregular units such as local militia and local populations for engineering construction efforts. All engineering efforts focus heavily on KDPs.

*See following page. Fig. 8-2 depicts reconnaissance and deployment activities occurring in the defensive zone.*

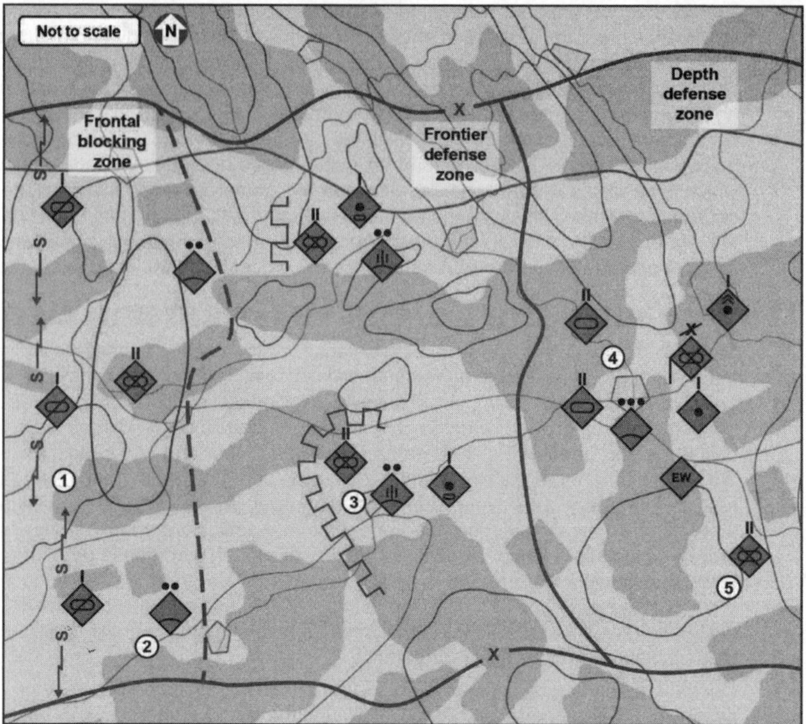


**(1) The cover group deploys a mix of reconnaissance and light infantry units into the frontal blocking zone. These units conduct reconnaissance and counterreconnaissance missions while screening the main body. (2) MANPADS sections establish ambush zones along potential aerial avenues of approach. (3) The frontier defense zone is occupied by the frontier defense group. The main line of defense consists of two key defense points, each defended by a mechanized CA-BN. SPG batteries and SPAAG sections are in direct support. (4) The depth defense group consists of two armored CA-BNs supported by a heavy towed battery, a rocket battery, and a SHORAD platoon. This group is charged with conducting the decisive counterattack against the enemy's main effort. (5) The reserve group consists of a mechanized CA-BN. It awaits orders to block enemy advances or to support the depth group's counterattack.**

| | | | | | |
|---|---|---|---|---|---|
| CA-BN | combined arms battalion | SHORAD | short-range air defense | S | screen |
| MANPADS | man-portable air defense system | SPG | self-propelled gun | SPAAG | self-propelled anti aircraft |
| N | north | EW | electronic warfare | | |

*Ref: ATP 7-100.3 (Aug '21), fig. 8-2. Reconnaissance and deployment in the defensive zone (example).*

# IV. Spoil the Enemy's Preparations

Once the reconnaissance efforts determine the enemy's presence and disposition, the defensive group commences efforts to spoil the enemy's offensive plan. In most cases, this phase involves two primary efforts: spoiling attacks and firepower assaults. Ideally, these efforts are conducted concurrently to achieve a combined-arms effect. Spoiling efforts are not intended to be decisive; instead, they are seen as enabling the main defensive efforts and eventual counterattacks.

Spoiling attacks, also called harassing attacks or harassing assaults, are limited-scope offensive actions that are designed to disrupt enemy movement and cohesion, reduce enemy morale, develop PLAA situational understanding, and manipulate the enemy commander's decision making. Spoiling attacks may take the form of offensive ground action, air attack, or information attack. Spoiling attacks target key enemy formations or capabilities, attempting to engage them when they are in a vulnerable state, such as during movement. Ground spoiling attacks often employ hit-and-run tactics—such as raids—and commanders are warned not to become decisively engaged. Enemy assets such as communications systems, high-value weapons systems, and LOCs are the most prized targets. Forces that conduct spoiling attacks are often expected to act independently, and thus they must be skilled and flexible. Objectives for spoiling attacks may include disrupting enemy command and communication to forward units, inflicting casualties, destroying or neutralizing key systems, and slowing the enemy's advance. Spoiling attacks occur throughout the defensive operation, and they are of particular importance in defeating any enemy breaches of the main defensive line.

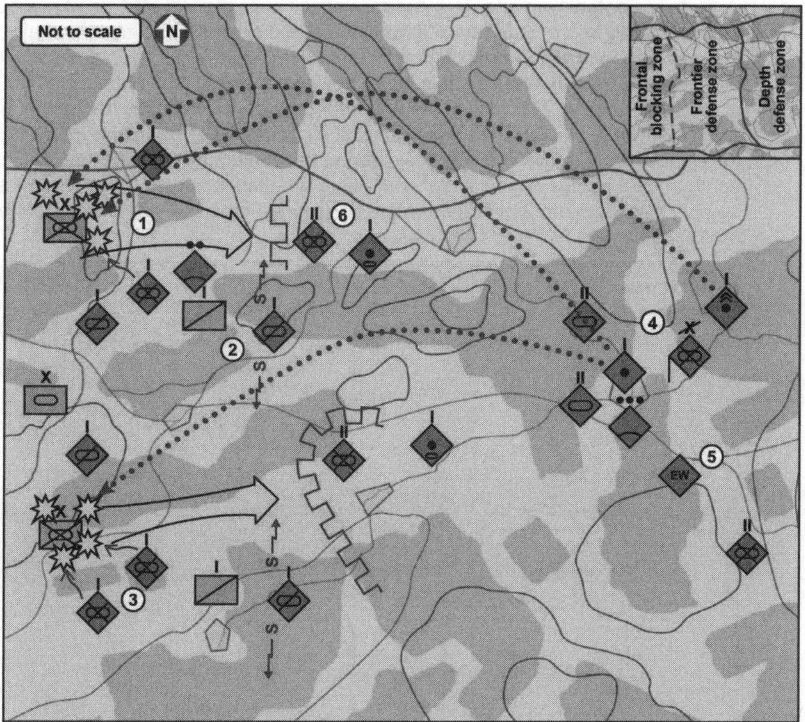


**(1) Two enemy mechanized brigades begin an attack on the defensive group's position. (2) Reconnaissance units conduct counterreconnaissance operations against enemy scouts. (3) Mechanized CA-BN units in the cover group conduct spoiling hit-and-run attacks against the enemy, forcing it to deploy early and slowing its progress through the frontal blocking zone. (4) The firepower group delivers firepower assaults against high-value enemy units. (5) The EW group simultaneously attempts to suppress enemy communications and deceive enemy collection systems. (6) The frontier defense group entrenches around the key defense points and prepares to conduct a blocking action.**

| CA-BN | combined arms battalion | EW | electronic warfare | N | north | S | screen |
|---|---|---|---|---|---|---|---|

***Note.*** Only select units shown for clarity purposes.

*Ref: ATP 7-100.3 (Aug '21), fig. 8-3. Spoil (example).*

Firepower assaults employ artillery groups or firepower strike groups to target enemy offensive forces with destructive massed fire. The firepower assault may employ a firepower assault zone, a predetermined and pretargeted area serving as an ambush point. If employed, firepower assault zones should be located along the enemy's expected avenue of approach or on terrain the enemy is likely to occupy. A firepower assault seeks to decimate the enemy formation when it is most vulnerable, either during movement or during a halt in open terrain. The best practice is to integrate the firepower assault with other attacks, particularly spoiling attacks by reconnaissance forces, to maximize the accuracy of fire and the combined-arms effect of multidimensional threats. Commanders must, however, ensure that the groups that conduct the firepower assault do not expose themselves to enemy counterfire unnecessarily, as it is likely that the enemy plans for a counterfire attack once friendly artillery forces are discovered. Commanders may employ mobile artillery in the firepower assault role or may rely on hardening or deception to protect their artillery force.

# V. Resist the Enemy's Assaults

The main defensive effort occurs when the enemy commences its main assault through the defensive zone. The enemy's attack consists of a mixture of maneuver, firepower, and information attack, coordinated and synchronized to overwhelm, destroy, or force the withdrawal of friendly units. The focus of the resistance phase is to blunt the enemy's attack; sap its combat power, cohesion, and morale; and put it in a vulnerable position. This allows the defender to seize the initiative through aggressive counterattacks and then transition to the offense. The PLAA anticipates fighting throughout the depth of the defensive zone, and it offers three guidelines to inform commanders during this critical phase: wage simultaneous resistance, prioritize combat strength, and use proper countermeasures. Figure 8-4 depicts the defensive group resisting the enemy's initial assault.

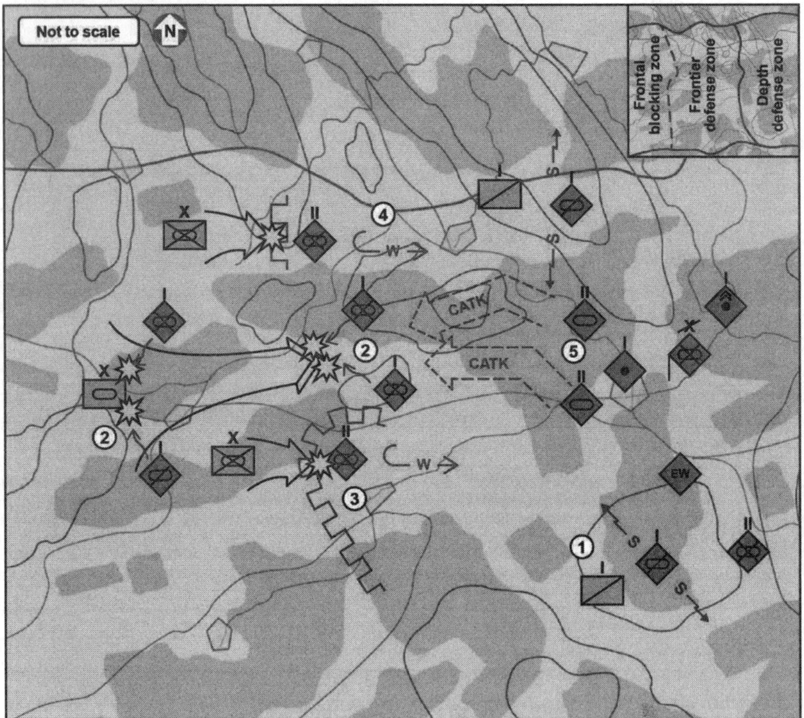


**(1) Reconnaissance units continue to conduct counterreconnaissance operations against advancing enemy scouts to prevent detection of the disposition and axis of the main counterattack. (2) The cover group commences spoiling attacks and raids against the enemy's main effort, causing disruption and forcing early deployment. These attacks are continuous and intended to canalize the enemy's main effort, directing it to the point in the frontier defense zone where the main counterattack is planned. (3) The frontier defense group conducts strong resistance against enemy attacks on key defensive positions. The enemy attempts to fix the two CA-BNs and destroy them in detail; entrenchments and firepower blunt the attack. (4) The CA-BNs conduct a retrograde action, falling back slowly and blocking enemy penetrations into rear areas. (5) The depth group begins to move toward the counterattack position. By this time, the enemy's position and axis of advance are well known, and the commander carefully chooses the time and place for the counterattack.**

*Ref: ATP 7-100.3 (Aug '21), fig. 8-4. Resist (example).*

# Resist the Enemy's Assaults (Guidelines)

*Ref: ATP 7-100.3, Chinese Tactics, (Aug '21), pp. 8-11 to 8-13.*

## Wage Simultaneous Resistance

The enemy's attack will likely target the full depth of the defensive zone with depth assaults, flanking assaults, and encircling maneuvers. These various actions must be resisted using a comprehensive and integrated approach. Resistance may take the form of symmetric, force-on-force actions—such as meeting an armored thrust with an armored reserve—or it may be asymmetric, such as using electromagnetic attack to confuse and neutralize an enemy air depth assault. Commanders must anticipate heavy enemy activity within the defensive zone and skillfully and flexibly move their own forces—taking advantage of interior LOCs and knowledge of terrain—to meet and defeat the various enemy actions. Commanders employ two primary actions during resistance: blocking actions and repositioning actions.

Blocking actions involve a group or unit standing fast and tenaciously defending a specified position, with the intent to stop or delay an enemy action, reduce enemy cohesion, and inflict casualties on enemy units. Blocking actions should only be conducted when necessary, or when terrain or other factors make such actions highly advantageous. Blocking actions are to be performed judiciously, as they make a unit or group vulnerable to encirclement, isolation, and firepower attack. Blocking actions should be supported with firepower, obstacles, and reinforcements. They are ultimately intended to enable follow-on actions, such as counterattacks.

Repositioning actions are movements conducted by units or groups during the course of the defense. Commanders mix repositioning and blocking actions to conduct a tenacious withdrawal, ceding unimportant territory to the enemy only after inflicting heavy casualties. Skillful defense alternates blocking and repositioning in such a way that resistance is continuous and no unit is ever exposed to enemy assault.

## Prioritize Combat Strength

Effective resistance is underpinned by effective prioritization of combat resources. This guideline references the broader People's War principle of prioritization, but with a focus on defensive action. As a defensive group is assaulted through multiple domains and from multiple directions, the commander makes decisions about how and where to commit reserves and reposition units. Enemy thrusts should be met with the least force believed necessary. This differs significantly from offensive actions, where commanders are encouraged to concentrate as much combat power as possible against key objectives. Commanders should seek to minimize movement—especially long-distance movement—during critical phases of resistance. If a unit must move long distances, it is essentially unavailable as a fighting force. Commanders must prioritize the value of moving a unit during a defensive action, carefully considering the loss of the unit's combat power during the movement period and the vulnerability of the unit to enemy action. Effective prioritization is largely underwritten by high-quality, comprehensive reconnaissance and intelligence, as quality information tells the commander how and where to prioritize.

## Use Proper Countermeasures

Even the best defensive action is inherently reactionary. The defensive group must assess and react to the enemy, rather than dictating the fight on its own terms. This means that defensive combat places a higher premium on employing the right tactical means to counter enemy capabilities. Commanders must have a clear picture of the enemy capabilities they are facing and build operational systems to effectively counter their anticipated opponent. So too must commanders rapidly and effectively react to the enemy's actions as the situation unfolds, deploying friendly combat power to counteract enemy penetrations and assaults.

# VI. Counterattack

The counterattack, also called the mobile assault, is typically the culmination of a defensive operation. It requires the defensive group to mass combat power and conduct an aggressive, decisive attack against one or more enemy units, with the intent of disintegrating the enemy attack and forcing the enemy to either rapidly retreat or face annihilation. Counterattacks happen throughout a defensive action on smaller scales, but the decisive counterattack at the tactical level likely involves either the depth defense group or the combat reserve group conducting multidimensional penetrations of an enemy assault force. These penetrations target weak or exposed flanks of the enemy formation.

The timing and axis of the counterattack is critical. It should be timed around the moment when the enemy attack has culminated, and enemy forces are possibly overextended, low on supplies, and beyond their supporting or security elements. Resolute defense throughout the frontal blocking zone and the frontier defense zone should create favorable conditions for the attack, magnifying the combat power of the counterattack force. A counterattack should include one or more of four phases, which may be concurrent or in varying order. Figure 8-5 depicts a defensive group counterattacking the enemy.

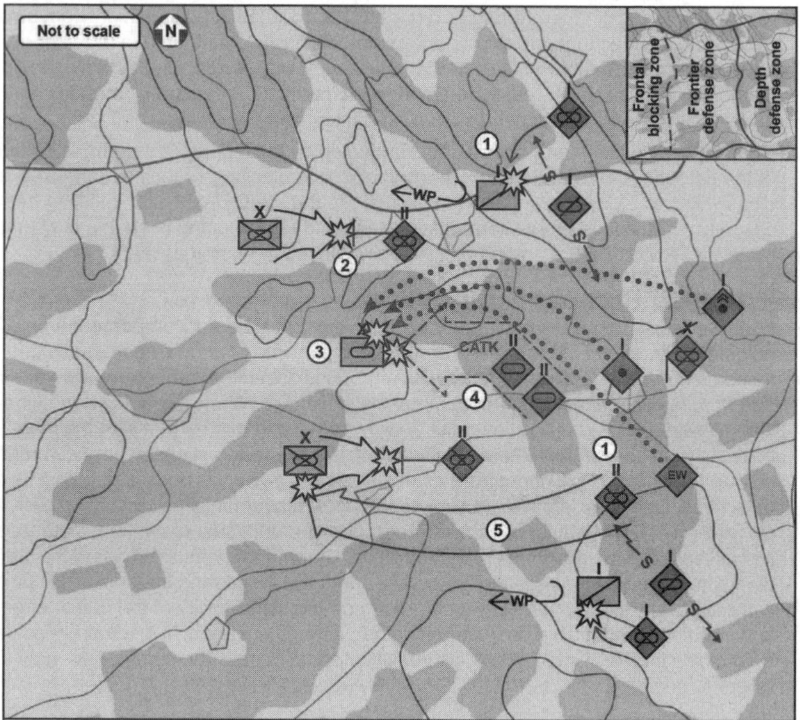


**(1) The cover group continues to conduct counterreconnaissance operations against enemy scouts. Mechanized infantry elements of the cover group conduct attacks against enemy scouts, forcing their withdrawal and exposing the flanks of the enemy assault group. (2) The frontier defense group continues to conduct blocking actions against the enemy's supporting attacks, preventing penetrations and slowing the enemy's advance. (3) Firepower is concentrated on the enemy's main effort. Tube and rocket artillery plus electronic attack are massed on the target in order to disrupt movement, reduce cohesion, and cause casualties. (4) The depth group conducts its counterattack assault. Despite being outnumbered, the depth group achieves local superiority through mobility, deception, and the effective use of firepower. The depth group assaults the enemy's main body head on, attempting to destroy the enemy's momentum and cohesion. (5) The combat reserve group is deployed in an attack on the enemy's exposed flank in an attempt to isolate the enemy's main body and force the enemy's withdrawal.**

*Ref: ATP 7-100.3 (Aug '21), fig. 8-5. Counterattack (example).*

# Counterattack (Phases)

*Ref: ATP 7-100.3, Chinese Tactics, (Aug '21), pp. 8-13 to 8-15.*

## Concentrate Fire

Concentration of firepower is often the touchstone that begins the counterattack in earnest. Commanders may use any combination of indirect fire—possibly in the form of a firepower assault—direct fire, and information attack, concentrated on the most powerful part of the enemy's formation. In many situations this will be the enemy's armored spearhead, and so the concentration of fire should involve a large number of antitank weapons. Concentrated fire disrupts the enemy's attack, creates casualties, and destroys key weapons systems, opening a window of opportunity for the counter-attack to break through the enemy's exposed flank.

## Seal Off Breakthroughs

Before a counterattack commences, any significant or dangerous enemy break-throughs in the counterattack zone must be neutralized or defeated. Enemy forces that have achieved significant penetrations in the defensive zone can threaten the counterattack with defeat or destruction. Breakthroughs can be sealed off using fire-power or direct attack, or the can be neutralized by attacking enemy LOCs. Sealing off breakthroughs does not mean that the enemy units must be annihilated; it only ensures that they cannot threaten the counterattack force as it conducts its mission. Once the counterattack is successful, the remaining enemy units will be isolated and encircled, enabling their destruction or forcing their withdrawal.

## Use Multi-domain or Multidirectional Assault

The decisive phase of the counterattack should take the form of a small-scale offensive operation, and it is informed by most of the principles and guidelines outlined in chapter 7. Ideally, a commander executes an attack that involves penetrations from multiple directions and through multiple domains, preventing the enemy from massing combat power and confusing it about the disposition and axis of the main effort. Direct assaults on the enemy's front are the preferred method of attack, as this fixes the enemy and immediately targets its most valuable units. Pincer movements are considered a secondary approach, and while they may be effective at targeting the enemy's exposed flank, they require longer movement and more time, which exposes the counterattack force to enemy artillery or direct attack. Firepower assaults can cut off or block access to the rear, while blocking units work to stop further penetrations. The assault targets the enemy's center of gravity, attempting to isolate enemy units throughout the defensive zone.

## Hold Key Defense Points

As the counterattack commences, commanders assess its effectiveness. If it is effective, they may order a continuation of the attack or consolidation of the gains. If it is ineffective, they may order the counterattacking force to assume a defensive posture and resist any further enemy advances. Regardless of effectiveness, commanders must immediately deploy the defensive group to retain KDPs, be they still held by friendly forces or recently retaken. Commanders should anticipate the enemy sending reinforcements or conducting supporting attacks promptly after it recognizes the threat of the counterattack. Having the defensive group deployed and in place to block these attacks ensures that the gains achieved by the counterattack are not lost.

# VII. Consolidate or Withdraw

During or after the counterattack commanders face a critical decision on how to proceed with the defensive battle. If the counterattack succeeded in decimating, blunting, or neutralizing the enemy assault, commanders may consider ordering a follow-on attack into the enemy's depth. Alternatively, they may consider holding fast, reinforcing and entrenching the current position to better resist any further enemy assaults. Commanders must consider casualties sustained, the overall readiness of available troops, the vulnerability of the enemy, and the advantages gained by each course of action when making a decision to continue advancing or hold in place.

If the counterattack was unsuccessful or the defensive group suffered serious casualties during the counterattack, the commander may order a withdrawal. The withdrawal should be orderly and decisive, with available groups conducting alternating retrograde operations and blocking actions. No unit should ever be exposed to an enemy pursuit attack. The deployment of air defense groups to deter or defeat attack by air is particularly important. Units are particularly vulnerable to firepower attack during withdrawals. Concealment and cover are important for the unit, and the firepower or artillery groups must prioritize counterfire operations if any unit is in the process of withdrawal. Figure 8-6 depicts the defensive group consolidating its position after a successful counterattack.

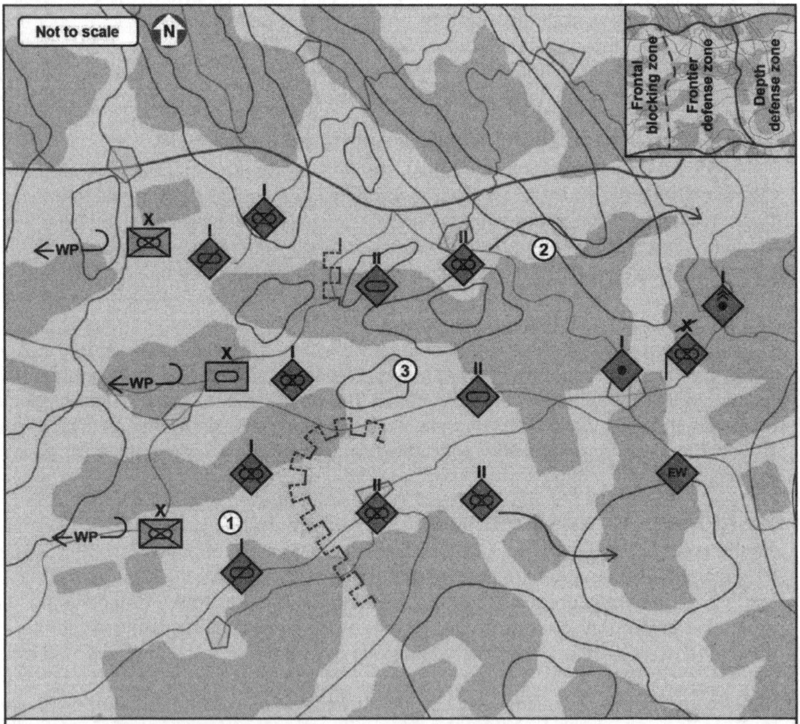


**(1) The cover group maintains contact with the retreating enemy, employing firepower to keep enemy forces off balance and ensure they cannot quickly reconsolidate. (2) Having suffered heavy casualties during its blocking actions against superior forces, the frontier defense group moves to rear areas for consolidation and refit, becoming the new reserve group. (3) The original depth and reserve groups occupy the key defensive positions and begin entrenching, preparing for the next enemy assault and providing a strong position from which to begin offensive operations in the sector.**

*Ref: ATP 7-100.3 (Aug '21), fig. 8-6. Consolidate (example).*

# III. Types of Defensive Operations

*Ref: ATP 7-100.3, Chinese Tactics, (Aug '21), pp. 8-17 to 8-30.*

The PLAA recognizes several types of defensive operations: positional, mobile, hasty, urban, diversionary, and specialized.

## Types of Defensive Operations

- I **Positional**
- II **Mobile**
- III **Hasty**
- IV **Urban**
- V **Diversionary**
- * **Specialized (Other)**

# I. Positional Defensive Operations

Positional defensive operations are defensive operations that rely on a strongly entrenched position, defensive terrain advantages, obstacles, and stubborn resistance. They are static in nature, and they are not designed to enable the rapid relocation of troops. Positional operations are not considered decisive, and they should always be executed as part of a wider defensive action. Positional actions defend a key piece of terrain or other target, delay or block enemy forces moving through an area, deplete an enemy's strength by forcing it to assault entrenched positions, or buy time and decision space for offensive operations in other areas.

*See following pages (pp. 7-18 to 7-19). Figures 8-7 through 8-10 depict a positional defensive operation.*

# Positional Defensive Operation (Example)

*Ref: ATP 7-100.3, Chinese Tactics, (Aug '21). (Note. Only select units shown for clarity.)*

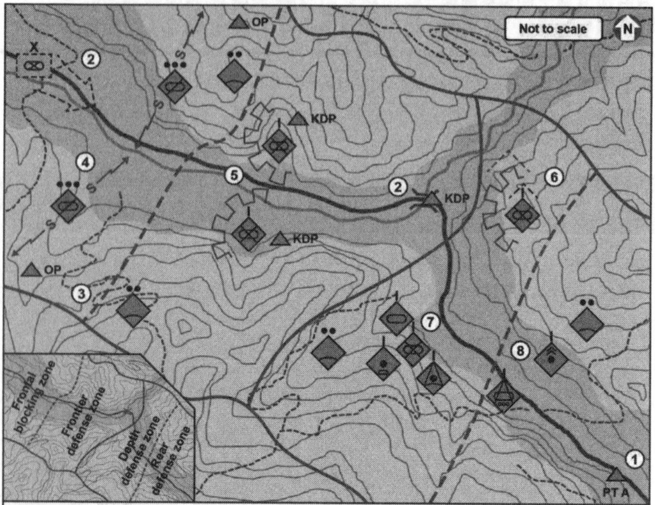


*Ref: ATP 7-100.3 (Aug '21), fig. 8-7. Positional defense (example part 1 of 4).*

**(1) A defensive group conducts a positional defensive operation in order to prevent the enemy from reaching Point A. (2) The defensive group commander expects the enemy to approach from the northwest and anticipates the enemy's main objective is the bridge. (3) OPs and MANPADS units are deployed throughout the defensive zone. (4) The cover group conducts reconnaissance and screens the main body. (5) The frontier defense group constructs fortified positions on either side of the river. (6) A deception group is deployed to the east of the bridge. (7) The depth defense group conceals itself along the main road south of the bridge. (8) Light rocket artillery and an antitank reserve occupy the far southeast.**

| | | | | | |
|---|---|---|---|---|---|
| KDP | key defense point | OP | observation post | S | screen |
| MANPADS | man-portable air defense system | PT | point | SS | service support |
| N | north | | | | |

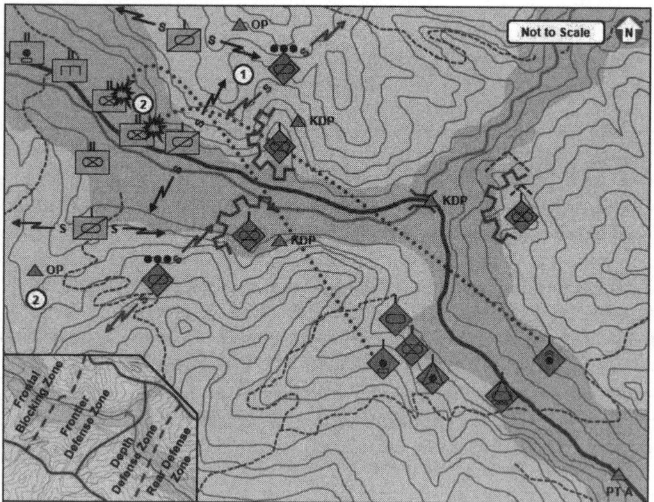


*Ref: ATP 7-100.3 (Aug '21), fig. 8-8. Positional defense (example part 2 of 4).*

**(1) As enemy scouts make contact with the cover group, scout platoons conduct reconnaissance and counterreconnaissance through the frontal blocking zone. (2) Forward observers occupying observation posts call for artillery fire against the enemy's main body, disrupting its movement and causing casualties.**

| | | | | | |
|---|---|---|---|---|---|
| KDP | key defense point | OP | observation post | PT | point |
| S | screen | | | | |

Ref: ATP 7-100.3 (Aug '21), fig. 8-9. Positional defense (example part 3 of 4).

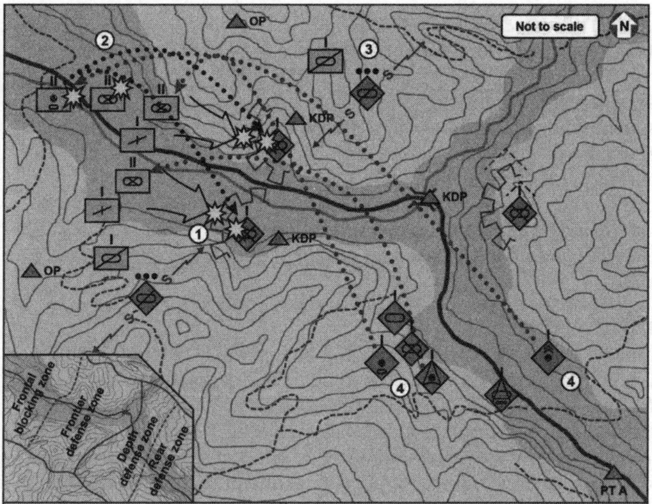


(1) Enemy mechanized battalions and engineers begin their assaults on the entrenched positions of the frontier defense group. Thanks to strong fortifications and favorable terrain, the frontier defense group can strongly resist the enemy's assault. (2) Enemy artillery supports the enemy assault. (3) Frontier defense group scouts continue to resist enemy reconnaissance efforts as they try to locate the friendly main body. (4) Tube artillery and mortars target the enemy's assault. Rocket artillery targets the enemy's self-propelled guns, whose location is revealed when they begin firing.

| | | | |
|---|---|---|---|
| KDP | key defense point | OP | observation post |
| N | north | PT | point |

Ref: ATP 7-100.3 (Aug '21), fig. 8-10. Positional defense (example part 4 of 4).

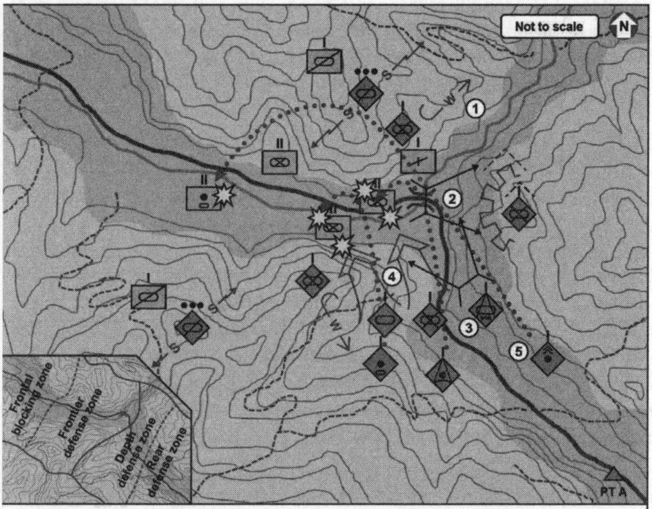


(1) Friendly frontline defense units withdraw after inflicting significant casualties and disrupting enemy movement. (2) The enemy main effort seizes the bridge and sets up a support-by-fire position, suppressing what it believes is the main defensive line overlooking the bridgehead. (3) Assault guns of the depth group open fire from concealed positions. (4) The remainder of the depth group emerges from its concealed position and conducts an assault against the exposed flank of the enemy main body. (5) Rocket artillery continues to suppress enemy self-propelled guns.

| | | | |
|---|---|---|---|
| N | north | S | screen |
| PT | point | W | withdraw |

Defensive Actions

When planning a positional defensive operation, three key principles should be considered: keep the defensive zone small, prepare sufficiently, and emphasize holding positions.

### Keep the Defensive Zone Small

Forces conducting a positional defensive operation are not expected to move with the same rapidity as those conducting more mobile forms of defense. This, in turn, means that concentrating combat power with a positional defensive group is more time consuming and difficult than with a mobile group. Positional defensive operations must ensure adequate concentration of combat power through a higher density of troops—rather than relying on mobility—which requires defensive zones to be smaller relative to defensive group size. This restricts the enemy's ability to maneuver, enhances the mutually supporting effects of defensive entrenchments, and enables defensive groups to operate with greater independence in the face of an enemy assault.

### Prepare Sufficiently

A positional defensive battle should be anticipated and carefully planned. Commanders must give subordinate units sufficient time to plan their defense and then build the necessary fortified positions on KDPs. Higher echelon defensive plans must account for the time necessary to prepare a defensive zone, and delaying actions may be needed to create the necessary time and decision space. Defensive group commanders should employ entrenchments, fortifications, and underground facilities, coupled with integrated firepower and information attack. Building these systems and creating these plans takes a considerable amount of time.

### Emphasize Holding Positions

A positional defensive operation should be built around holding strong pieces of terrain, even when outnumbered, under assault, or faced with intense firepower attack. Holding these pieces of terrain forces the enemy to commit and exhaust more and more troops in its attempt to break the defensive zone. With sufficient time and quality planning, a defensive group commander can create a comprehensive, in-depth plan that can hold strong positions, attrit the enemy, reduce the enemy's morale and cohesion, and buy other commanders in the area time and decision space.

## II. Mobile Defensive Operations

Mobile defensive operations are defensive operations that mix blocking and restraining actions with counterattacks, seeking to achieve a decisive defeat of an opponent's assault. Mobile defensive operations employ armored or mechanized forces that are able to maneuver and concentrate combat power decisively at key points throughout the defensive zone. Firepower, information warfare, and engineer support are all carefully integrated to support the main counterattack effort. The mobile defensive operation seeks to halt the enemy assault, isolate attacking enemy units, then either annihilate them or force their withdrawal.

*See following pages (pp. 7-22 to 7-33). Figures 8-11 through 8-14 depict a mobile defensive operation.*

When planning a mobile defensive operation, five key principles should be considered: plan and execute quickly and decisively, assume a large defensive zone, focus on the offense, emphasize flexibility and mobility, and assume command and communication challenges.

### Plan and Execute Quickly and Decisively

Unlike positional defensive operations, mobile defensive operations are often opportunistic or hasty in nature, as they are fundamentally reactionary to enemy actions.

This requires mobile defensive actions to be planned and executed on a much shorter timeline than positional defenses. Commanders must be flexible in their deployments and plan in advance as much as possible. They also must allow for changes to occur during the battle and allow the plan to adapt to the situation. Plans should, at a minimum, instruct lower-echelon units about key defensive positions, the main direction of defense, and how to employ key countermeasures to enemy actions.

## Assume a Large Defensive Zone

Mobile defensive operations should be conducted in a relatively wide geographic area. This forces the enemy to disperse its forces—making its intentions clearer earlier in the battle—and takes full advantage of the mobility of the defensive group. A larger relative area decreases the value of key defensive positions and makes defensive engineering more challenging and less effective. This, in turn, means that the bulk of the defense in a mobile operation falls to mobile units and their ability to concentrate combat power. The key competency of a mobile defense is the rapid transfer of combat strength, along with the ability to gracefully fall back or aggressively attack, as required.

## Focus on the Offense

The mobile defensive operation, though defensive at the operational level, is built around attack at the tactical level. An effective mobile defense emphasizes offensive operations, using the bulk of available combat power on attacks and counterattacks and seeking a decisive engagement with the enemy assault force. Other activities in the defensive zone should revolve around the decisive action by either manipulating enemy forces or providing direct support to the main counterattack. Mobile defensive operations must also employ a variety of combat actions, such as raids, ambushes, screens, firepower assaults, and information warfare to best support the counterattack effort.

## Emphasize Flexibility and Mobility

The success of a mobile defensive operation is contingent on the defensive group massing combat power at key times and places. Considering the large geographic area of the defensive zone and the assumption that the enemy possesses greater local combat power than the defensive group, a combination of flexibility and mobility are required to achieve decisive concentration. Commanders must be prepared to shift their center of gravity and conduct offensive actions across the flanks, depth, or rear areas of enemy formations, or anywhere else within their defensive zone. Defensive groups in mobile operations are often ad hoc or informal. Troops and units must be comfortable with the rapid forming and reforming of combat groups to fit the needs of the immediate situation.

## Assume Command and Communication Challenges

The combination of hasty action, a large defensive area, ad-hoc combat groups, and aggressive enemy actions make command and communication of a mobile defensive operation very difficult. Commanders must approach this issue with two different solutions: First, they must train their subordinates to act appropriately in the absence of orders or direct connection with their leaders. Second, they must maximize efforts to establish and sustain command and communication structure throughout the defensive area. Despite these efforts, commanders must assume that subordinate units will have to fight when isolated or out of contact, and they must provide their subordinates with the means to conduct operations without direct communication.

# Positional Defensive Operation (Example)

*Ref: ATP 7-100.3, Chinese Tactics, (Aug '21). (Note. Only select units shown for clarity.)*

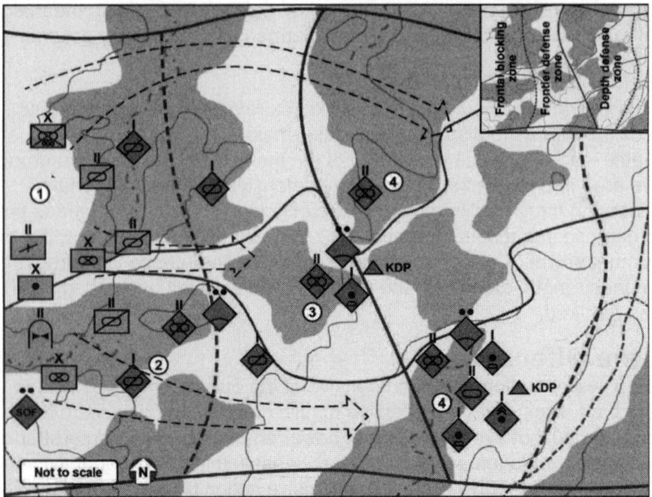


*Ref: ATP 7-100.3 (Aug '21), fig. 8-11. Mobile defense (example part 1 of 4).*

**(1) A defensive group conducts a mobile defense against a superior force, with orders to decisively defeat the enemy's attack through the area and annihilate as many enemy units as possible. (2) The cover group deploys into the frontal blocking zone. It focuses on concealment and mobility, preparing for the reconnaissance fight. (3) The frontier defense group deploys with a CA-BN and two reconnaissance companies, creating an additional layer of screening and reconnaissance. (4) The depth defense group consists of three CA-BNs that will rapidly maneuver in order to achieve local superiority over the larger enemy units, using firepower to compensate for their smaller numbers.**

| | | | |
|---|---|---|---|
| CA-BN | combined arms battalion | N | north |
| KDP | key defense point | SOF | special operations forces |

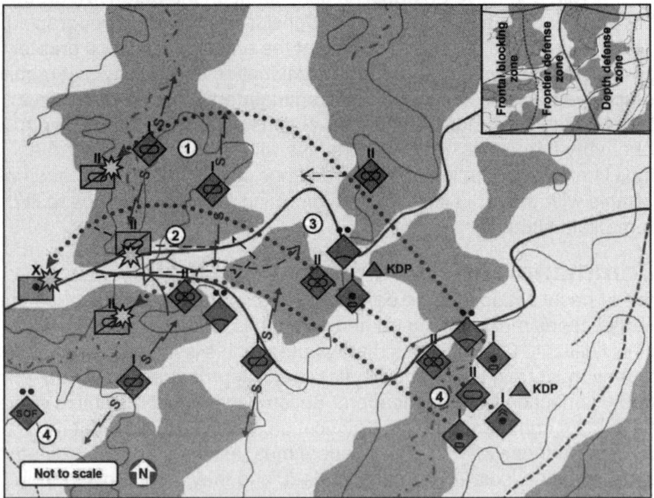


*Ref: ATP 7-100.3 (Aug '21), fig. 8-12. Mobile defense (example part 2 of 4).*

**(1) The reconnaissance/counterreconnaissance fight commences as cover group units screen the main body and obstruct enemy scouting efforts. Scout units conduct a defense-in-depth while firepower attacks against enemy scout forces disrupt their movements. (2) The cover group's medium CA-BN conducts a surprise raid against the enemy's center, badly damaging an enemy scout unit. (3) The mechanized CA-BN then moves to join an impromptu defensive line with two other CA-BNs. (4) Aided by SOF forward observers, the heavy rocket battalion targets the enemy's artillery brigade with counterfire. The enemy cannot respond, as the 300-mm rockets significantly outrange its lighter rocket systems.**

| | | | |
|---|---|---|---|
| CA-BN | combined arms battalion | S | screen |
| N | north | SOF | special operations forces |

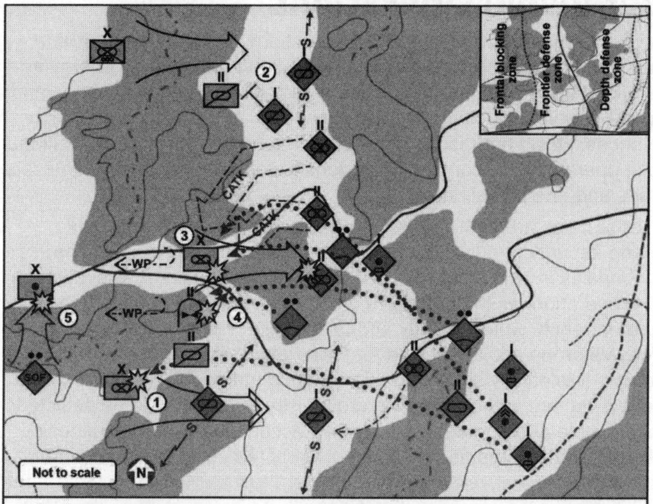


*Ref: ATP 7-100.3 (Aug '21), fig. 8-13. Mobile defense (example part 3 of 4).*

**(1) The enemy's main effort commences to the south. Cover group units continue to screen against a vastly superior enemy force, attempting to slow and disrupt the main assault through stubborn resistance and firepower. (2) In the north, the enemy's flank penetration is met with a combination of screening and blocking actions, designed to divert the enemy northward and protect the central counterattack. (3) In the center, the enemy, deprived of effective scouting, assaults what it believes to be an inferior force. The defensive group commander has concentrated combat power in the center, and conducts a decisive counterattack against the flank of the enemy brigade. The counterattack, coupled with a firepower attack, compels an enemy withdrawal. This isolates the two remaining wings of the enemy force. (4) Medium-range air defenses force the enemy's aerial attack to lower altitudes, where an ambush consisting of MANPADS and anti-aircraft guns is waiting. The ambush inflicts severe damage on the enemy aviation unit, compelling its withdrawal and leaving the main effort without its air component. (5) SOF elements in enemy rear areas conduct a raid against enemy artillery, disrupting its efforts to conduct fire support and counterfire.**

| | | | | | |
|---|---|---|---|---|---|
| CATK | counterattack | N | north | SOF | special operations forces |
| MANPADS | man-portable air defense system | S | screen | WP | withdrawal under pressure |

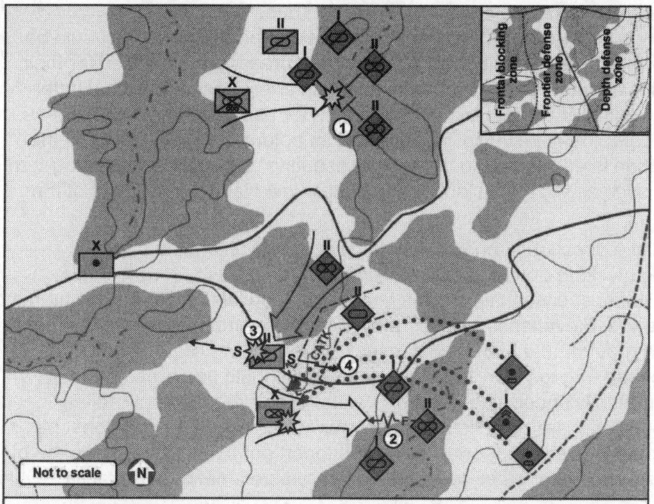


*Ref: ATP 7-100.3 (Aug '21), fig. 8-14. Mobile defense (example part 4 of 4).*

**(1) In the north, the remaining cover group units are reinforced by two CA-BNs. These units take up a blocking position and engage the enemy brigade, ensuring it cannot support the main effort to the south and maintaining its isolation. (2) The enemy's main effort continues its penetration, and is met by one of the depth group's CA-BNs, supported by cover group units. Though outnumbered, the CA-BN conducts a stubborn resistance, fixing the enemy brigade. (3) A mechanized CA-BN aggressively assaults the enemy's reconnaissance unit that is screening the main effort, putting pressure on the flank of the enemy's main body. (4) The depth group's heavy CA-BN conducts a decisive penetrating counterattack on the flank of the main body, aided by a concentration of all available firepower.**

| | | | | | |
|---|---|---|---|---|---|
| CA-BN | combined arms battalion | F | fix | S | screen |
| CATK | counterattack | N | north | | |

# III. Hasty Defensive Operations

Hasty defensive operations are those defensive operations conducted under emergent circumstances and on a short timeline, usually in direct response to enemy actions. Hasty operations attempt to make best use of terrain and reconnaissance and security operations—coupled with surprise and deception—to defeat, neutralize, or blunt enemy advances. Hasty operations may take the form of abbreviated positional operations, mobile operations, or contain characteristics of both, depending on available forces, terrain, and enemy actions.

Hasty defensive operations are highly dependent on two key actions: extensive reconnaissance and security and the seizing and fortification of key terrain. The importance of reconnaissance is enhanced due to the shortened timeline available for planning and preparation, as the defensive group has less time to plan for enemy contingencies and entrench, while security efforts can delay the enemy assault and buy the defensive group additional time for preparation. Seizing key terrain in the defensive zone is a relatively simple and time-saving method for increasing the overall combat effectiveness of the defensive group. Key terrain should be selected with a view toward simple and aggressive defensive actions, such as ambushes. Once terrain is taken, entrenchments should be constructed using whatever time and resources are available.

# IV. Urban Defensive Operations

Urban defensive operations are seen as optimal in many ways, but they also provide some unique challenges for the defensive group. The urban defense is conducted using the same principles as other forms of defense: building the defensive group and establishing the primary direction of defense and KDPs. Urban terrain is characterized by buildings, streets, and subsurface facilities; planners must account for and make best use of these terrain features when planning an urban defensive operation.

Reconnaissance in an urban defensive zone concentrates on mapping the characteristics of buildings, streets, and subsurface developments. Buildings should be assessed for strength, vulnerability, and defensibility. Streets are assessed for their effects on mobility and how they might be used by an enemy assault force. Underground facilities are carefully scouted and assessed for use as fortifications and as movement corridors; they must also be assessed as potential vulnerabilities. If the defensive operation is anticipated to be lengthy, scouting must also assess long-term support assets, such as water availability and supply-line viability in and out of the defensive area.

The main defensive direction in an urban defensive operation should prioritize the use of strong blocking positions that can control wide areas of terrain in the defensive zone. In most cases this implies large, strong buildings that can be further fortified or hardened. Elevation is seen as a key enabler in urban defense. If a force controls the high ground, that force can command a large physical area while being difficult to detect and engage. Key defensive positions should be spread out, but they should be mutually supporting. Support points, consisting of supply, fallback, and firepower positions, are established to the rear of key defensive positions. Multiple defensive positions are supported by each support point, and they are linked by communications trenches or other concealed and protected methods of transit.

Urban defensive operations are more likely to involve complications, such as the presence of civilians, militia, or security forces. Commanders must plan for the various problems associated with civilian presence while integrating militia and security forces into the defensive plan. They must also make clear to subordinates the rules governing the presence and treatment of civilians in the defensive zone. Security forces may be leveraged to assist with the removal or protection of civilians.

Obstructions and obstacles are often more effective in urban areas due to the canalizing effects of streets and the lack of available alternative routes. The defensive group should make best use of obstacles to manipulate enemy movement, encouraging the enemy to overextend past its support and leave itself vulnerable to counterattack. Obstacles should be carefully integrated with firepower. Firepower that is optimized to engage an enemy attempting to breach an obstacle creates a powerful combined arms effect. It should also be used to dominate key avenues of approach. Firepower in urban terrain is often magnified by the close proximity of the combatants and the canalizing nature of the terrain, but it may be degraded by the wide availability of cover and protection offered by strong buildings and underground facilities. Figure 8-15 depicts an urban defensive operation.

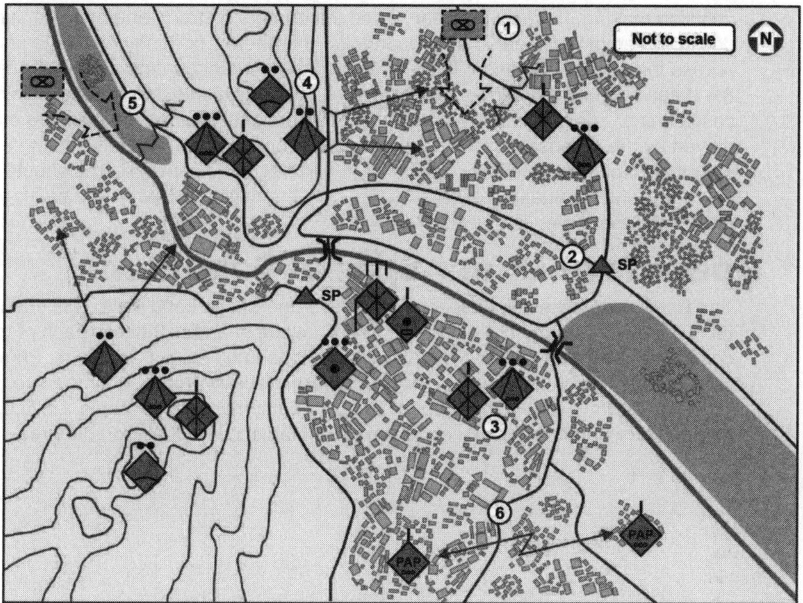


**(1) A defensive group built around a light CA-BN and augmented by antitank, engineer, PAP, and 122-mm SPG units, deploys to an urban area with orders to defend against an expected enemy attack from the north and northwest. (2) Two support points are established on main routes. (3) Four groups are built, each consisting of an infantry company and an antitank platoon, that are deployed to either elevated areas or key routes throughout the urban area. (4) Firepower teams consisting of antitank and MANPADS sections are deployed to elevated areas in support of maneuver groups. (5) Obstacles are deployed along main routes and integrated with direct and indirect firepower. (6) PAP units patrol rear areas and maintain control of the local civilian population.**

| | | | | | |
|---|---|---|---|---|---|
| CA-BN | combined arms battalion | MM | millimeter | SP | support point |
| M | meter | N | north | SPG | self-propelled gun |
| MANPADS | man-portable air defense system | PAP | People's Armed Police | | |

*Ref: ATP 7-100.3 (Aug '21), fig. 8-15. Urban defensive operation (example).*

# V. Diversionary Defensive Operations

Diversionary defensive operations include the multitude of different temporary or hasty missions conducted both in and out of the defensive zone to surprise, confuse, frustrate, or degrade enemy forces. Missions under the diversionary umbrella include screens and covers, counterreconnaissance efforts, ambushes, blocking actions, and raids. Diversionary operations are defined by their purpose. They are intended to force the enemy to deploy, deplete enemy supplies, inflict casualties, slow the enemy advance, suppress enemy reconnaissance, and provide early warning for the main defensive group. They are typically conducted far in front of the main defensive zone. The deeper the penetration into the enemy's territory, the more effective these operations tend to be.

Depth is the key principle during diversionary defensive operations. Deploying in depth enhances the effects of frustration and confusion upon the enemy, as it requires enemy forces to continually fight through stubborn resistance. Diversionary operations may be along the main direction of defense, or they may be deployed in alternative directions as a means of screening or covering the main defensive effort. Countermobility and firepower are key enablers of diversionary operations, as they magnify the combat power of the diversionary force without requiring additional troops.

Commanders of diversionary forces probably have significant latitude in their decision making. If the diversionary force encounters a strong enemy force, it may decide to remain concealed and reconnoiter the enemy, or it may decide to carefully harass the enemy formation. If the diversionary force encounters a weaker enemy, the defending unit may aggressively assault the enemy, seeking to inflict significant casualties or spoil the enemy's movement. Diversionary forces may also be employed by the defensive group to counterattack on the enemy's flanks as the enemy penetrates forward defensive positions. In general, diversionary forces should avoid being decisively engaged, always maintaining the ability to rapidly and safely withdraw to deeper defensive positions.

# * Operations Under Special Conditions

The PLAA recognizes eight special defensive operations conditions: hills, mountains, jungles, extreme cold, deserts, rivers, swamps, and nighttime. Each of these conditions requires modifications and adaptations to baseline operations. Primary considerations include ensuring that troops are properly prepared for the special condition—with cold weather gear or bridging equipment, for example. Units should also train for potential special conditions that may occur in their combat area.

# IV. Defensive Tactics

*Ref: ATP 7-100.3, Chinese Tactics, (Aug '21), pp. 8-30 to 8-41.*

The PLAA uses several different tactics during defensive operations, with their use informed by the principles discussed earlier in this chapter. Each of these may be employed by virtually any echelon, and each may be used independently, sequentially, or simultaneously. Figures in this section depict genericized units conducting simplified variants of the defensive tactics against a notional enemy.

## Defensive Tactics

- I Counterreconnaissance
- II Blocking Actions
- III Repositioning Actions
- IV Breakout
- V Withdrawal
- VI Firepower Defense
- VII Depth Defensive Actions
- VIII Rear Defense



# I. Counterreconnaissance

Counterreconnaissance as a tactic crosses operations of all types, and it is critical in each. Effective counterreconnaissance during defensive operations denies the enemy knowledge about the disposition and strength of friendly forces, the axis and timing of counterattacks, and the locations of key objectives. Counterreconnaissance may be active—attempting to aggressively defeat enemy reconnaissance efforts using direct action against reconnaissance elements—or it may be passive, using techniques such as camouflage, concealment, or decoys to fool or spoof enemy collection efforts. The best practice is to employ both active and passive measures in an integrated manner. Active measures deny the enemy knowledge of friendly forces, while passive measures feed the enemy false or misleading information, fooling it into making faulty decisions.

Much of the counterreconnaissance effort is conducted by the cover group, which is likely built around reconnaissance or other light, mobile forces. These units conduct screening and covering actions throughout the frontal blocking zone. PLAA reconnaissance units are generally lighter and less powerful than their Western equivalents, and so they rely more heavily on deception and integrated firepower to achieve their objectives rather than direct action. Electromagnetic attack and defense must be carefully integrated into the counterreconnaissance mission, and information warfare may be used to deceive the enemy. Counterreconnaissance must be practiced throughout the depth of the battlefield, as the enemy has a variety of reconnaissance and intelligence capabilities capable of collecting at any point and in any set of conditions.

# II. Blocking Actions

Blocking defensive actions are those actions that place a unit or group in a relatively static position, seeking to stop an enemy advance, attrit an enemy formation, or reduce enemy strength and cohesion through tenacious defending. Blocking actions are generally not decisive, but they are used to complement other actions on the battlefield by occupying enemy forces or slowing enemy advances.

Blocking actions make less use of depth than other defensive actions. Commanders executing a blocking action should seek to concentrate as much combat power as possible along the main line of defense, using terrain, obstacles, and entrenchments to enhance the strength of the position. Blocking actions are fundamentally less reliant on mobility and rapid reaction than other actions, and instead they rely heavily on terrain advantage, firepower, and brave defense to effectively resist the enemy. Commanders should generally consider blocking actions to be temporary in nature, as a static defense is always vulnerable to being overwhelmed, encircled, or isolated. Groups executing a blocking action should be prepared to retrograde if necessary, and commanders should make rear guards or reinforcements available if possible.

Blocking actions must make best use of firepower and obstacles to overcome the disadvantage of immobility. Artillery should be pre-registered and carefully integrated into the overall defensive plan. Firepower and obstacles should be employed cooperatively to create a protective boundary of firepower ambushes around vulnerable spots in the defensive line.

# III. Repositioning Actions

Repositioning actions are movements conducted to shift combat power during a defensive operation. These movements are intended to move combat power to blocking, counterattack, or fallback positions. As with blocking actions, commanders must use repositioning actions judiciously, as units on the move are more vulnerable to attack and firepower assault. Repositioning actions should be concealed if at all possible; movements at nighttime or during periods of bad weather are optimal. Repositioning actions should be performed rapidly with security provided to the main body, especially while it is in transit. Security must involve counterreconnaissance, counterfire, and screen or cover activity to counter direct action.

Repositioning actions are done for one of three reasons: to reposition a unit to a new, stronger, or more defensible blocking position; to attempt to shift the enemy's focus from one axis of defense to another; or to enable an adjacent unit to move, consolidate, or retreat. Repositioning actions are employed in careful coordination with blocking, screening, and covering actions to create an integrated, in-depth defensive effort. Commanders make use of repositioning actions to gracefully retrograde, ensuring that units on the move are well protected. Once a unit or group completes a repositioning action, it immediately converts to a strong defensive posture, enabling adjacent units or groups to conduct their own repositioning actions.

If the defense must retrograde into the deepest areas of the defensive zone, the defensive group must reposition its firepower and support units and its forward defensive groups. Commanders use the same principles when displacing rear area units as with forward units, while accounting for the former's lesser mobility and greater vulnerability. Air defense coverage becomes critical in rear areas, as rear area units on the move are highly vulnerable to air attack. Air defense coverage must be comprehensive and makes use of leapfrog or bounding overwatch techniques to maintain continuous coverage over moving units.

# IV. Breakout

Breakouts are conducted when a unit or group is encircled, flanked, or in danger of being encircled. The PLAA believes that most mechanized opponents will seek to encircle and isolate units or groups, using many of the same tactics and techniques the PLAA itself employs. This, in turn, has caused it to emphasize the importance of a unit being able to break out of an encircled or isolated situation while preserving manpower and avoiding destruction or surrender. Figure 8-16 on depicts a unit conducting a breakout action.

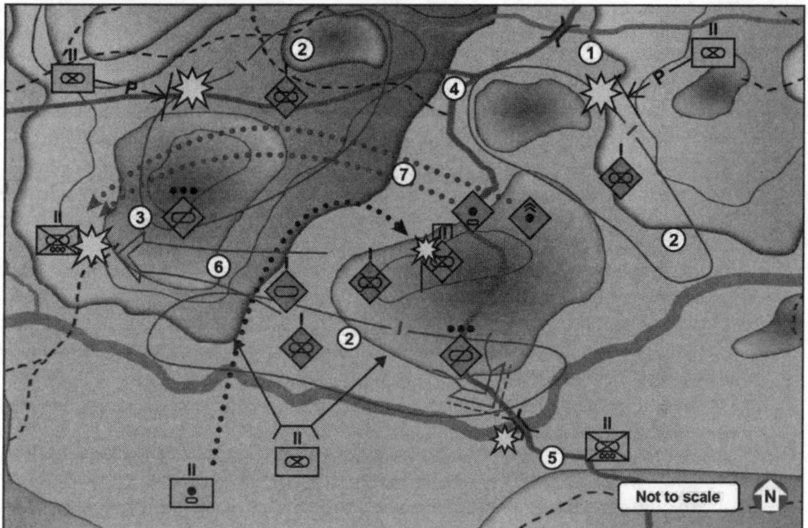


**(1) A defensive group is isolated and surrounded by successful enemy penetrations. The group commander orders a breakout. (2) The defensive group establishes a strong defensive perimeter, using terrain and entrenchments to amplify combat power and enable units to endure enemy firepower and attacks. (3) Careful but rapid reconnaissance reveals that the enemy's weakest position is a light armored unit occupying a blocking position to the west. (4) The enemy likely anticipates a breakout to the northeast or the south, making use of main roads. Instead, the commander chooses a more challenging breakout route, hoping to surprise the enemy. (5) A reconnaissance platoon conducts a feint operation along an expected route in order to deceive the enemy. (6) An ad-hoc assault group comprised of the defensive group's reserves conducts a violent assault along the designated axis of advance. Though outnumbered, the group attempts to achieve localized superiority using deception and maneuver. (7) A firepower attack with the limited available assets supports the breakout assault. If the assault is successful, one company will conduct a rear-guard action while the rest of the defensive group conducts a withdrawal to a new defensive position.**

| N | north | P | penetration |
|---|---|---|---|

*Ref: ATP 7-100.3 (Aug '21), fig. 8-16. Breakout (example).*

A breakout is as much a mental activity as it is a physical activity. Commanders must fight against the tendency of their troops to become passive, afraid, or despondent when facing long odds or a difficult situation. Encircled units will practically always be heavily outnumbered and outgunned. Through the use of skillful command, rapid maneuver, and careful deception, an encircled unit masses combat power at a weak spot in the enemy's position and, with the aid of other nearby units, groups, and systems, breaches the encirclement and moves quickly to a secure defensive position.

Encircled units have likely endured heavy multi-domain firepower assaults and attacks through various flanks on the ground. As such, an encircled unit is likely significantly weakened and damaged, which presents a major challenge to the commander. The nature of the enemy's attacks will be unpredictable and constant, making extensive planning an impossibility. Preparations for a breakout are thus necessarily hasty, and commanders must assume a higher degree of risk when planning deployments and maneuvers.

The most critical element of a successful breakout is coordination with supporting groups and units—particularly firepower groups. This too presents a challenge, both to the encircled commander and to any supporting units. The enemy will certainly attempt to target and suppress any communications between an encircled unit and its potential support to ensure its isolation and preclude any attempts at relief. Commanders may employ creative or subtle methods for coordinating a breakout action. They may also encourage units to act independently. While this may decrease the concentration of combat power, it may enable a subordinate commander to effectively seize an opportunity and the initiative that otherwise passes the unit by.

The breakout action begins with rapid planning. The encircled commander should attempt to contact all subordinate groups and units to achieve some level of integration and coordination of purpose. Plans should be simple and easy to understand and transmit. Commanders at all levels must place a high premium on concealing their actual intentions, as they are likely under intense reconnaissance and surveillance. Deceiving the enemy about the time and place of the breakout effort is a key enabler.

The encircled commander's first objective is to build a highly defensible position. A strong defensive base helps to resist enemy penetrations and provides the anchor for an eventual breakout assault. Commanders take advantage of interior lines and strong terrain to create an in-depth, 360-degree defensive perimeter. Hasty fortifications and entrenchments should be constructed by all units and groups within the defensive area. The commander organizes the encircled units into two primary groups: a frontier group that forms the main defensive line, and a reserve or assault group that conducts the actual breakout attack.

Coordination with external support should be attempted and employed wherever possible. The coordination priority for an encircled unit is firepower. A firepower assault may disrupt, disable, or blunt the enemy's assaults and create useable breaches in the enemy's encirclement. Firepower coordination is very difficult for an encircled unit; however, and commanders should strive to keep firepower planning as simple as possible. Intricate firepower schemes are of little use; instead, commanders of assisting units should employ massed area fires in simple phased schemes, placing the onus on the encircled unit to make best use of the firepower assault. Adjacent assault, security, or reconnaissance units may also assist the encircled unit by conducting demonstrations, feints, or penetrations of the enemy's assault force. Coordination with these units is likely difficult or impossible, so much responsibility lies with the individual group commanders to exercise their discretion in assisting the encircled unit.

After encircled commanders establish a strong central defensive position, they gather combat power from the reserve group and establishes the axis of the breakout assault. Deception and concealment are paramount during this phase, as the enemy will attempt to discover the disposition of any breakout attempt and destroy it with firepower or ground assault. A breakout attack should take the form of multiple simultaneous penetrations, each targeting a weak spot along the enemy's assault force.

These penetrations should be separate, but close enough to be mutually supporting. Commanders should strive to create a strong unity of purpose and strong morale in the assault group members. Though they are outnumbered, they can still achieve a victory and save their fellow soldiers if they succeed in breaking out of the encirclement. Once it is clear that one or more penetrations has succeeded, the commander should feed additional troops into the successful breach to secure it and enable the remainder of the encircled unit to withdraw.

As the remainder of the unit withdraws through the breach, the formerly encircled unit makes use of the reserve and assault groups, plus any outside assisting units, to secure the egress route. The enemy will react strongly to the breakout attempt, and it will attempt to counterattack. This counterattack must be anticipated and defeated with blocking actions and firepower. Movement out of the encirclement may be slow and arduous due to the condition of the encircled unit. Commanders must give these units as much time as possible to move to secure areas. The enemy will also enthusiastically pursue the formerly encircled unit with both air and ground forces. The higher echelon commander must establish a secure defensive position for the encircled unit to retrograde to, and commanders should launch spoiling attacks and employ air defense firepower to defeat the pursuit attack.

# V. Withdrawal

A withdrawal is a carefully organized and planned retreat from a combat situation. There are two types of withdrawal: active and compelled. Active withdrawal is when a unit or group conducts a retrograde operation to take up a more defensible position, better defend an adjacent unit under assault, avoid enemy firepower, or lure the enemy into overextending itself. A compelled withdrawal is a retrograde operation undertaken in direct response to an enemy attack, when a unit is unable to continue resistance in its current position or it is unwise to do so.

Withdrawals are among the riskiest tactical actions. They are nearly always done while in direct contact with an enemy force, and they are often executed in close proximity to a superior enemy. Well-executed withdrawals integrate a wide array of combat power and a variety of methods to ensure the safety of the withdrawing unit and prevent the enemy from taking advantage of the potential disorganization and vulnerability in the defensive zone. Active withdrawals tend to be relatively well planned and carefully executed, while compelled withdrawals may require very fast planning and a great deal of flexibility in execution.

The prescribed order for withdrawal ensures that vulnerable friendly units are protected as the withdrawing unit breaks contact with the enemy. First, the cover and screen forces in the forward zones withdraw and move quickly to the established assembly point to the rear to establish a new security zone. Next, the unit's rear security and logistics elements move, passing lines with the cover and screen forces and establishing rear security and logistics support at the new assembly area. Next, the main body withdraws, along with the commander. This is considered the key point in a withdrawal operation, and it should be timed to take best advantage of weather, time of day, and other environmental factors that help conceal the main-body movement. This leaves only the rear-guard defensive element and the firepower element; these groups withdraw last. The firepower element moves quickly to prepared positions in the new defensive area. The rear-guard element conducts a stubborn retrograde action to delay and disrupt the enemy's pursuit attacks.

Once at the new assembly area, the defensive group rapidly establishes a new defensive area while simultaneously evacuating casualties, reorganizing subordinate groups, and establishing contact with any new adjacent units.

*See p. 7-32. Figure 8-17 depicts a unit conducting a withdrawal action.*

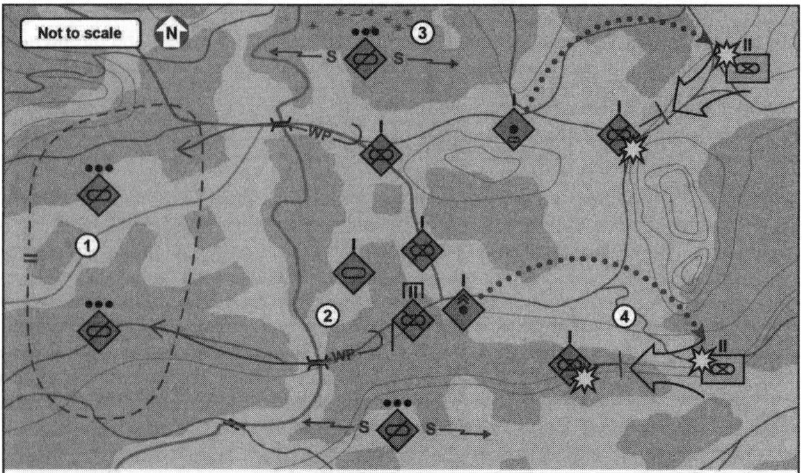


**(1) A defensive group conducts a compelled withdrawal from an untenable position. Cover group elements lead the repositioning action, finding a new, strong defensive position for the group to occupy while awaiting reinforcements. (2) The main body of the group moves rapidly in order to limit exposure to enemy assaults and firepower attacks. (3) Security operations along the flanks protect the main body during vulnerable movement periods. (4) Rear-guard units, supported by firepower attacks, conduct blocking actions against enemy forces, disrupting and delaying their attacks and enabling the group to set up a strong defense in the new position.**

| N north | S screen | WP withdrawal under pressure |
|---|---|---|

*Ref: ATP 7-100.3 (Aug '21), fig. 8-17. Withdrawal (example).*

# VI. Firepower Defense

The PLAA considers the application of firepower to be the primary tool for defeating enemy ground assaults. Firepower defense is often the trigger for decisive counterattacks, and it has an important amplifying function for blocking and harassment missions. Firepower is employed throughout the depth of the battlefield and in all phases of the battle to disrupt the enemy, inflict casualties, defeat penetrations, and isolate enemy formations.

Firepower in the defense is provided by the firepower system, which is a mixture of mobile and static artillery, air defense systems, antitank and other direct fire systems, and attack aircraft. Firepower is organized both vertically and horizontally. Horizontal firepower seeks to dominate a specific axis—usually the main direction of defense—with a mixture of artillery, antitank, and mobile firepower. Vertical firepower seeks to control the sky above the battlefield with air defense firepower and make use of the air domain through fixed- and rotary-wing firepower systems. Firepower is concentrated against key points at key times, and it is carefully synchronized with maneuver to mass combat power. Firepower system effectiveness is enhanced when it is integrated with obstacles.

*See following pages (pp. 7-34 to 7-35) for an overview and further discussion.*

# VII. Depth Defensive Actions

*Ref: ATP 7-100.3, Chinese Tactics, (Aug '21), pp. 8-35 to 8-37.*

Depth defensive actions are those that typically take place in the depth defense zone. These actions are typically calculated to be decisive, as this is where the enemy's assault force is often the most vulnerable and friendly defensive measures are strongest. Depth defensive actions seek to definitively turn back the enemy assault or—ideally—annihilate the enemy's assault force. These actions are built around the depth defensive group, which typically contains a high proportion of firepower, mobile, or armored forces able to move quickly and deliver decisive attacks to the flanks or front of an enemy assault force. This effort is aided by comprehensive security throughout the depth defense zone, spoiling attacks, and blocking actions.

## Depth Security

Security throughout the depth defense zone may be provided by a number of different units or groups that either deliberately or dynamically conduct security operations. The cover group, having withdrawn from the frontal blocking zone, may be called upon to screen or cover the depth group. Additional security can be provided by the reserve group, scouting groups, or elements of the frontier defense group. It is critical that the commander make best use of available forces to maximize security. This will often involve reorganizing or reconstituting disparate groups into useful security forces in an ad hoc manner. Deliberate security measures may also be employed in the depth defense zone, but planning must account for the unpredictability of the enemy's axis of advance into the zone. If the enemy approaches from an unanticipated direction, security deployments may be useless unless they can be rapidly adapted.

Security forces may also take on the mission of shaping the enemy's actions and disposition in anticipation of the decisive counterattack. Shaping and deception operations attempt to orient the enemy's attention away from the axis of the main assault while fixing, deceiving, or blocking enemy forces that might react to the counterattack. Effects of demonstrations and feints are amplified by information warfare operations.

Security operations in the depth defense zone are broadly similar to those in other areas. Commanders of security groups, however, should be encouraged to defend more stubbornly and attack more aggressively than commanders of other groups, as the enemy should be prevented from entering the rear defense zone if at all possible. This may involve security groups being ordered to perform impromptu or hasty blocking actions, making best use of terrain, firepower, and hasty fortifications to prevent enemy penetration into rear areas.

## Spoiling Attack

Spoiling attacks are small-scale offensive actions that target an enemy formation's cohesion and morale. In the depth defense zone, spoiling attacks are primarily intended to defeat, blunt, or disrupt enemy maneuvers, such as flanking movements and penetrations, that may affect or threaten the primary counterattack effort. Spoiling attacks may be conducted by any type of group, but commanders should seek to avoid siphoning off power from the depth group for nondecisive actions, if at all possible.

## Counterattack

A counterattack is typically the decisive action within the depth defense zone. Ideally, it consists of an aggressive, concentrated assault by the depth defense group against a vulnerable portion of the enemy's assault force. The counterattack should be heavily supported by firepower, information warfare, and adjacent or supporting units. The counterattack seeks to break the enemy's cohesion and will, ending its thrust or penetration, and forcing it to choose between retreat and annihilation.

# Firepower Defense Purposes

*Ref: ATP 7-100.3, Chinese Tactics, (Aug '21), pp. 8-35 to 8-37.*

The PLAA prescribes six primary reasons for employing firepower in the defense:

## Target Enemy Information Networks and Firepower Systems

A mixture of artillery, attack aviation, and electronic warfare targets the enemy's communications networks and artillery systems. This action seeks to either destroy or disrupt these key enemy systems, undermining the enemy's overall offensive action and enabling the decisive counterattack.

## Strike Assembling or Deploying Troops

Enemy troops are at their most vulnerable when deploying and massing for an attack. (The preferred PLAA term is "unfolding" to mean deploying, but this publication uses deploying to reduce confusion.) Commanders should seek to target enemy forces during this phase of their offensive action by judiciously employing blocking forces and carefully timing firepower assaults. A mixture of aerial firepower and rocket and tube artillery is best suited for this firepower mission.

## Resist Enemy Assaults

Directly supporting friendly units that are resisting enemy assaults throughout the frontier defense zone is the backbone of the firepower defense mission. These attacks seek to destroy or disrupt enemy formations before and during contact and provide agile, adaptive support to defensive groups as the battle evolves. This firepower is provided mainly by massed rocket and tube artillery.

## Defeat Airborne, Infiltration, and Flanking Maneuvers

If mobile reserve units are unavailable, firepower may be used to disrupt or blunt enemy infiltrations and flanking maneuvers. These enemy units are often exposed, unarmored, or otherwise vulnerable, and targeting them with massed firepower may defeat them without the need for ground intervention.

## Defeat Penetrations

As the enemy's assaults begin to penetrate through the frontier defense zone, enemy units will encounter continued resistance and counterattacks from the depth defense zone. Firepower defense is employed throughout the depth defense zone to support counterattacks and cause casualties in enemy assault groups. Much of this firepower is provided by antitank and mobile groups.

## Target the Enemy's Depth

As the defensive action proceeds, the enemy's rear area is an important target. Enemy reinforcements, resupply efforts, fallback positions, command and communication nodes, and security forces are all commonly located in the enemy's depth area. A mixture of long-range rocket artillery and fixed-wing aircraft provides most of the firepower for this mission.

## Firepower Defense Capabilities

The bulk of the firepower system is comprised of four capabilities: artillery, antitank, aerial, and air defense.

## Artillery Firepower

Artillery firepower consists of indirect firepower systems that deliver either precision or massed fire throughout the depth of the battlefield. This includes rocket artillery, static and mobile tube artillery, and mounted and dismounted mortars. Artillery systems are the backbone of the firepower system. The PLAA envisions a high density of artillery systems to be a decisive capability in most defensive actions. Artillery systems perform every mission inherent in a firepower defense mission, including ground targeting, counterreconnaissance, antitank, information attack, counterfire, and air defense via targeting enemy aircraft support systems and aircraft on the ground.

## Antitank Firepower

The deployment of antitank firepower is a critical component of any defensive action against an opponent equipped with armored or mechanized forces. The PLAA employs an extensive array of antitank weapons, including assault guns, antitank guided missiles, short-range antitank weapons, and specialized artillery munitions. The PLAA employs a comparatively low concentration of antitank guided missiles, relying more heavily on massed short-range antitank weapons than its Western counterparts.

The employment of antitank firepower is one of the few ways in which the PLAA seeks to confront an enemy armored force. Commanders are encouraged to mass as much antitank firepower as possible along the expected avenues of approach for enemy armored forces and integrate the antitank groups carefully with obstacles and artillery. Concentrated antitank firepower seeks to destroy as many enemy armored vehicles as possible, with the understanding that the antitank group may suffer heavy casualties in the process. However, without this concentration of firepower and its associated risks, the enemy's armored thrust would likely succeed.

## Aerial Firepower

Aerial firepower is the combined capabilities of fixed-wing attack aircraft and attack helicopters. These two capabilities are somewhat different but largely complementary. Fixed-wing aircraft can deliver huge destructive power, but they are typically not well-coordinated with the ground element in contact, and they are often somewhat limited in their precision and accuracy. Thus, fixed-wing attacks are best employed targeting enemy rear areas: halting reinforcements, destroying supplies, and disrupting command and communication. Conversely, attack helicopter assaults are carefully coordinated with ground forces, and they are precise in their application of firepower. They are, however, far more vulnerable to ground and air defense firepower. Helicopters are best employed as a highly mobile reserve—targeting enemy flanking actions, encirclements, and infiltrations, as these are least likely to have substantial air defense capabilities or defensive counterair support. Commanders may also choose to employ attack helicopters to blunt the enemy's decisive attack. While this approach adds to the value of the decisive counterattack, it exposes the helicopters to more-concentrated air defense firepower.

## Air Defense Firepower

The PLAA is keenly aware of the massive destructive potential present in enemy air attack, and it places heavy emphasis on defending ground units from enemy aircraft. Air defense firepower is deployed throughout the defensive area using a tiered and layered approach, and it is enhanced by more-powerful, longer-range systems that are deployed across the theater. The PLAA employs a wide variety of air defense systems. Man-portable air defense systems (MANPADS) are widely fielded across formations and groups of all types. Self-propelled antiaircraft guns (SPAAGs) are widely employed by mechanized and motorized units, and short- and medium-range missile systems are present at the CA-BDE level. These systems are employed using a combined arms approach, wherein the capabilities of one system offset the vulnerabilities of another.

The PLAA divides target types by altitude. Ultralow-altitude air defense is provided by a combination of small arms, crew-served weapons, and SPAAGs. Low and medium air defense is provided by a mixture of SPAAGs, MANPADS, and missiles. High-altitude air defense is provided by theater assets, such as long-range missiles and fighter aircraft. Tactical echelons are most concerned with ultralow- and low-altitude threats, such as unmanned aircraft (UA) and helicopters, and seek to defeat these systems through a mixture of concentrated firepower that defends key assets and air defense ambushes that strike at enemy aircraft as they move along air routes. Air defense systems are oriented by section, while the coverage of the defensive zone is ideally all-directional.

# VIII. Rear Defense

The rear defense zone is the deepest and most vulnerable section of the defensive battlefield. Commanders and defensive groups should strive to prevent significant enemy penetrations into the rear defense zone. Decisive actions should be planned for more-forward areas. Rear defense operations seek to blunt, delay, or defeat enemy penetrations, protect critical assets in the rear area, and—if necessary— enable the retreat or retrograde of vulnerable forces in the rear defense zone. Operations in the rear defense zone are undertaken by the reserve group, by rear security units—possibly including militia or paramilitary security units—and by any other groups that have moved into the rear area due to the maneuvers of the battle.

The rear defense zone typically houses important and potentially vulnerable systems and units, such as artillery groups, air defense groups, supply and logistics groups, command groups, and aviation groups. These groups are necessarily more vulnerable to direct attack. Protecting them or enabling their withdrawal is a critical mission in the rear defense zone.

Rear defense also includes the mission and associated tasks of defeating smaller enemy penetrations and infiltrations. The enemy seeks to establish a combat presence throughout the depth of the battlefield, and it will use aerial insertions and ground infiltrations to move small, mobile units into the rear area. These enemy units are charged with conducting covert direct operations, such as sabotage and raids, along with reconnaissance and surveillance activities. Rear defense units must be capable of detecting and neutralizing these enemy forces, either through direct action or deception. Commanders must assess the threat level in rear areas and ensure that the rear defense or reserve groups have the necessary combat power available to counter these threats.

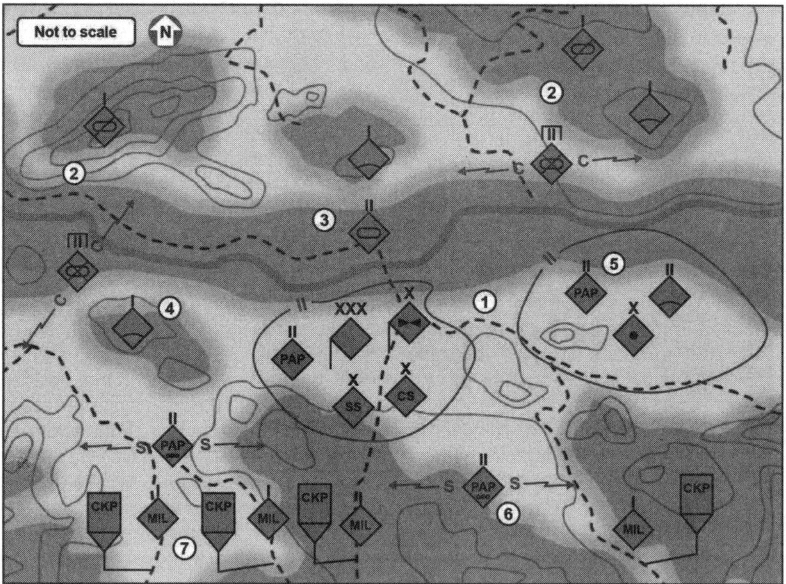


**(1) A rear-area security zone is established around a group army headquarters element, combat aviation brigade headquarters, combat support and service support brigades, an HQ-9 SAM battalion, and an artillery brigade. (2) The cover group deploys reconnaissance units and two defensive groups to the frontal blocking zone, creating the main line of defense against enemy penetrations. (3) A heavy CA-BN serves as the depth group and combat reserve. (4) Medium-range HQ-16 SAM batteries are deployed through the frontal blocking zone, integrated digitally with the HQ-9 battalion. (5) Dismounted PAP battalions serve as security groups for the high-value targets. (6) Mounted PAP battalions conduct screen missions across southern avenues of approach, deterring infiltrations, raids, or surveillance by enemy units. (7) China militia units establish checkpoints along supply routes, conducting traffic and local route security.**

*Ref: ATP 7-100.3 (Aug '21), fig. 8-18. Rear area defense (example).*

## Rear Security

Security in rear areas is a critical enabler of the overall defense. Numerous key groups and capabilities reside in the rear defense zone, and the enemy will seek to target these high-value assets with a mixture of infiltration, penetration—particularly by aerial forces—firepower, and electromagnetic attack. Security forces in rear areas consist of a mixture of static groups that resolutely defend the most valuable assets and mobile forces that react rapidly and aggressively to any intrusions.

Commanders determine the responsibilities of the rear security groups according to the priorities of the battle and available forces. First, the commander must establish a clear priority of rear area assets. Those assets most critical to the overall prosecution of the defensive battle are placed highest. Then, a rear defense zone plan is developed which assigns available troops to defend or secure assets based on priority. Static or less-mobile groups conduct positional defensive actions. This cohort consists of militia, armed police, and dismounted soldiers. These groups carefully entrench around their specified defended asset and work in concert with the commander of the asset to harden the entirety of the critical area. Simultaneously, more-mobile forces, such as mounted police, mechanized infantry, or armor (if available), are formed into mobile defensive groups. These groups are placed carefully to best counter anticipated enemy axes of intrusion and to best support the more-static units in their positional defenses.

Airborne infiltrations and penetrations should be deterred, neutralized, or attritted by air defense firepower which is carefully placed along anticipated axes of advance. Ground infiltrations are met aggressively by mobile defensive groups, then isolated and defeated. Static defensive groups should seek to fix any enemy attack forces, enabling the mobile defensive groups to attack them along flanks or from the rear. Enemy firepower attacks in the rear area are countered by concealment, hardening, and aggressive counterfire operations. A key enabler of counterfire operations in rear areas is breaking the enemy's linkage with reconnaissance and surveillance assets as they conduct targeting operations. This can be accomplished by destroying the platform or neutralizing the communications link between sensor and shooter.

## Cover

Cover missions are of particular importance in the rear defense zone. Commanders employ a mixture of mobile armored, mounted, and reconnaissance assets using a 360-degree approach, always assuming that the rear defense zone is vulnerable from any direction. Cover units deploy a sufficient distance from the main body (between three to five kilometers [km] for a CA-BDE) to give adequate early warning about enemy activity and grant adequate space for stubborn resistance. Deployments of covering units focus on anticipated primary axes of advance. Units are carefully integrated with mobile defensive groups. As a cover unit encounters an enemy penetration, it resists while falling back to defensible terrain, seeking to fix the enemy intrusion and enable the mobile group to decisively counterattack.

Cover missions in rear areas must place a high priority on air defense firepower. It is highly likely that enemy air intrusions will overfly areas occupied by cover forces; this provides an excellent opportunity to conduct aerial ambushes. Cover groups in rear areas should, if possible, be augmented by MANPADS or anti-aircraft guns to make best use of this tactic.

If units are adjacent to one another, commanders between the units should coordinate their respective rear area cover missions. It may be unnecessary to deploy a cover force facing a friendly unit. Commanders must be aware of what is to the left and right of their rear defense zone and weigh their deployments accordingly. Commanders must also deploy cover forces along any seams or borders between units that may be exploited by enemy infiltration or penetration.

## Sabotage and Countermobility

As rear defense groups may be both outmanned and outgunned in the event of a major enemy penetration, rear defense zone operations must make best use of both countermobility and sabotage missions. Countermobility efforts focus on building integrated obstacle systems that slow the enemy, force it to spread combat power or change axis of advance, or canalize it into ambush or counterattack zones. Rear defense groups typically have more time and assets to build obstacle systems, and they should rely heavily upon them to multiply available combat power. Particular attention should be paid to slowing the enemy as it enters engagement areas for firepower systems—both direct and indirect—and manipulating enemy movement to make it more susceptible to spoiling attacks or counterattacks.

Sabotage is the deliberate, covert targeting and destruction of enemy assets to negatively affect the enemy's operations. While sabotage has its place everywhere on the battlefield, its usefulness is most pronounced in the rear defense zone. Militia, SOF, or paramilitary units are highly effective sabotage groups, as they can blend into the surrounding environment and subtly sabotage critical assets without provoking a strong enemy response. Sabotage efforts work in concert with countermobility efforts to slow the enemy advance through the rear defense zone. These efforts include, but are not limited to destroying bridges, roads, tunnels, or other mobility corridors; ambushing enemy supply elements as they move to resupply the enemy attack group; or conducting small ambushes of isolated enemy units. Sabotage teams should be kept small to minimize their signature, and sabotage efforts should be conducted in a decentralized manner.

Chap 8

# I. Antiterrorism Actions

*Ref: ATP 7-100.3, Chinese Tactics, (Aug '21), pp. 9-1 to 9-9.*

Antiterrorism and stability actions are conducted by a wide variety of different organizations within the Chinese government, ranging from both the regular and special operations (SOF) forces of the People's Liberation Army (PLA), to the People's Armed Police (PAP), to national and local law enforcement. These actions include domestic law enforcement, stability operations, maritime security, and international stability operations. The PLA broadly categorizes antiterrorism actions as part of maintaining both domestic and international stability, while PAP stability actions are largely domestic in nature and focus on internal threats and dissent. This chapter discusses national-level antiterrorism and stability operations, focusing predominantly on the mission types undertaken by People's Liberation Army Army (PLAA) and PAP units under the control and direction of the national government.

## I. Overview of PLAA Antiterrorism Operations

Terrorism in the modern sense is a relatively new phenomenon in China, and the Chinese government is still working to establish a clear understanding of what constitutes both domestic and international acts of terror. The PLA defines terrorism as acts of violence that disrupt national unity and societal stability by creating casualties or damaging property. This definition is very broad, far more sweeping and ambiguous—possibly deliberately so—than most Western definitions. This helps to explain why the PLA views antiterrorist actions as supporting national stability. Instead of only acts of violence perpetrated to achieve a political or social objective, the Chinese definition also includes acts that threaten peace and stability within the country. This, in turn, gives Chinese authorities a very broad set of circumstances in which antiterrorism laws and operations can be applied. The growing Chinese interest in antiterrorism operations coincides with rising religious and sectarian tensions in specific regions of China and the rise of global terrorism through the latter half of the 20th century.

## II. The Informationized Battlefield and Antiterrorism Operations

The PLA recognizes several characteristics that shape antiterrorism activities: urgency, complexity, significant effects, joint nature, and asymmetry. As with all other military operations, China is acutely aware of the changes the informationized battlefield has brought to its antiterrorism mission. Unlike offensive and defensive actions, however, many of the effects the informationized battlefield has on antiterrorism operations are related to media, public opinion, public policy, and public perception.

### Urgency

Antiterrorism missions are urgent in nature. Generally speaking, all of the initiative during a terrorist attack or campaign rests with the terrorist organization. Proactive antiterrorism efforts are possible, but they are limited in scope and effectiveness by the difficulties associated with identifying and neutralizing potential terrorists before they strike. This makes most antiterrorism activities reactionary and highly time sensitive. Constant vigilance is an important consideration for any unit or organization charged with conducting an antiterrorism mission, and rapid response to a terrorist event is one of the most important elements in preventing casualties, maintaining stability, and suppressing further terrorist acts.

### Complexity

Antiterrorism missions are complex. Terrorist activities are often sophisticated and complex, making use of media, religious or political factionalism, civilians, and international actors to manipulate the situation to the terrorists' advantage. By framing an activity as defending religious freedom or democracy, terrorists can gain sympathy from other parts of the world. By coercing local civilians, they can conceal their activities and disrupt efforts to find and suppress them. This requires antiterrorism forces to be equally sophisticated and highly aware of both the effects their actions have on the global political dialogue and how they can push back against attempts to manipulate local civilian populations by building trust and maintaining a strong visible presence.

### Significant Effects

The effects of terrorism can be significant. One of the primary features of terrorism is that the psychological and political effects of terrorist acts can often far exceed the actual physical effects. Terrorism is designed to amplify the effects of violence by disrupting lives, bringing anxiety to a society, and pressuring those conducting antiterrorism efforts into overreacting to real or perceived threats. These effects may be realized even when a terrorist operation fails. The simple threat may be enough to achieve the desired effect. Antiterrorism efforts must be valued according to their ability to maintain the long-term peace, prosperity, and stability of a society, not just in terms of their ability to directly root out and defeat terrorist elements.

### Joint Nature

Antiterrorism operations must be joint in nature. Antiterrorism missions require cooperation between different elements, perhaps more than for any other security activity. The complexity of terrorist activities, along with the fact that they often cross international borders, causes a high premium to be placed on open and effective cooperation. In this context, joint refers to three different forms of cooperation: between the military, police, and civilian antiterrorism units domestically; between the political and military parts of the government; and between different international actors.

### Asymmetry

Terrorist actions are nearly always asymmetric. Terrorists seldom enjoy any significant parity in equipment or training to that of their opponents, and they seek to offset these disadvantages by attacking vulnerable targets from unforeseen directions. Most antiterrorist organizations are also asymmetric in nature. While they may be formally trained and equipped, most are drawn from police or militia organizations, and so they do not habitually employ aggressive military tactics. Operational methods used by both sides are described as irregular. Terrorists employ ambushes, raids, and sabotage, while antiterrorists employ their own ambushes, along with careful counterattacks, search and annihilate, and infiltration methods.

## III. Antiterrorism Missions and Tactics

Unlike most other military operations, antiterrorism operations are prescribed to be tightly controlled and commanded by a central authority. This ensures that the antiterrorism operation is tightly integrated and that subordinate groups conduct their own operations in accordance with the principles of antiterrorism operations. Freedom of action by lower echelon groups is viewed as less critical, as the scope of antiterrorism operations is typically much smaller than major military operations, and the consequences for a subordinate's mistake are often more significant.

There are five primary antiterrorism missions; each has numerous subtypes and permutations based on the operational environment and the nature of the threat.

# IV. Principles of Antiterrorism Operations

*Ref: ATP 7-100.3, Chinese Tactics, (Aug '21), pp. 9-2 to 9-3.*

China recognizes the significant challenges that antiterrorism operations face and their significant differences from more-traditional military operations. As a result, the principles that the PLA and the rest of the Chinese Security Apparatus apply to antiterrorism operations are unique and much more subtle than those that apply to more-symmetric kinds of actions.

## Active Prevention

One of the most insidious elements of terrorism is that terrorist actions can still achieve much or all of their desired effects even when the mission itself fails. Terrorist acts can also target a virtually infinite number of different elements of a society, from the military, to education, to infrastructure, to government, to simple random targets. The only sure way to eliminate the terrorist threat is to suppress it when it is still in the embryonic stage. This requires antiterrorism forces to take constant, targeted preventative measures in order to identify and suppress terrorism before it reaches a more formal stage. Authorities must also take measures to reduce the intensity of terrorist actions by hardening vulnerable targets, keeping the local population informed, and ensuring that emergency response capabilities are robust and well maintained.

## Rapid Reaction

In spite of the best efforts to actively prevent them, terrorism events typically occur suddenly; often with little or no widespread warning. This requires the forces and organizations that respond to them to plan carefully for many different contingencies and react with great rapidity when a terrorist attack does occur. These organizations should be localized and know the population, terrain, and other characteristics of their assigned area. They must war-game many different possible courses of action to save time when actual events occur. When an event does occur, the planning is rapidly revised to meet the needs of the real-world situation. The priorities in these situations are (1) to secure the immediate area, (2) to reduce or contain civilian casualties, and (3) to rapidly counterattack and attempt to kill or capture the terrorist actors.

## Joint Actions

The complex legal framework of antiterrorism activities requires a unique joint approach to antiterrorism operations. PLAA, PAP, China Militia, national police, and local police must all work jointly under a unified command structure in order to achieve the best results. Because terrorism activities can take on an incredibly wide variety of different forms, structures, and capabilities, antiterrorism forces must be able to adopt any potential posture, from passive and cooperative to high intensity. This approach requires a true joint command that has the authority to unify and coordinate actions across numerous different organizations. This principle is inherently political, and deciding who actually commands an antiterrorism group is likely to be a decision made at higher levels of government. Leaders of subordinate organizations must put aside their own agendas and those of their commands in order to achieve harmony and unity of purpose within a joint group.

## Legality

One of the most important elements of a terrorist action is the undermining of the rule of law and the customs and observances of a people. Terrorists deliberately attack these important institutions to magnify the effects of their violent actions by eliciting reactions of anger and fear from their victims. Antiterrorist groups must take great care to act in a legal manner, observing all relevant laws and regulations to make clear to the population that they have the moral high ground. Similarly, antiterrorism groups must carefully observe all important local customs and culture to gain the trust of the people.

Many of the same terms, techniques, and tactics carry over from full-scale military operations.

## Antiterrorism Missions

| | |
|---|---|
| A | Active Prevention |
| B | Riot Control |
| C | Active Antiterrorism Attack |
| D | Hostage Rescue |
| E | Border Control and Security |

## A. Active Prevention

One of the most insidious elements of terrorism is that terrorist actions can still achieve much or all of their desired effects even when the mission itself fails. Terrorist acts can also target a virtually infinite number of different elements of a society, from the military, to education, to infrastructure, to government, to simple random targets. The only sure way to eliminate the terrorist threat is to suppress it when it is still in the embryonic stage. This requires antiterrorism forces to take constant, targeted preventative measures in order to identify and suppress terrorism before it reaches a more formal stage. Authorities must also take measures to reduce the intensity of terrorist actions by hardening vulnerable targets, keeping the local population informed, and ensuring that emergency response capabilities are robust and well maintained.

## B. Riot Control

Riot control is seen as one of the most important elements in maintaining domestic stability. Rioting is viewed as a direct threat to Chinese national security, and control efforts are centered on ensuring the safety of both people and property. The methods used, however, are aggressive—employing many of the same tactics and techniques used during high-intensity military operations. Riot-control operations consist of four phases: control and blocking; divide and encircle; pursue and destroy; and eliminate enemy remnants and suppress follow-on rioting.

*Note. PLA doctrine does not make much differentiation between demonstrations and large-scale destructive riots. The methods prescribed are essentially high-intensity military tactics. It may be inferred that riot control in this context is referring to a large-scale insurrection or highly destructive riot. It should also be noted that all rioters are referred to as terrorists, independent of actual affiliation with a terrorist group or its motivations.*

*See following pages (pp. 8-6 to 8-7) for discussion of phases of riot control operations and examples.*

## C. Active Antiterrorism Attack

Active antiterrorist attacks are those attacks that proactively target terrorist bases of operations and leaders. They may be independent missions, or they may be conducted concurrently with or in support of other antiterrorist activities. The general approach to these attacks is much the same as for a riot-control or ground offensive operation. Decisive attacks penetrate the enemy defenses and isolate enemy units, allowing their destruction in detail. There are, however, two important activities that must be emphasized during a decisive pre-emptive attack. First, reconnaissance and search must be comprehensive and meticulous, ensuring that the entire battlefield is carefully assessed and all terrorist targets identified. This may be a lengthy process and commanders must be patient, not launching their attack until it is complete. Second, antiterrorist groups must emphasize the blocking actions or cordon that surrounds the terrorist target. Terrorists likely have preplanned escape routes in place. They will rapidly flee the area if threatened, so a substantial cordon force may be required in order to fully contain the area. A careful search should reveal all major possible egress routes.

## D. Hostage Rescue

Hostage rescue missions integrate multiple different competencies in such a way that decisive application of both physical and psychological force frees the hostages and eliminates the terrorist group. Hostage rescue missions are arguably the highest profile of all antiterrorism mission types. They often are carefully followed by the world at large, and they have outsized significance and consequences for both security forces and society at large.

A hostage rescue mission employs five different groups: a blocking group, an assault group, a reconnaissance group, a psychological warfare group, and a rescue group. These groups work in concert with one another to break terrorist resistance while protecting the hostages. The rescue mission commences after hostages are taken and antiterrorist forces are organized and deployed. The first phase of the operation is the unfold phase, where the blocking unit deploys around the periphery of the combat area and secures land, sea, and air routes from either ingress or egress. The other groups spend this period organizing themselves and developing the plan of action.

Psychological warfare operations commence in the next phase of the mission. This naturally centers on the psychological warfare group, although a variety of other organizations may have roles to play. The psychological warfare phase typically consists of negotiating with the terrorist group. Negotiations are never used to concede to the political demands of the terrorists, but rather to begin the process of isolating them, dividing their forces, and disintegrating their strength. Negotiations should be deliberately difficult, causing psychological stress to the terrorists, but they should not be so aggressive as to cause a threat to the hostages. They may also serve to distract terrorists from ongoing assault preparations, enabling the assault group and reconnaissance group to conduct their missions without detection. If it is possible to resolve the situation through negotiation without making political concessions to the terrorists, this should always be the first choice.

The final phase of the hostage rescue mission is the raid or attack. This consists of the reconnaissance and assault groups carefully assessing the situation, developing an assault plan, and then bravely and aggressively executing the plan. Reconnaissance efforts should ascertain the exact position of the terrorists and hostages, identify all of the potential ambush points or strongpoints in and around the objective, and provide the assault team with a comprehensive plan of attack against the heart of the terrorist group. The psychological warfare effort should have caused distress and division among the terrorists, possibly compromising their defenses and isolating the different terrorist groups. Careful concealment during the approach phase, coupled with surprise and precision in the attack, are the best ways to ensure

# Riot Control Operational Phases

*Ref: ATP 7-100.3, Chinese Tactics, (Aug '21), pp. 9-6 to 9-7.*

## Control and Blocking

The control and blocking phase always initiates a riot-control action. This phase consists of blocking groups who focus on denying possible rioter escape or reinforcement routes deploying around the riot area. Blocking operations should include both military and police forces, ideally working in concert with one another to properly leverage the capabilities of each. Blocking should also include the denial of communications and the suppression of ports, subterranean facilities, and other possible ways of escape or reinforcement. Blocking is continuous throughout the operation, but blocking forces should maintain a flexible approach to their deployment, ready to reinforce key areas if and when they come under direct assault. During this phase, comprehensive reconnaissance is conducted throughout the riot area to ascertain the positions of the leaders, key rioter defensive positions, and any other important pieces of intelligence to inform the upcoming decisive phases.

## Divide and Encircle

The divide and encircle phase is the decisive phase of the riot-control operation. It employs a mixture of surprise attacks and storming attacks to overwhelm the rioters, with a focus on dividing the various groups, isolating them from their leadership, and then annihilating them. Some groups are highly vulnerable to isolation and will cease resistance once cut off from their leaders, while others will fight determinedly—even when completely cut off and surrounded. Antiterrorist forces must be aware of the tendencies of their opponents as they face destruction. Some riot participants will surrender willingly and without further resistance, while others will respond with suicidal violence, even when facing certain defeat. Antiterrorist force leaders must employ force measures appropriate to their opponent. Less-fanatical adversaries may be willing to peacefully surrender if offered the opportunity, while suicidal, fanatical individuals pose a lethal threat until they are captured or killed.

## Pursue and Destroy

The next phase of the riot-control operation is the pursuit and destruction of any remaining remnants. It is critical that as many terrorists as possible be captured or killed, as any who escape will remain a threat and can easily consolidate and rally new recruits in a short period of time. As most riots occur in urban environments, pursuit will likely be on city roads and through populated areas. Pursuing forces integrate both land and air assets to keep careful track of the positions of those retreating, canalizing them into less densely populated areas and blocking their escape to safe havens. Once the groups are pushed to a favorable position, they are blocked from all sides and attacked until captured or destroyed. Antiterrorist groups may also employ techniques such as ambushes to surprise and defeat retreating rioters.

## Eliminate Enemy Remnants and Suppress Follow-on Rioting

Once most rioting forces are captured or killed, the operation transitions from pursuit to search. Antiterrorist groups conduct comprehensive searches of the combat area in order to find any remaining isolated remnants. This search can be either immediately after the successful destruction of the main group or some time afterward, giving remaining enemy forces time to relax and believe themselves safe from reprisal. Searches must be careful, leaving no blind spots or areas unchecked, but the antiterrorist group also must not disrupt the lives of peaceful citizens in the search area.

*Ref: ATP 7-100.3 (Aug '21), fig. 9-1. Riot control (example; part 1 of 3).*

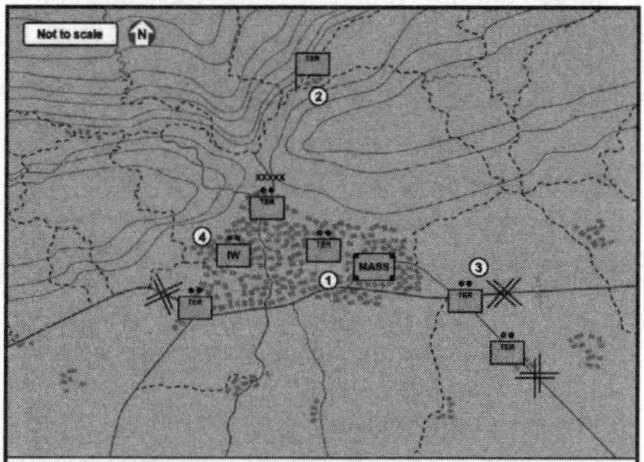


(1) A terrorist organization creates a civil disturbance in an urban area. Using a mixture of violence and information warfare in the form of deceptive propaganda, local civilians are fooled or coerced into supporting the disturbance. (2) Terrorist leadership elements establish a headquarters in the nearby hills and communicate to their followers with cellular phones, computers, and simple radios. (3) Roads into the area are blockaded and manned by small terrorist cells. (4) The civil disturbance is carefully filmed and broadcast over social media, ensuring that international news organizations receive the footage.

| | | | |
|---|---|---|---|
| IW | information warfare | N | north |
| MASS | mass demonstration | TER | terrorist |

*Ref: ATP 7-100.3 (Aug '21), fig. 9-2. Riot control (example; part 2 of 3).*

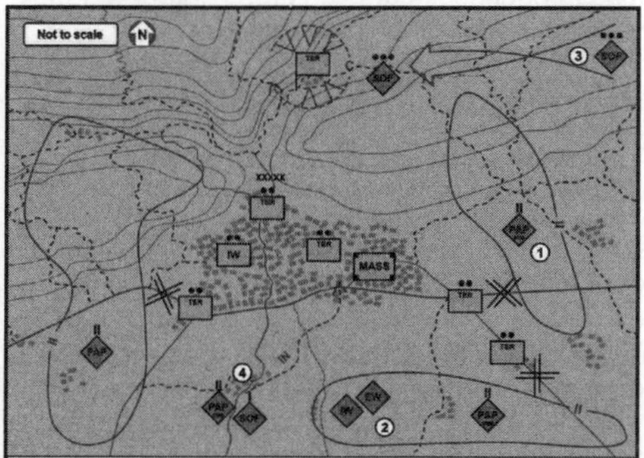


A PAP mobile detachment deploys to control the riot. (1) Three PAP battalions deploy to blocking positions around the area, cordoning it off and ensuring no travel in or out. (2) An EW group deploys to the area, suppressing all over-the-air communications, including cellular phones and radios—isolating distant terrorist cells from one another. An IW group begins targeting terrorist networks and computers and ensures that the area is cut off from internet access. (3) A SOF platoon, receiving a mix of electronic intelligence information and satellite imagery, identifies the location of the terrorist leadership cell. It conducts an air insertion and establish a cordon around the cell, isolating it physically from the rest of the terrorist cells. (4) A PAP battalion and SOF company comprise the assault group; they begin their intensive reconnaissance of the main riot area.

| | | | | | |
|---|---|---|---|---|---|
| C | cordon | IW | information warfare | PAP | People's Armed Police |
| EW | electronic warfare | MASS | mass demonstration | SOF | special operations forces |
| IN | infiltration | N | north | TER | terrorist |

*Ref: ATP 7-100.3 (Aug '21), fig. 9-3. Riot control (example; part 3 of 3).*

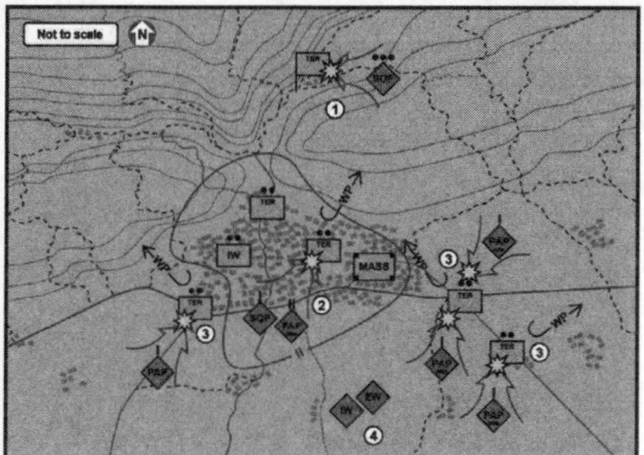


(1) A SOF platoon conducts an assault on the isolated and cordoned terrorist leadership cell, killing or capturing the leaders. (2) A SOF company conducts a carefully measured surprise attack on the heart of the riot, targeting terrorist leaders and carefully avoiding civilian casualties. As terrorists flee the area, the PAP cordon captures them. (3) PAP companies detached from the outer blocking force engage the now-isolated terrorist groups, who quickly cease resistance and either disperse or surrender. Mounted units pursue terrorists as they attempt to escape the area. (4) The EW and IW groups continue to suppress terrorist communications, while ensuring that the operation is recorded and rebroadcast as an antiterrorist propaganda tool, dissuading future riots.

| | | | | | |
|---|---|---|---|---|---|
| EW | electronic warfare | N | north | TER | terrorist |
| IW | information warfare | PAP | People's Armed Police | WP | withdrawal under pressure |
| MASS | mass demonstration | SOF | special operations forces | | |

that the terrorists are killed and the hostages unhurt. Nonlethal munitions—such as smoke, flash, and flare rounds— may be used to temporarily disorient the terrorists without harming hostages. Sniper teams should also be employed, both as reconnaissance platforms and as overwatch. Multiple sniper teams are deployed in a variety of different positions around the combat area, creating a variety of threats to the terrorists. Snipers may also be employed as ambush teams to engage terrorists attempting to escape the area.

## E. Border Control and Security

China places a high priority on border security in general, and in particular where border regions are unstable or dangerous. Border security falls to a number of different organizations, from local and national police active units of the PLAA. The territory designated for enhanced security measures is called a border control area.

Borders cross three domains: land, sea, and air. In order to effectively control the border, the border-control force must control traffic through all three domains. Land borders are controlled by a mixture of firepower and physical occupation. Forces align along widely used corridors and critical areas, and they make use of mobile forces to react to unexpected intrusions through less-patrolled areas. Firepower may be used to augment these forces or to close areas of the border that are not physically manned. Border-control areas do not necessarily need to be placed directly on the border. If terrain or other conditions dictate, they may extend some distance into Chinese territory.

Border control areas must be carefully cordoned through the use of blocking forces and checkpoints. Blocking forces employ defense-in-depth techniques to ensure that the area is secure from both ingress and egress, and checkpoints control the flow of traffic in and out of the area. As border-control efforts become harsher, the negative effects on the civilian population and economy increase. Local officials must carefully balance the need for security with the negative effects on the lawful population and cross-border trade. Ports are another area of emphasis: while they are not likely to be targeted by offensive action, they provide ingress routes for undesirable forces and cargo. In a high-security border zone, inspection efforts through ports must be significantly enhanced.

Air and sea border control typically involve the use of exclusion zones to deny the use of airspace or waterways to unwanted traffic. A no-fly zone or restricted airspace denies opponents the use of airspace over the border-control zone, while a naval exclusion zone denies use of waterways. These measures must be widely publicized to ensure that domestic and international aircraft or ships do not inadvertently violate them, and sufficient air and naval and coast guard forces must be present to ensure that the measures can actually be enforced.

# II. Stability Actions

*Ref: ATP 7-100.3, Chinese Tactics, (Aug '21), pp. 9-9 to 9-14.*

## I. The Informationized Battlefield and Stability and Security Actions

Security and stability missions are among the highest priorities for the Chinese Security Apparatus. In past generations, much of the domestic security mission fell to the PLA. Today, most of the burden is shouldered by the PAP, along with a mixture of national and domestic law enforcement agencies. In addition to peacetime stability and security missions, the PAP has a significant role in supporting wartime missions. The PAP should not be underestimated as a military and paramilitary force; it is comparable to a regional power's national army in size and combat power. The PLA recognizes several characteristics of stability and security actions: political nature, hybrid warfare, threat strength, and international peacekeeping.

The PAP command structure and control mechanisms are still somewhat opaque to outsiders. It appears that PAP units are organized into brigade-sized formations called mobile detachments that are subsequently assigned to a specific region. In wartime, group armies may take over operational control of PAP units to support group army operations, primarily in rear area and security roles.

Figure 9-4 depicts a genericized PAP mobile detachment as it might be organized for wartime support operations.

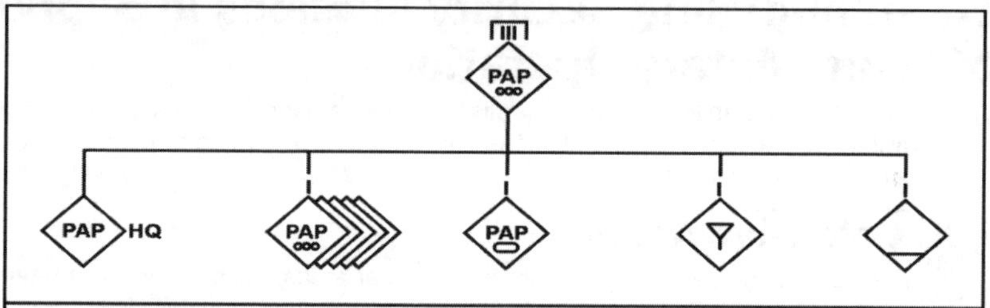


| HQ | headquarters | PAP | People's Armed Police |
|---|---|---|---|

*Ref: ATP 7-100.3 (Aug '21), fig. 9-4. Task-organized People's Armed Police mobile detachment (example).*

### Political Nature

While all military activities are political to a certain extent, security and stability operations are more heavily influenced by politics than are force-against-force operations. Security and stability forces must coexist with local populations, and they may be required to protect them from attack or to suppress civil unrest. Social media is more readily available in security zones removed from the front line. Security-group activities are often high-profile and likely to be seen and publicized. As a result, security and stability forces are given more leeway and more responsibility for implementing domestic policy than other forces as they conduct traditional law enforcement alongside their military duties.

### Hybrid Warfare

Hybrid warfare seeks to significantly influence behind-the-lines operations, where security and stability forces typically operate. Activities of a hybrid opponent may include the use of either civilians or soldiers in civilian attire to conduct raids, ambushes, or other direct actions in rear areas; the use of firepower or electromagnetic attack to target key assets in rear areas; the use of information attack to influence civilian populations throughout the security zone; or the use of special operations or irregular forces to conduct deep direct action throughout the security zone. The clash between security and stability forces and the enemy's offensive hybrid warfare elements will likely be a key factor on the informationized battlefield.

### Threat Strength

The ability to threaten rear areas with a variety of different attack methods has expanded as firepower and irregular threats have proliferated and grown in capability. Security areas now face many of the same levels of threats as forward defensive areas, but security and stability forces do not enjoy the same combat power as most of their military counterparts. Security forces must ensure they can effectively neutralize the anticipated threats throughout their assigned security area, and they must carefully allocate combat power to ensure the right balance of forces in accordance with their higher echelon's priorities.

### International Peacekeeping

The multiple disciplines required of stability and security forces place them in an ideal position to function as international peacekeepers. As China seeks to expand its international role through participation in peacekeeping missions, it is likely that the government will lean heavily on its PAP units to function in an expeditionary capacity. As with any other stability-type mission, peacekeeping missions are often complex and ambiguous, requiring mentally agile leaders and versatile units.

# II. Stability and Security Missions in Support of Joint Military Operations

As with other stability and security missions, those performed in support of joint military operations are primarily defensive in nature. They can range across many combinations of domains.

## A. Defensive Actions

The PLA recognizes that assets throughout rear areas, such as air and seaports, bridges, railheads, command and communication nodes, and missile operation sites will be high-value targets for the enemy. Much of the responsibility for defending these assets from land, sea, and air attack falls to security forces.

## B. Ground Defense

Ground defense actions by security forces employ a defense-in-depth approach to make it difficult or impossible for the enemy to successfully attack key assets. Enemy strike methods against rear areas include direct attack by ground forces, secret attacks by special or irregular forces, and firepower attacks. Security forces employ a mixture of blocking actions along key avenues of approach, strong defenses around key assets, and the use of ambushes and raids against enemy intrusions.

Blocking forces employ either static defense methods or checkpoints. Static defense—allowing no one to pass through the defended area—is employed when the chances of enemy action are high, and the movement of friendly forces or civilians through the area is not critical to wider operations. Checkpoints— allowing carefully screened personnel to pass through—are employed in more-secure areas or where friendly movement is required. Static defense techniques are more secure

# III. Principles of Stability & Security Operations

*Ref: ATP 7-100.3, Chinese Tactics, (Aug '21), pp. 9-10 to 9-11.*

Stability and security operations are conducted in accordance with a fixed set of principles, as are all other PLA activities. The principles governing stability and security operations, however, are referred to specifically as guiding thoughts, likely inferring that they are meant to be broader in scope and less restrictive than other Chinese military principles. This is in keeping with the view that stability and security missions are more complex and ambiguous than other operations, and thus they require greater latitude for leaders to act in accordance with their judgment.

## Situational Awareness

While situational awareness is required for military units of all types, it is of particular importance for security units operating in concert with PLA units on an active campaign. Security force commanders must be acutely aware of the general objectives and strategy of the military forces they are supporting, and they may be required to operate with little or no direct oversight in securing rear areas, supply and communications lines, and key assets. Security forces may also be required, in dire situations, to face off against much more powerful opponents in order to delay the enemy or buy time for the supported force to reposition or retreat. Security forces must be prepared to make this sacrifice either when called upon or when they recognize the situation requires it.

## Key-Point Concentration

Security forces operate with far less density than do regular army units. Contiguous deployments throughout the security zone are likely impossible, considering available forces and the number of different assets that must be defended. Commanders must carefully prioritize what they want to defend and from what type of threat. This enables the security force to properly allocate its subordinate forces to concentrate greater security on key points throughout the security zone. At the same time, security forces must be able to continue operations even when enemy forces are moving through the security zone, provided that they are moving through unimportant or indefensible areas.

## Defense Focus

Security forces will most likely be outnumbered when facing conventional opponents, and they may be outnumbered even by irregular opponents. This makes offensive actions difficult, as security forces will seldom be able to concentrate sufficient combat power to attack and destroy an opponent. Instead, they must focus on the defense, building their operations around steadfast protection of key points, enabled by extensive entrenchments and a defense-in-depth approach. The doctrine of active defense still applies; however, when a security unit can achieve local superiority, it must attack and spoil the enemy's preparations. Security units contribute significantly to the victory of the larger military force by delaying enemy forces, attritting enemy formations, and disaggregating enemy capabilities in the security zone, setting the enemy up for decisive defeat by a counterattack.

than checkpoints, but they create more hardship for the local population and friendly forces. Commanders must weigh the risk of threat to an asset with the needs of friendly forces and the local population when choosing which method to employ.

Security units should make extensive use of entrenchments, hardening, and fortifications to enhance the defensive effort. This is particularly true for static assets: if an asset cannot be moved, it is far easier for the enemy to target and destroy it. Hardening and fortifying these assets forces the enemy to commit more resources to their destruction while protecting security forces from enemy attack and multiplying the effects of their defensive efforts.

As enemy forces move into the security zone, security forces engage the intruding forces through a mixture of stubborn defenses, counterattacks, raids, and ambushes. If the enemy force is relatively small and weak, a decisive counterattack or ambush may seek to annihilate it or force its withdrawal. If the enemy force is powerful, security forces conduct stubborn retrograde operations, delaying the enemy's advance, inflicting casualties, and disrupting enemy cohesion in anticipation of a powerful counterattack from a main-body defensive group. Retrograde operations cease once the security group reaches a key asset. At this time, the security unit must bravely defend without further retreat, using all of its remaining combat power to resist the enemy's assault.

## C. Coastal Defense

Security-force responsibilities revolve primarily around neutralizing shore raids and harassments that target wide areas of shoreline. Most of these duties fall to the Chinese Coast Guard, a subsidiary of the PAP. Due to the vast distances of vulnerable coast in China, this necessarily requires security units to be widely dispersed, limiting their combat power. Security forces are not intended to resist major amphibious operations; this responsibility falls to the PLAA. Instead, they are to remain vigilant against small-scale enemy intrusions over the shore, mainly reconnaissance raids, sabotage raids, and insertion or extraction of enemy agents. They also take on numerous civil missions, including coastal patrol and policing, regulation and protection of trade, and interdiction of smuggling.

## D. Air Defense

Security-force responsibility for air defense is similar to coastal defense. Security forces have little to no active air defense capability, but they are instead responsible for responding to aerial insertions of ground forces into the security zone. China fears the use of helicopters to conduct air insertions of troops into rear areas, and it relies heavily on security forces to counter this threat. Security forces will be widely dispersed and cannot resist a major air assault operation, but they are expected to detect and neutralize smaller aerial insertions of commandos, raiders, or enemy agents. The best method is to ambush helicopters as they descend in altitude and consolidate. An alternative method is to target enemy forces while they are vulnerable immediately after landing. This requires security forces to detect and track enemy air elements as they move into the security zone, then rapidly mass combat power in and around the landing zone. Security units should make extensive use of observation posts integrated with radar and the wider air defense network.

## E. Defense of Social Stability

China places a high priority on security and stability at all times; this emphasis also extends to combat theaters. Security forces play an important role in maintaining social stability in rear areas. The policy focus is almost entirely on social stability in Chinese domestic territory, but this emphasis likely extends to any foreign territory that has been occupied, conquered, or annexed during the course of a conflict. There are three primary components to the defense of social stability mission: guard socially vital assets, maintain social order, and address the effects of enemy attacks.

# Defense of Social Stability (Components)

*Ref: ATP 7-100.3, Chinese Tactics, (Aug '21), pp. 9-12 to 9-13.*

## Guard Socially Vital Assets

A subcomponent of guarding key points throughout the security area, security forces have responsibility for guarding those assets that are vital to social stability. In a wartime theater, these assets include distribution stations for food, water, and fuel; civilian transportation nodes; civilian government facilities; and public utilities. Unlike more-aggressive security missions that may face enemy regular forces, security groups charged with guarding these assets should focus on reducing their visibility and either concealing the target or deceiving the enemy as their primary courses of action. The most likely enemy actions against targets of this type are irregular forces and recruited criminal elements operating in a secret or concealed fashion. As such, measures to guard these assets are as much police work as military security work, and integration between civilian police forces and security forces is vital.

## Maintain Social Order

The component of maintaining social order reflects the ongoing Chinese fear of counter-revolutionary activities that may leverage propaganda and politics to undermine the Chinese government and military, seeding discord through rear areas. This mission type necessarily involves interaction between security forces and the population at large as the former tries to locate and neutralize enemy or other elements attempting to sow discord, cause riots, harass military units, or sabotage infrastructure. Commanders must work together with civilian law enforcement units when conducting this mission, and they should carefully weigh the risk to friendly operations against the risk of alienating either friendly or occupied peaceful civilian populations through the use of excessive force.

## Address the Effects of Enemy Attacks

Enemy efforts in rear areas will sometimes succeed, whether they are from firepower attacks, ground attacks, or attacks by irregular forces. Security forces have an important role to play in dealing with the effects of these attacks. If the local population does not see national leaders reacting strongly and quickly to minimize the effects of a successful enemy attack, its morale will suffer. In particular, casualties suffered in rear areas must be quickly and efficiently cared for, either treated on site or evacuated to proper medical facilities. Destroyed buildings, roads, bridges, or other infrastructure must be first secured, then rehabilitated as quickly as possible not only for the strategic benefit, but also to show the local population the strength of the government. The primary role of security forces in this mission is care of casualties, followed by the securing of destroyed or damaged areas.

## Defense of Supply Routes and Traffic

The most important mission for security forces in the security area of an active combat theater is the management of military traffic and the defense of transportation and supply lines. The PLA rightly fears rear area attacks on its transportation infrastructure as a means to cutting off its powerful frontline army units. In areas not near the front line, the bulk of this mission falls to security units. There are three primary components to defending supply routes and traffic: route security, traffic management, and emergency repairs. Route security requires constant, active vigilance against a variety of enemy attack methods that might seek to destroy friendly convoys or bridges, tunnels, and other key transportation assets. Active defense is conducted along main routes to discover enemy activity and neutralize enemy efforts.

Traffic management involves the careful routing of military traffic to ensure that the most valuable convoys use the most secure routes, and that cargo is effectively prioritized in accordance with the needs of the theater. Civilian traffic must be carefully controlled to ensure that priority military traffic has the right of way.

## F. Security for the Main Force

The PLA envisions its security forces in an expanded role, providing direct security for the PLAA. This likely involves defense of the rear and flanks of a main body, along with defending critical assets and routes of march within the main body's combat area. It also includes management of the military region, including the wider geographic area occupied by the main body. Security forces integrate with the reconnaissance and security elements of the main body when conducting this mission, with armed police supporting military operations by securing rear areas.

This mission requires the security force to carefully integrate with the main body, keeping close track of its activities. If the main body advances, the security force provides rear area security and route security. If the main body retreats, the security force may be required to provide flank security or, in dire situations, conduct rear-guard actions against enemy forces. Main-force security operations also contain a counterreconnaissance element. Security forces must ensure that enemy scouts cannot infiltrate rear areas and conduct reconnaissance or targeting of the main body or key assets in the security area. This may involve direct action against enemy scouting elements, some of which may be powerful armored units. Armed police units have only minimal antitank capabilities; these must be enhanced if the security unit anticipates contact with enemy heavy forces.

## G. Deception Actions

The PLA also envisions its security units contributing to ground actions through deception, primarily through the use of feint actions. Security forces—particularly armed police—are well suited to missions of this type due to their high density of light, mobile armored vehicles and the ability to move quickly through a variety of different terrain types. Five different forms of feint are envisioned for security forces:

- Feint movement: a security force moves in such a way that it deceives the enemy into thinking a larger army movement is occurring.
- Feint assembly: a security force simulates a more-powerful ground force assembling on the ground for an offensive action.
- Feint attack: a security force conducts an actual limited attack on an enemy position.
- Feint defense: a security force conducts a limited or simulated defense against an attacking enemy.
- Feint withdrawal: a security force pretends to be a larger force withdrawing from the battlefield.

All of these measures are supported with false intelligence, including fake radio traffic. Their general intent is to fool enemy reconnaissance and surveillance assets into thinking that the security unit is actually a mechanized or armored friendly unit and to manipulate them into acting a certain way due to this misbelief. This, in turn, puts actual main body forces on the offensive, knowing that the enemy has been fooled and is pursuing an action against a deception unit. Commanders must remember, however, that security forces are not equipped to operate in close quarters with enemy heavy forces, and they should strive to protect their assigned security forces during deception actions.

# Chap 9 Equipment & Capabilities

*Ref: ATP 7-100.3, Chinese Tactics, (Aug '21).*

The PRC is a near-peer competitor to the United States, with significant political and strategic interests worldwide, though the PRC's primary geographic focus is Eastern and Southeastern Asia and the Western Pacific. The PRC has developed significant capabilities in all domains: land, air, maritime, space, and cyberspace. Capabilities in specific operational environments include a number of different unit and system combinations able to challenge the combat power of U.S. and allied forces. The PLA represents a significant competitor to U.S. and allied forces, especially in the Western Pacific.

## PLA Capabilities Overview



Appendix A describes the capabilities and limitations of PLA maneuver forces: infantry, armor, marine, and airborne.

Appendix B describes PLA indirect fire capabilities, including mortars, tube artillery, and rocket artillery in support of group army and tactical operations; this includes an overview of integrated fires systems.

Appendix C describes PLA tactical air defense forces controlled by its land forces, including an overview of integrated air defense systems up to the group army echelon.

Appendix D describes how the PLA aviation assets—primarily rotary-wing manned aircraft and unmanned aircraft—support tactical operations.

Appendix E describes the use of engineers and chemical defense forces in support of tactical operations.

Appendix F describes the PLA's communications and network architecture at the tactical level.

Appendix G describes special operations forces both at tactical and national/strategic echelons.

# Worldwide Equipment Guide (WEG)

The Worldwide Equipment Guide (WEG) was developed to support the TC 7- 100 series and all OPFOR portrayal in training simulations (live, virtual, constructive, and gaming). The equipment portrayed in the WEG represents military systems, variants, and upgrades that US forces may encounter now and in the foreseeable future. The authors continually analyze real world developments, capabilities, and trends to guarantee the OPFOR remains relevant.

Published in three volumes, (Ground; Airspace & Air Defense Systems; and Naval & Littoral Systems) the WEG is the approved document for OPFOR equipment data used in U.S. Army training. Annual updates are posted on the ATN website. Therefore it is available for downloading and local distribution. Distribution restriction is unlimited. This issue replaces all previous issues.

## Worldwide Equipment Guide (Dec 2016) Volume 1: Ground Systems

OPFOR equipment is broken into four "tiers" in order to portray systems for adversaries with differing levels of force capabilities for use as representative examples of a rational force developer's systems mix. Equipment is listed in convenient tier tables for use as a tool for trainers to reflect different levels of modernity. Each tier provides an equivalent level of capability for systems across different functional areas.

## Worldwide Equipment Guide (Dec 2016) Volume 2: Air and Air Defense Systems

Chap 1 provides the basic characteristics of selected rotary-wing aircraft readily available to the OPFOR. Because of the increasingly large numbers of variants of each aircraft, only the most common variants produced in significant numbers were addressed. Chap 2 provides the basic characteristics of selected fixed-wing aircraft readily available to COE OPFOR across the spectrum of joint operations. Chap 6 covers theater missiles and chap 7 covers air defense systems.

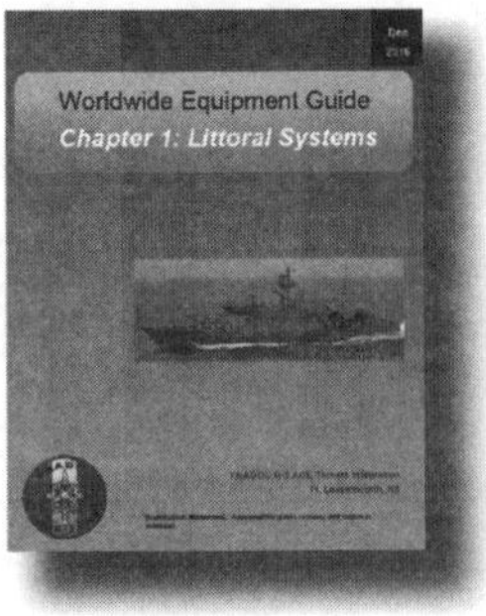


## Worldwide Equipment Guide (Dec 2016) Volume 3: Naval Systems

Naval surface vessels vary widely, from aircraft carriers displacing tens of thousands of metric tons to meters-long rubber inflatable boats for littoral use. Chap 1 focuses on vessels for use in littoral ("near the shore") operations. Chap 2 covers aircraft carriers and surface combatants. These vessels are designed to conduct overt attacks against naval, air, and/or ground targets or to perform patrol functions. Chap 3 focuses primarily on undersea vessels capable of blue-water operations, and is currently limited to submarines.

# Chap 9 A. Maneuver Capabilities

*Ref: ATP 7-100.3, Chinese Tactics, (Aug '21), app. A.*

People's Liberation Army (PLA) maneuver forces include the People's Liberation Army Army's (PLAA's) combined arms brigades (CA-BDEs), special operations forces (SOF), army aviation brigades, and the handful of remaining infantry divisions, along with the airborne brigades of the People's Liberation Army Air Force (PLAAF) and the marine brigades of the people's Liberation Army Navy (PLAN). The paramilitary formations of the People's Liberation Army Reserve infantry units, People's Armed Police (PAP), and China Militia units may be used for rear area or flank security or economy of force and deception (feint) missions. While the entirety of the PLA is undergoing substantial reform and modernization, significant but decreasing numbers of older systems are expected to be fielded throughout frontline formations through 2035—the scheduled date for PLA equipment modernization to be complete. Unlike Western armies that tend to field very homogenous formations from an equipment perspective, the PLAA employs a very wide variety of vehicles with varying capabilities and readiness levels. CA-BDEs with more-critical missions or in more strategically important areas receive newer and better equipment, with older equipment gradually trickling down to less-important commands. The PAP employs a wide variety of equipment types, generally tending toward light, motorized forces along with inexpensive mortars, recoilless rifles, armored personnel carriers (APCs), and limited numbers of helicopters.

The core of the PLAA's maneuver component consists of around 73 CA-BDEs of three different types: light (motorized), medium (mechanized), and heavy (armored), along with a handful of legacy division structures. In addition, there are approximately 18 divisions in the reserves and militia, though their equipment and readiness levels are much lower than those of the active-duty force. All or nearly all of the active component is either motorized or mechanized. CA-BDEs regularly conduct large-scale training maneuvers against a full-time opposing force brigade. The PLAA brigade structure is described in detail in chapter 2.

## I. Maneuver Capabilities Equipment Overview

The PLAA employs a wide variety of tanks, infantry fighting vehicles (IFVs), and APCs, ranging from high-end modern systems to Cold-War relics. Historically, the PLAA has been reticent to throw anything away. As new equipment arrived, older equipment was handed down to lower-readiness units, then to reserves, then militia, and finally to massive storage facilities. The recent set of PLAA reforms, however, moves away from this model, seeking greater homogenization through a force equipped entirely with modern equipment. This necessarily reduces the overall size of the army, and it promises significant cost savings in the maintenance and operation of obsolete equipment. At the present time, however, the PLAA still employs a variety of armored vehicles, with most CA-BDEs still equipped with older or obsolescent vehicles. Due to the vast differences in capability between in-use systems, it is critical that PLAA units' equipment be carefully assessed on an individual basis, rather than generically.

*Note. The equipment overview provided in this appendix is not intended to be exhaustive, but rather to introduce the most widely employed variants of major equipment types used by the PLA.*

## A. Tanks

The PLAA fields a massive number of tanks, most of which are main battle tanks (MBTs), with a handful of light and amphibious tanks. The MBT force currently has more than 6,500 vehicles—more than twice the number fielded by any other country. Many of these, however, are early Cold-War vintage and of dubious utility on the modern battlefield. Early Chinese tanks were either direct copies or heavily influenced by Soviet designs, but the more-modern tanks are entirely indigenous and show more design similarities with Western tanks than with Russian ones.

Despite its age, the Type 59 tank is still fielded in limited numbers in active units, and large numbers are maintained in reserve or mothball status. The Type 59 was derived from the ubiquitous Soviet T-54/55 tank and shares most of its design features. Its armor is homogenous steel, and its early 100-mm gun is inadequate against modern MBTs. The PLAA has, however, periodically upgraded its Type 59 fleet, adding modern thermal sights, a larger gun, improved gun stabilizers, explosive-reactive armor (ERA), and improved crew survivability systems. Modernized Type 59 tanks are lightweight, reasonably mobile, possess relatively modern fire control systems, and are simple to operate and maintain. However, their protection—even when upgraded with ERA—is poor by modern standards. Though the tanks employ a copy of the once-ubiquitous L7 105-mm main gun that can fire both tungsten and uranium penetrators, main gun performance is poor compared to modern larger guns.

The Type 80/Type 88 tank was the first Chinese indigenous tank design. The PLAA mixed Western and Soviet design philosophies in the Type 80, using a Soviet-style chassis and turret design with a Western gun and crew configuration. The result was an MBT that was competitive with the T-72, T-80, and early M1 Abrams designs. The first variants entered service in the early 1980s, and they have been continuously upgraded. Those currently in service are essentially comparable to the much-later Type 96. Upgrade versions feature advanced computerized fire control, a mixture of homogenous and applique composite armor, a 125 millimeter (mm) smoothbore gun capable of firing penetrators, high explosives (HE) and missiles, ERA, and a significantly upgraded engine. While an older design, upgraded Type 80/88 tanks are competitive with older Western MBTs, and they can threaten even the most modern designs.

The Type 96 tank is the most widespread modern MBT in the PLAA inventory. A fully indigenous design derived from the Type 80, the Type 96 was the first Chinese tank to feature full composite armor and a Western-style arrow-shaped turret, in contrast to the bowl-shaped turrets of the Type 80/88. The Type 96 is a relatively light modern MBT with excellent mobility, leading the PLAA to deploy most of its Type 96s in hilly regions and other areas of difficult terrain. Composite ERA and laser jammers have been added to the original Type 96, improving its survivability against antitank guided missiles (ATGMs). The Type 96 was originally only an iterative upgrade, and it was not truly competitive with either its Western or Russian counterparts when it was first deployed. It was built in large numbers, however, and upgraded variants—including Types 96A1 and 96B—have improved lethality and protection.

The Type 99 tank is China's newest MBT. It represents the PLAA's new commitment to modernization and investment in technology. Using the same design approach as the Type 96, the Type 99 employs a Russian-style chassis with a Western-style turret. It features a full suite of modern protective systems, including spaced composite armor enhanced with ERA; full chemical, biological, radiological, and nuclear protection; crew survivability upgrades; and an active protection system. Its 125-mm main gun can fire sabot, HE, and missiles, and its advanced suspension and powerful engine give it excellent cross-country mobility. An upgraded variant, the Type 99A, enhances the protection scheme with thicker armor and multi-layer ERA, a more powerful engine, and improved electronic warfare and network capabilities. Due to its high cost, it is not anticipated that the Type 99 will fully replace the Type 96, and the Type 99A will only see service in elite "digitized" units.

The Type 62 is an evolution of a Cold-War tank that has been modified and modernized into a light tank platform. Developed specifically to operate in the challenging mountains and jungles of southern China, the Type 62 was essentially a scaled-down Type 59, armed with an 85-mm gun and a lightweight armor package. At around 30 metric tons, the Type 62 filled the gap between heavy MBTs and very light airborne and amphibious tanks. Modernized variants feature upgraded armor and targeting packages, but they are badly undergunned and have virtually no offensive capability against modern MBTs. These tanks are still in limited use in specific geographic areas, but they are being phased out in favor of the Type 15 light tank.

The Type 15 is a brand-new light tank developed to take over the role filled by the Type 62. It features the ubiquitous 105-mm rifled gun and composite armor, along with an advanced remote weapons station and autoloader. The Type 15 is likely not intended to compete directly with modern MBTs, but rather to enhance the firepower of mountain and jungle units. It is also likely tailored for the export market—numerous lower-tier militaries around the world are seeking to replace their Soviet-era equipment, and a modern tank in the 30-metric-ton class meets many of their requirements. The Type 15 was just recently accepted by the PLAA, and it is in only limited use by operational units.

The Type 63 is a widely proliferated amphibious light tank. Early Type 63 models were built primarily to provide PLAA forces with an armored platform capable of crossing rivers and other small bodies of water. These vehicles are still in limited service in lowland regions. The upgraded Type 63A was built specifically for the People's Liberation Army Navy Marine Corps (PLANMC), and it is capable of crossing open-ocean areas. Both types are capable of firing while embarked, though the Type 63A has a greatly enhanced on-the-move capability. Both tanks have only light homogenous armor, and they are not intended to withstand attacks from modern ATGMs or enemy tanks. While still in use by both the PLAA and PLAN, Type 63s are being phased out in favor of more modern amphibious tank systems.

The ZLT-05 amphibious tank is a hybrid IFV, tank, and assault-gun platform with impressive performance across even large bodies of water. A modern 105-mm gun employs sabot, HE, and ATGMs from land or water, using a modern fire control computer and extensive stabilization. While heavier than the Type 63, the ZLT-05 still features only light armor, and it is intended to protect only against fragmentation and light weapons. The ZLT-05 is intended to be the standard amphibious tank throughout the PLANMC and, as such, it is the backbone of the PLA's short-range amphibious capability.

## B. Infantry Fighting Vehicles

Despite being one of the early adopters of IFVs, the PLAA employs a lower density of them than do other large militaries, particularly as compared to Russia. There are numerous varieties and configurations of Chinese IFVs, and categorizing them can be complex: weight, configuration, and firepower are not conclusive, as they are with the U.S. military. Instead, one must look at the organization of the unit in question: a mechanized (medium) CA-BDE employs its armored vehicles as both IFVs and APCs, while motorized (light) CA-BDEs employ their vehicles almost exclusively as APCs. Chinese IFV designs have generally mirrored Soviet and Russian designs going back to the BMP-1 of the 1960s, although PLAA preferences now seem to be trending toward favoring wheeled as opposed to tracked vehicles. The PLAA actively fields roughly 4,000 IFVs of all types, plus a large number of older vehicles in reserve status.

The Type 86 IFV is a copy of the ubiquitous BMP-1, and it is still widely fielded by active PLAA formations. The original Type 86 featured a 100-mm gun, while the newer Type 86A was rebuilt with a 30-mm autocannon. Variations include troop transport (with an eight-soldier capacity), scouting, engineering, ATGM carrier, air defense, amphibious, and command vehicles. The Type 86 is tracked, with good cross- country mobility and considerable firepower for its weight. Its protection is poor, however, consisting only of thin homogenous steel and minimal survivability improvements.

The ZBD-04 IFV is a heavy tracked vehicle that strongly resembles the Russian BMP-3. It features a powerful engine, improved protection, and impressive firepower, though it weighs nearly twice as much as the Types 86 and 86A that it was built to replace. The ZBD-04 is broadly similar to the M2 Bradley in configuration, size, and capability. It employs both a high-velocity 100-mm main gun and a 30-mm chain gun, coupled with ATGMs and machine guns—a remarkable amount of firepower for an IFV platform. It can carry seven soldiers, and it is fully amphibious, though it cannot cross ocean waters. Its armor is modular—ranging from the homogenous steel hull to applique composite and ERA add-ons—giving it frontal protection from cannon fire in addition to all-around machine gun and fragmentation protection. Other variants of the ZBD- 04 include tank destroyer (replaces passengers with additional missiles), heavy tank destroyer (replaces normal HJ-8 ATGMs with heavy HJ-10 ATGMs), command, and reconnaissance vehicles. The reconnaissance variant is the heaviest and most-powerful scouting platform in the PLAA. An improved variant, the ZBD-04A, adds additional armor protection and improved fire control and network capability, though at the cost of additional weight.

The ZBD-03 IFV is the PLAAF's primary airborne armored platform. The ZBD-03 is very small compared to its contemporaries—it is only eight metric tons, as compared to 24-plus metric tons for the ZBD-04 and nearly 30 metric tons for the M2 Bradley. Despite its small size, it enjoys excellent firepower. A 30-mm autocannon and a lightweight ATGM launcher give the ZBD-03 punch similar to much-larger IFVs. Its protection, however, is very light, and it is likely only fully resistant to small arms and fragmentation. The IFV's mobility is outstanding, thanks to its light weight, high power-to-weight ratio, and advanced suspension. The ZBD-03 is both air mobile and air droppable, and it forms the primary armored component of the PLAAF's airborne corps. It is envisioned as the main offensive capability of the airborne corps, and it is intended to take on enemy rear area forces, garrison forces, and other soft targets as part of airborne operations. This IFV was not designed to confront heavy mechanized forces. Variants include a command vehicle and a tank destroyer variant with a larger ATGM launcher instead of a gun.

The ZBD-05 IFV is the PLANMC's amphibious fighting vehicle. It shares a chassis and machinery with the ZLT-05 light amphibious tank, the primary difference being that the IFV carries eight passengers and employs a 30-mm autocannon, rather than the 105-mm main gun. The ZBD-05 can cross open ocean with relative speed and safety, and it can engage targets while embarked due to advanced stabilization of the gun and missile launchers. The ZBD-05's protection is similar to that of the ZLT-05. It is necessarily light and only proof against small arms and fragmentation. The ZBD-05 is in wide service with the PLAA's amphibious brigades and the PLANMC; proliferation will likely increase significantly as the PLANMC expansion continues.

The ZBL-08 is an IFV built on the relatively new Type 08 8 x 8 wheeled vehicle chassis. The Type 08 chassis is broadly similar to the U.S. Stryker or the Russian BTR, though it is an indigenous Chinese design, and it is somewhat heavier than either of its contemporaries. The ZBL-08 variant carries seven passengers, and it employs a 30-mm autocannon plus ATGMs. It has good cross-country mobility, and it is notably fast on improved roads, making it a first choice for urban combat situations. Its armor is likely proof against heavy machine-gun fire and fragmentation, though it may have limited protection against light cannon from some directions. The IFV variants are employed in much the same way as their tracked counterparts—as firepower support for dismounted infantry. A reconnaissance variant is equipped with a sophisticated sensor suite including radar, thermal sensors, electro-optical sensors, and a laser designator, in addition to the gun and missile package of the IFV.

## C. Armored Personnel Carriers

As with IFVs, the PLAA employs a relatively low density of APCs as compared to other major militaries. APCs comprise the main mobility element of light CA-BDEs, providing light infantry with enhanced mobility and protection but contributing only limited offensive firepower. Most PLAA APCs are tracked, but a growing number are 8 x 8 wheeled vehicles. In addition to the PLAA's APC force, the PAP employs a significant number of APCs in support of its security, domestic stability, and antiterrorism roles.

The Type 63 APC is one of the first Chinese indigenous armored fighting vehicle (AFV) designs, and it is very similar to the U.S. M113 in size, capability, and role. Simple, lightweight, and amphibious, it has enjoyed a long and useful lifespan—just like the M113. It employs lightweight steel armor to defend against small arms and fragmentation, and it can carry up to 10 passengers. The Type 63's design has lent itself to a number of different variants, including command vehicle, ambulance, and mortar or artillery carrier. It has been periodically upgraded with more-powerful engines and better suspension, and it will continue in PLAA service for the foreseeable future.

The ZSD-89 APC was developed as an offshoot of an export product. The PLAA was impressed by its carrying capacity and mobility and ordered it to serve alongside the Type 63s already in service. The ZSD-89 is one of the world's largest APCs in terms of capacity, able to carry 13 passengers or a large number of ATGMs, artillery rounds, or mines in other configurations. It is amphibious and protected against fragmentation and heavy machine-gun rounds. Like the Type 63, the ZSD-89 has been developed into a large number of different variants, including mortar and artillery carriers, command vehicles, scouting vehicles, supply and recovery vehicles, tank destroyers, and IFVs.

The Type 08 is the APC variant of the ZBL-08 8 x 8 wheeled AFV. It can carry up to 10 passengers, and it enjoys a substantial advantage in both speed and range as compared to tracked vehicles. The Type 08 employs a number of advanced features, such as a remote weapons station and advanced suspension, and it is likely to be the PLAA's APC of choice as older vehicles are retired, particularly in the medium CA-BDE. The Type 08's chassis is large and well-suited to a variety of roles, including command, communication, artillery and mortar carrier, antiaircraft gun, reconnaissance, ambulance, electronic warfare, engineering support, and assault gun. The PAP also employs Type 08s in domestic security and antiterrorism roles. This is the largest and most-powerful vehicle employed by the PAP.

The ZSL-92 is a light APC that employs two different 6 x 6 wheeled chassis that are collectively referred to as the WZ-551. Despite a long and troubled development period, the ZSL-92 is widely proliferated in the PLAA, the PAP, and internationally. An IFV variant is equipped with a 25-mm autocannon, and it can carry nine passengers, while the APC variant employs a heavy machine gun, and it can carry up to 11 passengers. Due to its light armor, however, it is unlikely that the PLAA employs the ZSL-92 in an offensive role. The ZSL- 92's light weight and road mobility make it a natural choice for the PAP, and large numbers are employed in engineering, crowd control, and security roles.

## D. Tank Destroyers, Assault Guns, and Antitank Guided Missile Vehicles

The PLAA places a very high priority on antitank firepower in tactical formations. As a result, it extensively modifies virtually all of its AFV platforms to enhance their antitank, antipersonnel, and anti-obstacle firepower. The two most common approaches to these modifications are either to replace a machine gun or cannon with a tank gun (usually in the 105-mm class), or to add additional ATGM launchers and replace the crew compartment with additional ATGM storage. These systems provide enhanced firepower to light and medium CA-BDEs, typically operating as one company within a battalion.

Tank destroyers and assault guns are generally similar in capability and appearance. Nearly all employ an APC or IFV chassis, with the crew compartment occupied by the machinery and ammunition to support a 105-mm main gun and turret. Nearly all of the PLAA's 105-mm guns are derived from the widely proliferated British L7 gun of the 1960s. These guns still pose a threat to even modern armor because of upgrades to their muzzle velocity and ammunition. In addition to their mobile antitank role, these guns also have useful antipersonnel and anti-obstacle capabilities, and they are employed by battalion or group commanders whenever enhanced firepower is required. Among the most common assault-gun types are the ZTL-11 (a modified Type 08 APC) and the PTL-02, based on the ZSL-92 APC.

ATGM vehicles are widespread in PLAA formations of all types. Virtually every battalion-level organization employs at least a platoon of ATGM vehicles. For lighter units, this represents the bulk of their mobile antitank firepower. Due to their thin armor, ATGM vehicles are best employed in ambushes or standoff attacks where they cannot be targeted by enemy tanks or IFVs. They are almost exclusively a defensive weapon, though they can employ their missiles to target bunkers or other fortifications if required. Virtually every PLAA APC model has been modified to be an ATGM vehicle.

## E. Light Armored Vehicles and Wheeled Vehicles

Most of the PLAA's light tactical vehicles are in the Mengshi family, a group of vehicles derived from the U.S. H1 Hummer, which is the civilian version of the U.S. M998 high mobility multipurpose wheeled vehicle. Like the M998, the Mengshi family of vehicles has been periodically upgraded, starting primarily as soft-sided utility vehicles and now featuring extensive armor, mine-resistant chassis, and remote weapons stations. These vehicles fill a wide variety of roles, from simple transport, to ambulance, to command, and to maintenance, fulfilling practically every role available for a light wheeled tactical truck. The PLAA, carefully following coalition activities in Iraq and Afghanistan, also developed the Mengshi into capable patrol vehicles, scouting platforms, and local fire support in the form of the CSK series of vehicles. The latest Mengshi vehicles are significantly larger than their Hummer predecessors, strongly resembling U.S. joint light tactical vehicles in both size and capability. These vehicles are ubiquitous throughout the PLAA and the PAP, providing much of the motorization support for ground forces along with a growing protection and firepower suite.

The PLA employs a variety of heavy wheeled vehicles, varying widely in size and payload. The MV3 is the generic medium truck, usually operated in a 6 x 6 configuration. It is largely comparable to the U.S. M939 or light medium tactical vehicle, though it is a newer design than either. Numerous different models of larger trucks provide progressively greater payload. The largest varieties are up to 10 x 10 in configuration and serve as transporter-erector-launchers for intercontinental ballistic missiles (ICBMs) and long-range surface-to-air missile (SAM) systems. The density of cargo trucks—particularly the medium-lift cargo trucks that are the backbone of tactical logistics—is relatively low in the PLA, and it affects China's ability to conduct expeditionary operations or operations a significant distance from its internal logistics structure.

## F. Small Arms and Direct Fire Weapons

China's military independence from the Soviet Union can be tracked with the growth of its small arms industry. While the early PLA relied entirely upon Soviet and Soviet-derived weapons, years of investment built the Chinese small-arms industry into one of the world's largest and most sophisticated. This development was paralleled by the advancement of the PLAA's service rifle: over the last 50 years it transitioned from a simple copy of the Soviet SKS, to the development of variants based on the AK 47, AKM, and AK-74, to completely indigenous contemporary designs.

A-30. The QBZ-95 is the PLAA's primary service rifle today. A bullpup design, the QBZ-95 is optimized for ease of use and volume of fire. It is well regarded as a simple, sturdy, and comfortable weapon. Its 5.8-mm x 42-mm round offers more power and better ballistic performance than the standard 5.56-mm North Atlantic Treaty Organization (NATO) round, though at slightly increased weight and recoil. It is less accurate than most of its Western competitors, and it lacks the modular upgrade capabilities—such as rail systems— enjoyed by most modern assault rifles. The next generation of the QBZ-95 is thought to include a flat-top rail system instead of simple iron sights. The QBZ-95 has been modified into grenade launcher, carbine, and light support-weapon variants. Older service rifles—mostly AKM and AK-74 variants—are still in wide use across militia and reserve formations.

Like most Western Armies, PLAA formations employ three classes of machine gun: squad automatic weapons, primarily the QBB-95 light machine-gun variant of the QBZ-95; general purpose machine guns— a mixture of older Soviet-derived weapons and newer Western-style general-purpose machine guns; and heavy machine guns in the 12.7-mm caliber class, including a direct copy of the U.S. M2HB. In addition, 14.5-mm heavy machine guns are employed as antiaircraft weapons throughout tactical formations. General- purpose machine guns are operated by machine-gun teams and mounted on vehicles of all types; heavy machine guns are typically mounted on armored vehicles. Heavy antiaircraft machine guns are fielded in each combined arms battalion (CA-BN) heavy-weapon company. Generally speaking, PLAA machine guns can be considered comparable—and in many cases practically identical—to their Western equivalents.

## G. Man-Portable Missiles and Rockets

The PLAA employs a very high density of man-portable antitank weapons throughout its formations, echoing a general emphasis on antitank capability that is visible throughout the organization. The PLAA's historical approach to man-portable anti-tank warfare was simple mass, equipping formations of all types with huge numbers of unguided rockets—mostly based on the ubiquitous Soviet RPG-7—and simple ATGMs, predominantly the HJ-73, a copy of the Soviet AT-3 Sagger. These systems are still in use and still pose a threat to many armored vehicles due to periodic upgrades and the density of their deployment. The PLAA, however, is beginning to take a quality-over-quantity approach with regard to its ATGMs, seen most easily in the development and fielding of the HJ-12. Developed specifically to compete with the U.S. FGM-148 Javelin, the HJ-12 employs most of the latest ATGM developments, including a top-down attack profile, tandem warheads, soft launch, and a fire-and-forget targeting capability. These systems are expensive and complex, however, and their density is limited by their cost.

The PLAA still employs a high density of unguided man-portable rockets throughout its formations, particularly in light infantry units. These weapons employ simple HE, antitank, fuel and air explosive, incendiary, and demolition warheads, filling a variety of different roles and providing significant enhancement to the firepower of the infantry squad. The PF-89 is the most widely proliferated shoulder-fired antitank rocket system, and it is very similar to the NATO AT-4 in both size and capability. The FHJ-84 is a lightweight twin- rocket system that fires an incendiary warhead for anti-personnel or demolition use, and the PF-97 is a heavy fuel and air explosive rocket system. The PF-98 does not have a Western equivalent. It is a powerful 120-mm antitank and antipersonnel rocket system, usually fired from a bipod. This weapon gives the light infantry squad a lethal capability against a modern MBT, though its weight, size, and cost limit its proliferation.

# II. Maneuver Capabilities and Limitations

In just over 20 years, the PLAA's maneuver capabilities evolved from a force mixture of light infantry conscripts and obsolete tanks to a nearly fully motorized and mechanized force that employs a variety of cutting-edge armored vehicles and advanced guided munitions. This evolution is incomplete, however. A significant number of CA-BDEs are still equipped with older systems, and they are likely to remain so for the foreseeable future. So too has the movement towards professionalization within the maneuver force produced mixed results. While the quality of both recruits and non-commissioned officers (NCOs) has increased significantly, the PLAA still struggles with decentralizing planning and operations at lower tactical echelons. The maneuver force has also struggled with the complexities of the informationized battlefield and modern mechanized forces.

The most significant capability enhancement in the PLA is the development of a viable expeditionary capability. Previous PLA forces planned only to operate in Chinese territory, in neighboring territories, and in Chinese territorial waters. In order to establish itself as a great power, however, China believes that a powerful ground expeditionary capability is required. This is the primary driver behind the expansion of the PLANMC. It is likely the most significant effort in the PLA to meet the new expeditionary requirements. Despite these efforts, the PLA still cannot deploy or sustain a heavy mechanized force outside of shared land borders and its territorial waters, limiting the combat power of an expeditionary force.

PLAA leaders at tactical echelons are technically and tactically competent, with fairly regular training in both staff and command functions. The officer corps is professionalized and well educated, with many officers completing both advanced degrees and advanced military educations. Many PLAA officers are commissioned enlisted personnel. The PLAA relies very heavily on its officer corps for leadership functions. A professionalized NCO corps is a relatively new development.

Despite these reforms and a greater emphasis on independence and initiative at lower tactical echelons, PLAA formations still struggle with independent action. Small unit leaders are afraid to make decisions due to a zero-tolerance culture, and officers remain hesitant to empower subordinates lest they underperform and thus reflect poorly on the officer. Except in elite units such as SOF or reconnaissance, the battalion is likely to be the lowest echelon capable of operating independently for any significant period of time, and even the battalion's independence is limited by a small staff and a low density of support assets. Significant resources are being devoted to mitigating these gaps, however, such as the establishment of training facilities that mimic U.S. Army Combat Training Centers and the development of a professionalized, dedicated opposing force.

The Chinese soldier is tough, physically fit, and deeply loyal to the country and the Communist Party of China (CPC). China's growing urbanization and improving social infrastructure are increasing the health, education, and general quality of its conscripts across the board. Training quality, however, varies wildly. Some units receive substantial training resources, such as maneuver exercises against a professional opposing force, while others may not even receive adequate ammunition for regular marksmanship practice. It is challenging for outsiders to effectively assess a unit's training readiness level due to the wide variations between brigades, even within the same group armies. Readiness levels vary even more among reserve and militia units. Some maintain regular training schedules and musters, while others exist practically in name only. Reservists are nearly all taken from former active-duty soldiers and, as such, enjoy a baseline level of training, but a reserve or militia unit, if activated, will take an extended period to reach a useful training readiness level. Conversely, the elite elements of the PLA—SOF, prestigious CA-BDEs, airborne, and marine forces, among others—enjoy substantial training resources and are likely comparable in competence to well-trained Western formations.

# Chap 9 B. Fire Support Capabilities

*Ref: ATP 7-100.3, Chinese Tactics, (Aug '21), app. B.*

No other capability area within the People's Liberation Army Army (PLAA) has received greater emphasis during the recent period of reform than fire support. In just over two decades, PLAA fire support evolved from a collection of aging Soviet-derived equipment to a largely indigenous, sophisticated, widely varied, and numerous collection of gun and rocket systems. The PLAA employs a mixture of mortars, howitzers, and rockets as the backbone of its fire support, including a notably high density of modernized rocket systems capable of employing precision and near-precision guided munitions.

The PLAA's emphasis on firepower manifests most clearly in its investment in modernizing its fire support capabilities. Experiences and observation led PLAA leaders to conclude that their maneuver units faced significant challenges when matching up against their approximate equivalents in the U.S. and Russian armies, so the decision was made to pursue enhanced firepower as a means of offsetting that capability gap. Joint fire support—from fixed-wing or rotary-wing aircraft, or from ships or submarines—is much less well developed. Close air support is virtually nonexistent as a capability, and the PLAA employs a very low density of attack helicopters, though this number is set to increase. Naval firepower is similarly largely absent from the People's Liberation Army's (PLA's) portfolio, though amphibious forces can provide much of their own firepower support through amphibious tanks, assault guns, and mortars.

The PLAA fields ground-based fire support in very high densities throughout its formations. Battalions of all types operate organic medium mortars, while combined arms brigades (CA-BDEs) employ heavy mortars, howitzers, and rocket artillery (multiple rocket launchers—MRLs) in their organic fire support battalions. The group army's artillery brigade also features a mixture of guns and rockets, adding in a heavy rocket battalion. These units may be retained by the group army to mass fires on a critical target, or they may be task-organized down to the CA-BDE in order to enhance the CA-BDE's firepower. The CA-BDE's organic firepower is roughly equivalent to that of a U.S. brigade combat team (BCT): 27 guns, along with a light rocket battery. Most of the CA-BDE's guns are 122 millimeter (mm) versus the predominantly 155-mm guns of the BCTs, but the BCT lacks an organic rocket capability. The CA-BDE possesses a higher density of guns and better mobility of firepower systems, while the BCT enjoys the greater range and destructive power of larger, higher-caliber gun systems. The density and destructive firepower of reinforcing fires is strongly in the PLAA's favor. The artillery brigade includes up to 72 guns, plus up to 40 MRLs of various caliber. These systems give the group army a significant advantage in both range and firepower over equivalent Western formations.

Fire direction, targeting, and forward observation have also been major areas of investment for the PLAA. China is the world leader in unmanned aircraft system production and development, and the PLAA was an early adopter of unmanned aircraft as intelligence, surveillance, and reconnaissance platforms in support of its artillery—particularly its long-range rocket artillery. CA-BDE artillery battalions include fire-finding radars, battlefield surveillance radars, long-range electro-optical and infrared sensors, and sound- ranging equipment, in addition to well-equipped forward observers, both mounted and dismounted. The PLAA devotes significant training resources to the battlefield surveillance and targeting mission. Observers of all types are likely well trained, and integration between sensors and shooters is likely agile, redundant, and reliable.

# I. Fire Support Equipment Overview

The PLAA employs a huge variety of guns, mortars, and rockets. China manufactures a variety of these systems for export. This makes assessing PLAA equipment challenging. Much like with armored vehicles, units of the same type may possess systems with very different capabilities. One notable trend is how the PLAA prioritized the modernization of its artillery systems. Self-propelled guns (SPGs) and MRLs received enormous investments and have largely completely modernized systems, while many towed and mortar systems can be dated back to at least the 1960s.

*Note. The equipment overview provided in this appendix is not intended to be exhaustive, but rather to introduce the most widely employed variants of major equipment types used by the PLA.*

## A. Mortars

The Type 87 mortar is the standard light mortar employed throughout the PLAA, reserves, and militia units. A simple design similar to U.S. and Soviet or Russian light mortars, it fires a variety of ammunition, including high explosives (HE), smoke, and illumination, out to roughly 4,000 meters. These mortars are the basic indirect firepower of the light combined arms battalion (CA-BN), with each battalion employing a battery's worth of mortars (12 to 16 tubes) in the firepower company. The Type 87 is man-portable, though it is rather heavy by 81 or 82-mm standards.

The W-99 mortar looks like a hybrid between a mortar and a light howitzer. A towed system, it is referred to by the PLAA as a rapid-fire mortar, and it is capable of an impressive rate of fire, far surpassing that of man-packed 81 or 82-mm mortar types. It does, however, require significantly more support, including a prime mover and more ammunition. It is not man-portable, but it represents a significant increase in the firepower available to the infantry battalion. This capability is not replicated in U.S. formations.

## B. Towed Howitzers

The Type 54 howitzer is a simple 122-mm pack gun derived from the World War II-era Soviet M-30. Despite its advanced age, the M-30/Type 54 is still widely fielded and is almost certainly the most numerous light howitzer design in the world today. Chinese designs were periodically upgraded over the years, becoming lighter, increasing reliability and rate of fire, and simplifying march-order and emplacement procedures. While old, the Type 54's performance is roughly in line with modern 105-mm guns, though it is somewhat heavier than most contemporary light howitzers. It provides the bulk of the artillery firepower for the light CA-BDE, and it still equips many of the towed howitzer battalions of the artillery brigades. The most common evolved versions of the Type 54 are the Type 83 and Type 86, both redevelopments from the 1980s that feature superior chassis technology and increased rates of fire. The Type 54 family is a versatile system. It can provide both indirect or direct fire, and PLAA forces train to employ these systems as direct-fire antitank guns in addition to their role as artillery.

The Type 66 howitzer is another Soviet design, based around the D-20 152-mm howitzer. This system dates from the 1960s, and it still comprises most of the PLAA's heavy tube artillery capability. The Type 66 is a large and heavy gun with middling-to-poor performance by modern standards, but when employed competently and in mass, it can still deliver decisive firepower. The system is simple and robust, able to operate in austere environments with minimal maintenance support. Its rate of fire is similar to that of more modern guns, but its range and accuracy are significantly worse when firing conventional ammunition. Rocket-assisted ammunition and precision-guided rounds offset these limitations somewhat, but their availability is limited. The Type 66 equips the heavy towed batteries of PLAA artillery brigades, and it is used to provide heavy fire support for maneuver operations.

## C. Self-Propelled Howitzers

The PLZ-89 is the PLAA's most widely used SPG. It mates the gun from the ubiquitous Type 54/83/86 122-mm towed howitzers with the chassis of the Type 63 light tank, creating a simple, lightweight, mobile self-propelled artillery system. The performance and protection of the PLZ-89 are both unremarkable, but the system is robust and easy to employ. The PLAA employs around 700 PLZ-89s throughout its formations; nearly all are in the artillery battalions of medium or heavy CA-BDEs. These systems provide close-range fire support for CA-BDE maneuver actions, typically rolling just behind the maneuver battalions. Though outranged and outgunned by more-modern 155-mm SPGs, the PLZ-89's mobility and simplicity make it indispensable to the PLAA. The SH-3 is a recent modernized variant, offering improved fire control and mobility.

The PLZ-07 is the PLAA's newest 122-mm SPG. It employs a modernized version of the PLZ-89's gun, itself derived from the obsolete Soviet D-30, though its performance is improved with modernized fire control and advanced ammunition. The primary advance of the PLZ-07 is the use of the modern ZBD-04 armored personnel carrier (APC) as its chassis, giving it improved cross-country and amphibious performance, greater range, and improved reliability versus older SPGs. The PLZ-07 will likely be produced in large numbers, eventually replacing all of the older PLZ-89 subtypes throughout the PLAA. Like the older system, the PLZ-07 is outgunned by 155-mm systems, but its mobility and simplicity make it very well adapted for closely supporting maneuver formations.

The PLZ-83 was the PLAA's first heavy tracked SPG, mating the gun from the Type 66 howitzer with a generic Soviet heavy armored fighting vehicle (AFV) chassis. Though the 152-mm gun provided enhanced firepower compared to the widely fielded 122-mm models, the PLAA was dissatisfied with the PLZ-83's mobility and reliability. The ballistic performance of the gun was also poor. Relatively small numbers were fielded, with the PLAA initially equipping division artillery units with them, and today equipping individual batteries within artillery brigades with them. These systems are still in service, though they are set to be replaced by more-modern systems.

The PLZ-05 is the PLAA's new heavy SPG, mating a new 155-mm gun with a specially developed light AFV chassis. The PLZ-05 can employ a wide variety of munitions and enjoys the impressive range and accuracy of a fully modernized 155-mm gun system. In addition, the PLZ-05 employs an autoloader, increasing its sustained rate of fire to 10 rounds per minute, with a 3-rounds-in-15-seconds burst capability. The PLZ-05 is comparable to other advanced SPGs in most respects, and it represents a major increase in indirect-fire capability for the PLAA. It is likely that all artillery brigades will eventually employ a battalion's worth of PLZ-05s as the older PLZ-83s are phased out. The PLZ-05 is a very large and heavy system with substantial support requirements, but its cross-country mobility remains excellent.

## D. Multiple Rocket Launchers

The PHL-81 is the PLAA's most widely distributed 122-mm MRL. A simple truck-mounted design derived from the Soviet BM-21, the PHL-81 provides both CA-BDE artillery battalions and artillery brigades with long-range destructive firepower. Though the PHL-81 is derived from a 1960s-era design, it has been periodically upgraded with digital fire control and a wide variety of munitions, including HE, antitank submunitions, fuel-air, smoke, and mine scattering. More-specialized types of ammunition, including radio-frequency jamming, incendiary, and chemical, have been produced in the past or in other nations, and they may be available to PLAA forces.

Precision-guided 122-mm rounds may also be available, though their cost and complexity limit their proliferation. Nominal range for the modernized 122-mm rounds is around 50 kilometers (kms). The PHZ-89 is essentially a tracked version of the PHL-81, developed to operate in support of PLAA mechanized and armored forces. It fires the same suite of 122-mm rocket rounds while improving protection and mobility at the expense of weight and cost. These systems feature a rapid-reload capability, enabling a second rocket pod to be fired within minutes and essentially doubling a battery's firepower against a time-sensitive target. Both the PHL-81 and the PHZ-89 are set to be replaced with domestically produced systems in the relatively near future. The PHL-11 is the new wheeled variant, and the PHZ-11 as the tracked.

The PHL-03 is the PLAA's version of the Soviet or Russian BM-30 Smerch, a widely proliferated heavy (300-mm) MRL. PHL-03 batteries provide long-range striking power for the PLAA's artillery brigades. Advancements in the 300-mm class of rockets make these among the most powerful and versatile artillery pieces fielded today. The PLAA operates a wide variety of 300-mm rocket ammunition types, including HE, antitank submunitions, fuel-air, incendiary, radio-frequency jamming, mine scattering, near- precision, and precision. New long-range variants push engagement ranges out more than 150 kms while maintaining reasonable accuracy. Reload of the PHL-03 is relatively slow. Each tube must be reloaded separately, rather than exchanging a pod. The PLAA likely views its 300-mm rocket class as the backbone of its operational fires capability of the future, bridging the gap between its gun and light rocket artillery and its ballistic missile force, and it is the primary offset for the lack of a meaningful close air support capability.

# II. Fire Support Capabilities and Limitations

The PLAA relies heavily on fire support to offset capability shortcomings in its maneuver forces and air power. Artillery of all types—mortars, tubes, and rockets—is deployed in significant density across every type of PLAA formation. Fire support is integrated effectively with maneuver operations. In particular, SPGs train to maneuver closely with armored and mechanized forces. Targeting can be decentralized, with lower echelon commanders integrating artillery spotters, unmanned aircraft (UA), and other surveillance assets. Massed fires are still considered the primary objective of artillery forces, and artillery commanders will generally seek to mass fires on the most valuable targets within range of their systems.

PLAA towed systems are much older and heavier than their Western counterparts, but they still have adequate tactical mobility to accomplish their fire support tasks. The PLAA relies on density and mass to accomplish these tasks, and logistics support for artillery is likely one of its primary limitations. The PLAA has a lower density of logistics support than do Western militaries—particularly the U.S. Army—and so sustaining ammunition requirements for operational artillery units may prove a significant challenge. PLAA SPGs are considerably lighter than their Western counterparts, and they are both outranged and outgunned, but they have fantastic cross-country mobility and easier sustainment requirements. The ability of these lightweight SPGs to maneuver closely with armored units is a key element of PLAA tactical operations.

PLAA rocket artillery is versatile and effective. It mixes a wide variety of different effects through the use of different munitions, integrated on simple, robust, mobile chassis. Lighter systems lack the destructive power of heavier Western MRLs, but their simplicity and versatility make them indispensable. Heavy MRLs comprise a huge part of the artillery brigade's firepower. These systems are carefully integrated with forward observers and other surveillance assets to make the most of their destructive potential.

While the PLAA fields a variety of precision and near-precision munitions for all of its fire support systems, the cost and complexity of these munitions limit their usage to the highest-priority missions. Precision munitions use a variety of guidance methods,

# E. Ballistic Missiles

*Ref: ATP 7-100.3, Chinese Tactics, (Aug '21), pp. 9-2 to 9-3.*

The DF-11 is the People's Liberation Army Rocket Force's (PLARF's) most widely employed short-range ballistic missile (SRBM). Originally designed to provide operational-level area fires out to 300 kms to theater commanders, subsequent versions have expanded the range out past 700 kms. Accuracy has also increased, reducing circular error probable to only 30 meters, giving theater commanders a long-range precision strike capability. The DF-11 can employ both conventional and nuclear warheads; conventional warheads may be either HE submunition or fuel-air explosive. The PLARF is capable of structured attacks with its DF-11 fleet, and the solid-fuel rocket and mobile transporter-erector-launchers enable rapid launch and reload operations.

The DF-15/16 is a newer family of systems that provides the bulk of the PLARF's operational-level fire support capability. These systems range between 600 kms and 1,000 kms, and they employ advanced antiballistic missile countermeasures such as terminal maneuvers and decoys. Early variants were not accurate enough for precision strikes, but modernized variants enjoy a circular error probable of 30 meters or less. These missiles can employ nuclear or conventional warheads, and have a significantly larger payload (over 600 kilograms) than do most SRBMs.

The B-611 is a new SRBM system that likely seeks to replicate the capabilities of the Soviet SS-26 Iskander. Its capabilities include a 500-km range, advanced penetration aids, good mobility, rapid reload, and precision targeting. The B-611 is likely intended to complement the 300-mm rocket class with an enhanced penetration weapon, enabling the targeting and destruction of antiballistic missile systems prior to more general targeting.

*Note. This section only discusses PLARF ballistic missiles that may be employed in support of tactical formations; it does not include long-range or intercontinental ballistic missiles (ICBMs).*

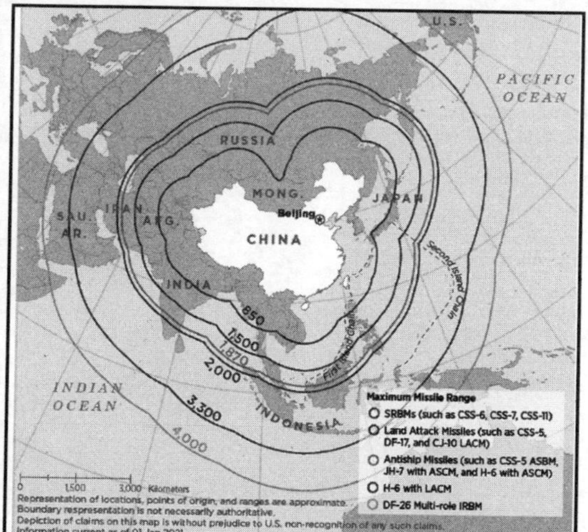


*Conventional strike ranges (Annual Report to Congress: Military and Security Developments Involving the People's Republic of China 2021, p. 62.)*

*See pp. 2-54 to 2-55 for more complete discussion of PLA Rocket Force (PLARF) missile capabilities, to include long-range and intercontinental ballistic missiles (ICBMs).*

including laser targeting, inertial, satellite, imagery, and radar. High-end ballistic missiles employ multiple methods to enhance accuracy at extreme ranges. The PLAA process for mensurating a target is not well known, but is likely less precise than U.S. processes. The decision-making process supporting targeting, however, is likely well-rehearsed and meticulous—in part due to reactions from the bombing of the Chinese embassy in Belgrade by American aircraft in 1999.

Like their Russian counterparts, PLAA fire support elements are very concerned with the counterfire threat, and significant resources are given over to counterfire operations. PLAA artillery units generally have good tactical mobility and practice the simple "shoot-and-scoot" methodology to avoid counterfire, while carefully concealing their positions from aerial attack and observation. The PLAA employs a variety of counterfire systems including sounding, radars, and visual means, and counterfire missions are highly prioritized.

Chap 9

# C. Air Defense Capabilities

*Ref: ATP 7-100.3, Chinese Tactics, (Aug '21), app. C.*

Ground-based air defenses are ubiquitous throughout the People's Liberation Army Army (PLAA), attempting to offset shortfalls in joint airpower. Experiences from Allied air power during the Korean War— and careful observation of the effects of joint airpower during Operation Desert Storm—profoundly influenced the People's Liberation Army's (PLA's) belief in air defense as a critical battlefield capability. Every maneuver formation larger than a company fields an organic air defense capability, and larger organizations employ a mixture of air defense systems to create a combined arms effect. Mass and depth are the operative words for PLAA air defense. Most of its short-range systems are relatively primitive and lack capability against modern air threats. They are, however, deployed very densely and broadly across the battlefield, creating a deterrent effect against low-altitude air threats. Medium- and long-range systems are significantly more lethal, capable of standing off against advanced air threats at any altitude.

Man-portable air defense systems (MANPADS) and heavy machine guns are fielded in the battalion's firepower company, and a mixture of self-propelled antiaircraft guns (SPAAGs), towed guns, and short-range infrared surface-to-air missile (SAM) systems, some medium-range radar systems, and a sensor and command component comprise the brigade's air defense battalion. Group army air defense brigades mix gun systems, short-range infrared missile systems, and medium-range radar missile systems. Air defense brigades also add an electronic air defense battalion that employs a variety of electronic warfare systems focused on defeating or neutralizing the enemy air threat. PLAA systems may or may not be fully digitally integrated into a wider integrated air defense system (IADS). Army units have the capability for digital integration all the way down to individual gun or missile teams, but it is not clear if this capability is fielded. Procedural controls such as airspace control measures and weapon control statuses are likely the dominant methods of fire control for most PLAA air defense units, excepting the medium-range SAM units.

Longer-range SAM systems are operated by the People's Liberation Army Air Force (PLAAF) as part of a nationwide IADS. PLAAF long-range SAM battalions are fully digitally integrated and likely centrally controlled by theater commands in cooperation with fixed-wing defensive counterair operations. Medium- range SAM units are also operated by the PLAAF, used either to defend long-range systems from oblique or pop-up attacks or to bridge gaps in SAM coverage. The PLAAF envisions its long-range SAM systems as wide area denial weapons, able to deny the use of airspace to enemy aircraft and create air superiority, or at least air parity, from the ground. These systems employ advanced counter-precision-guided munitions and electronic counter-countermeasures along with a variety of interceptor types and advanced, networked sensors.

## I. Air Defense Equipment Overview

PLAA air defense capabilities are primarily comprised of guns—towed, self-propelled, and hybrid— and missile systems that range from man portable to long range.

*Note. The equipment overview provided in this appendix is not intended to be exhaustive, but rather to introduce the most widely employed variants of major equipment types used by the PLA.*

## A. Towed Guns

The PG-87 is the most widespread towed antiaircraft gun in the PLAA's inventory, with several thousand units equipping light combined arms brigades (CA-BDEs), reserve units, and militia units. The PG- 87 is an up-gunned variant of the widely proliferated Soviet ZU-23, first fielded in 1960. PLAA variants employ two barrels firing 25-millimeter (mm) ammunition—in contrast to the Soviet 23 mm— giving Chinese weapons increased range and lethality. The PG-87 was further upgraded with enhanced targeting (including an infrared sight and fire control radar), improved training and direction mechanisms, and larger ammunition magazines. The PG-87 can fire a variety of ammunition types, including high explosive, armor piercing, and tracer. The PG-87's range is limited to two kilometers (kms), but it is lethal against aerial targets of all types that enter its engagement envelope. The PLAA also employs its PG-87 guns in direct-fire mode against ground targets. Its 25-mm ammunition has similar performance to infantry fighting vehicle (IFV) autocannons in this role.

The PG-99 is a newer towed twin 35-mm system likely derived from the Oerlikon GDF family of guns. The PG-99 makes use of a variety of different search and tracking sensors, including an electro-optical visual sensor, a laser range finder, and digital integration with supporting radars. Traverse and elevation are rapid and mechanically assisted, and the gun has excellent mobility. Its emplacement time is less than two minutes. The PG-99 can fire high explosive, incendiary, and armor piercing rounds past four kms, and it can also engage ground targets, including light armored vehicles. The PG-99's gun system is also the primary weapon for many of the PLAA's SPAAGs and hybrid systems, simplifying sustainment and crew training requirements.

## B. Self-Propelled Guns and Hybrid Systems

The PGZ-04/04A marries the ubiquitous ZBD-04 armored personnel carrier (APC) chassis with four 25-mm autocannons taken from the PG-87 towed gun and four QW-2 short-range infrared missiles, creating a mobile, protected short-range air defense (SHORAD) platform. The PGZ-04 enjoys the same protection as its APC and IFV counterparts, and it likely has comparable mobility. Its sensor suite includes the same electro-optical, infrared, and laser rangefinder as the PLAA's towed guns and it adds a short-range, low-altitude search radar. The 25-mm guns have a range of around two kms, while the QW-2 missiles can reach out to six kms. Firing doctrine likely instructs crews to employ guns against low and slow targets such as unmanned aircraft, reserving the missiles for standing off more-dangerous high-performance threats such as helicopters. Crews train to employ the 25-mm guns against ground targets as well. The combination of the advanced electro-optical, infrared sensor and the four autocannons likely make the PGZ-04 a particularly effective ground support system.

The PGZ-07/09 is a heavy SPAAG built on the PLZ-05 chassis. It mates the 35-mm gun from the PG- 99 towed system with a short-range radar system. The PGZ-09 may also employ short-range infrared missiles (likely the same QW-2 as the PGZ-04), though this is unconfirmed. The 35-mm guns employ a variety of ammunition types out past four kms, and they can also engage ground targets: the firepower of the PGZ-09's guns in a direct-fire role surpasses nearly every existing IFV. The PGZ-07s complement the PGZ-04s in the heavy CA- BDE's air defense battalion, and they are also fielded in the air defense brigade.

## C. Man-Portable Missile Systems

The HN-5 is China's first MANPADS, developed from the early variants of the Soviet 9K32 Strela-2. China was late in developing MANPADS as compared to the U.S. and the Soviets, and by the time the HN-5 reached operational status it was already obsolete. Upgraded variants were produced throughout the 1980s and 1990s, including an all-aspect version, though these too are now largely ineffective against modern countermeasures. The HN-5 claims a range of four kms, but this distance is significantly limited by aspect angle. HN-5 variants are employed as man-pack weapons and mounted on tactical vehicles of all types—particularly light wheeled vehicles—mated with a digital electro-optical fire control sensors. The remaining HN-5/5A/5B rounds are employed by reserve and militia units, and they may still effectively engage larger unmanned platforms or older military aircraft that lack modernized MANPADS countermeasures.

The FN-6 is a reverse-engineered copy of the European Mistral system. Though technically still a MANPADS, the FN-6 is a heavier and more-powerful missile than most MANPADS, and it is best suited for use on vehicle-mounted platforms or from stabilized ground positions. The PLAA employs FN-6 variants on a wide variety of vehicles, most notably the FN-6A. The FN-6A combines the infrared missile with a heavy machine gun on a light tactical vehicle chassis, mimicking the U.S. Army's AN/TWQ-1 Avenger SHORAD platform. It is likely that most future SHORAD vehicles will employ the FN-6 and evolved variants—such as the FN-16—as either a dedicated SHORAD system or a bolt-on option, as required. The FN-6 is an all- aspect weapon with a range of more than six kms.

The QW-1/2 are the PLAA's newest and most advanced MANPADSs. They are likely reverse-engineered or rough copies of the Russian 9K310 Igla-1 system. Each is a lightweight, shoulder-fired system with an advanced infrared seeker and continuously upgraded counter-countermeasures. The newer QW-2 has been upgraded to increase its performance against very low and fast-moving targets, potentially giving it a counter-cruise missile capability. The QW series is also mounted on a vast array of vehicles, ranging from light wheeled systems to heavy armored systems. Most of these systems employ only visual and infrared sensors, though some integrate short-range radars. QW-series missiles have a range of roughly six kms.

Chinese MANPADS systems are enabled by an advanced command and control system called SmartHunter. This system includes an advanced short-range radar, identification of friend and foe capability, and network backbone to support decentralized operations by MANPADS teams. SmartHunter is vehicle-mounted and tactically mobile, able to provide a high-quality air defense sensor and fire control support in advanced or austere areas.

## D. Short-Range Missile Systems

The HQ-6D is a short-range system that combines a semi-active radar-homing interceptor—derived from the U.S. AIM-7 Sparrow—and a truck-mounted transporter-erector-launcher (TEL). An HQ-6D battery includes a medium-range surveillance radar, tracking radars, and up to eight TELs, each with four ready missiles. The HQ-6D is less lethal and less survivable than more-modern active radar-homing systems, but it is still used for point defense missions against low-flying aircraft and cruise missiles. Evolved versions can be centrally controlled at the battalion echelon, and they can be integrated into a theater IADS. The HQ-6D's interceptor has a slant range of roughly 18 kms. The HQ-6D is the primary short-range system employed by the PLAAF.

The HQ-7 is a widely proliferated short-range infrared system based on the French Crotale. Though relatively old by SAM standards—it was first fielded in the early 1980s—the HQ-7's simplicity, mobility, and lethality have contributed to a long and ongoing service life in the PLAA. HQ-7 interceptors are deployed on mobile trucks, at fixed sites, and on ships. A battery consists of a short-range search radar and three launchers, each with either four or eight ready missiles. Launchers can be cued by the radar or engage targets visually. The interceptor's slant range is roughly eight kms against a fast-moving target and up to 15 kms against a slower target. The HQ-7 is typically the primary missile system of the medium CA-BDE, with one or two battery's worth of equipment organic to that organization. Other batteries are employed in static defense roles of critical assets. The HQ-7 is employed by both the PLAA as a mobile SAM platform and by the PLAAF as a lower-tier system.

## E. Medium-Range Missile Systems

The HQ-16/16A is a highly modified variant of the Russian 9K37 Buk. The HQ-16 combines a modernized semi-active radar-homing interceptor with a mobile tactical six wheel drive truck. Its interceptor was originally procured as a naval system, then adapted to land operations once its potential was recognized. A battery consists of a command post, two multi-mode (surveillance/tracking/illumination) radars, and four to six TELs, each with six ready missiles. The HQ-16 found a useful home within the PLAA. Its mobility makes it difficult to target when part of joint suppression of enemy air defenses, and its range/lethality occupy the space between the HQ-6/7 and the HQ-9. The HQ-16 is used primarily as a point defense system for critical assets, especially those vulnerable to attack by cruise missiles. Its interceptors are relatively simple and inexpensive, making it an attractive option for wide proliferation. The original interceptor's slant range was around 40 kms. This has been extended past 70 kms in the newest HQ-16B/C variants. The PLAA operates the HQ-16, but it likely integrates it fully into theater IADS along with the long-range PLAAF SAM systems.

The HQ-17 is a reverse-engineered copy of the Russian 9K331 Tor-M1 SAM system. The HQ-17 marries a simple command-guided interceptor with a robust tracked chassis and a powerful electronically scanned radar array, creating a mobile and protected medium-range SAM system. HQ-17s are primarily housed within heavy CA-BDEs, providing a uniquely lethal and long-range air defense capability within the CA-BDE's air defense battalion. HQ-17 batteries consist of a medium-range radar and four launchers, each with eight ready missiles. Though it uses an older command guidance system, the HQ-17's accuracy and lethality are thought to be excellent. Some analysists claim that the HQ-17 is capable of challenging targets such as cruise missiles and precision munitions. Its slant range is quoted as anywhere between 12 kms and 20 kms; the detection range of the radar is past 30 kms.

## F. Long-Range Missile Systems

The HQ-2/A/B is an evolved version of the aging Soviet S-75/SA-2, one of the earliest effective SAMs. Modernized HQ-2 batteries employ a multi-mission radar instead of the numerous single-purpose radars of earlier versions, but the command-guided interceptor remains largely unchanged. The slant range of PLAAF HQ- 2 interceptors is roughly 45 kms. The HQ-2 lacks the maneuverability and precision of more-modern systems, but its speed and range still pose a threat to larger, less-maneuverable aircraft. It is likely that the large numbers of HQ-2 systems still in use are envisioned as deception or decoy platforms, forcing the enemy to alter plans or expend expensive and rare joint suppression of enemy air defense resources to defeat or neutralize obsolete, inexpensive equipment. Each HQ-2 battery includes a multi-mission radar and up to six launchers, each with one ready missile.

The HQ-12/22 was originally intended as an incremental improvement to the HQ-2, but due to a lengthy development timeline it entered service at about the same time as the next generation of SAM systems. It employs a semi-active radar-homing interceptor and a multi-mission, phased-array sensor. It is truck mounted and significantly more mobile than the HQ-2—though not sufficiently so to support maneuver forces. Initial variants had a slant range of around 70 kms. The newer HQ-22 pushes this out to around 170 kms. An HQ 12/22 battery consists of a multi-mission radar and four launchers, each with two ready missiles. While the HQ 12/22 does not have performance comparable to that of the HQ-9 and newer SAM systems, it is far less expensive, and it is used to reinforce more-advanced SAM deployments with adequate, less-expensive interceptors and radars.

The HQ-9/HQ-15/HQ-18 is a family of long-range SAM systems based around the Russian S-300. Beginning with the HQ-9, these systems represent one of the largest and highest-profile military programs in Chinese history, helping to form the backbone of the PLA's anti-access and area denial strategy. The S-300 was built using a system-of-systems concept, enabling a battery or battalion to employ a mixture of radars and interceptors to meet operational needs. The PLA began development of the HQ-9—a reverse-engineered S-300—after carefully observing the devastating effects of allied air power during OPERATION DESERT STORM. The HQ-9 was specifically designed to deny large areas of airspace to fourth-generation Western aircraft through a combination of advanced sensors; large, long-range interceptors; and modernized electronic counter-countermeasures. While initial variants lacked the kinematic performance of contemporary Russian and U.S. SAM systems, periodic upgrades dramatically improved both the speed and maneuverability of interceptors. The HQ-15 was the first major modernization, adding an antiballistic missile capability and extending both detection and engagement ranges past 200 kms. The HQ-18 was the second major upgrade, adding a new pair of high-performance interceptors along with a larger, more-powerful radar. The HQ-18's interceptor ranges are shorter, but its performance against the most challenging targets—ballistic and cruise missiles, other precision-guided munitions, and maneuvering aircraft—is likely superior to the HQ-15. An HQ-9/15/18 battery typically consists of a long-range search radar, a powerful tracking radar, and up to eight TELs, each with four ready interceptors. Interceptors and radars can be mixed to a certain extent, either within a battalion or even within a battery. China recently purchased several S-400 systems from Russia. If the country follows its previous practices, these systems will be used to develop reverse-engineered upgrades to existing systems. All HQ-9/15/18 systems are integrated into theater IADS.

# II. Air Defense Capabilities and Limitations

Chinese air defense capabilities are comprehensive and numerous, mixing mobile, short-range systems under the control of the PLAA with powerful, modernized long-range systems operated by the PLAAF. Maneuver units will typically not operate without at least local SHORAD systems in support. MANPADS are widely distributed throughout PLAA formations of all types, and they are also employed by militia and reserve units throughout rear areas. This creates a tiered and layered air defense capability, using the combined arms effect of multiple systems to offset the vulnerabilities of one system with the strengths of another.

Long-range and many medium-range systems operate as part of the theater IADS. This integration amplifies the effectiveness of individual batteries by enabling them to participate in data sharing and centralized fire control. It is not known how effectively hardened the Chinese IADS communications network is, but it is likely among the highest-priority network systems in the PLA. Lower echelon units may also have the capability to participate in a wider air defense network construct, but they are not dependent on it. Battalions and batteries operate in a decentralized fashion, employing localized networks that integrate sensors and shooters at tactical echelons. This enables even the smallest units, such as a MANPADS team, to receive data and fire control support from their higher echelon headquarters.

The PLAA's continued interest in gun-based systems is unique among the world's advanced militaries: no other modern army employs such a density of gun systems. While guns have significant limitations in range and accuracy as compared to missiles, their simplicity, versatility, and capability against low-altitude air threats validate the continued PLAA investment. Indeed, several other militaries— the U.S. among them— are revisiting heavy antiaircraft guns, decades after abandoning them, as offsets to unmanned aircraft and attack helicopters.

Chap 9

# D. Aviation Capabilities

*Ref: ATP 7-100.3, Chinese Tactics, (Aug '21), app. D.*

People's Liberation Army Army (PLAA) aviation capabilities consist of a mixture of attack, multirole (medium), and transport helicopters, plus a number of different unmanned aircraft systems (UASs). Army aviation was a low priority of the PLAA for most of its history. This is reflected in the very low density of Chinese rotary-wing systems compared to other first-order militaries. As a basis for comparison, the U.S. Army's density of helicopters per capita is over eight times that of the PLAA.

## I. Aviation Equipment Overview

PLAA aviation capabilities consist of attack, multirole, transport, and light helicopters and unmanned aircraft (UAs).

*Note. The equipment overview provided in this appendix is not intended to be exhaustive, but rather to introduce the most widely employed variants of major equipment types used by the People's Liberation Army (PLA).*

### A. Attack Helicopters

The Z-10 is China's first modern attack helicopter. Pursued after the PLA studied the effectiveness of the U.S. Apache during Operation Desert Storm, the Z-10 is the result of collaboration between the Russian Kamov design bureau and the PLA. It strongly resembles other modern attack helicopters both physically and in capabilities, and its specifications are competitive with any contemporary design. The Z-10 has a crew of two and employs a chin-mounted chain gun—likely a copy of the U.S. M242 25-millimeter (mm) system—along with a versatile mixture of missiles and rockets. Z-10 missiles include both precision air-to-air and air-to-surface weapons. The Z-10 can carry up to 16 missiles of various types, four rocket pods, or some combination thereof, as required. The Z-10 also employs a modern sensor suite, including fire-control radars, infrared optics, and an optional targeting pod with an electro-optical targeting system. The Z-10 has an integrated electronic warfare capability that can be augmented by an add-on pod.

The Z-19 is a lightweight attack and reconnaissance helicopter that was developed in concert with the Z-10 and is derived from the Z-9 utility helicopter. It also employs a two-person crew, though it is substantially smaller and lighter than the Z-10. Unlike virtually every other attack helicopter, the Z-19 does not use a machine gun or cannon; its offensive capability comes either from eight missiles (air-to-air or air-to-surface) or two rocket pods. Its sensor suite is similar to that of the WZ-10, with a radar system and infrared targeting. The Z-19 features an integrated targeting pod under the nose, enhancing its targeting and reconnaissance capabilities at the expense of a gun system. Z-19s are fielded in the reconnaissance battalions of AABs; their primary tactical role is reconnaissance and targeting in support of heavier attack platforms. They may also be used more extensively in the rotary-wing air-to-air role; their small size, smaller infrared signature, and increased agility makes them well-suited to the low-level counter-air mission.

### B. Multirole Helicopters

The Mi-17 is a long-serving Russian medium-lift and -utility design still in limited use in the PLAA. An evolved version of the Mi-8, the Mi-17 is slightly larger and more powerful than the U.S. UH-60 Black Hawk. It is a simple utility design able to carry or sling a useful payload or up to two squads' worth of troops. Winglet-equipped variants can carry a substantial offensive payload, including rocket and gun pods, though they lack the sophisticated targeting capabilities and protection schemes of dedicated attack helicopters. The PLAA uses several different Mi-17 variants, though their performance and roles vary little. Mi- 17s serve mostly in the utility helicopter battalions of AABs.

The Z-9 is a light multirole helicopter derived from a European design. It is versatile and lightweight, able to serve in light transport, attack, reconnaissance, search-and-rescue, or maritime roles. In attack roles, it employs a mixture of light cannons and up to four missiles, both air-to-air and air-to-surface. The Z-9 may be considered a compromise platform. It was fielded as one of China's first modern military helicopters. It lacks the endurance and payload of a transport helicopter and the protection and sensors of an attack helicopter. Nonetheless, Z-9s are fielded widely throughout the PLAA, the People's Liberation Army Air Force (PLAAF) and People's Liberation Army Navy (PLAN). Z9s will see widespread use in multiple roles for the foreseeable future.

## C. Transport Helicopters

The Z-8 is a modernized version of a 1960s-era French heavy-lift helicopter, the SA-321. China reverse-engineered and modernized the French design, making it the standard heavy-lift platform for both the PLAA and the PLAN. The Z-8 is both fast and long-ranged, and it can carry a substantial payload. The basic design is very old; however, and it is both inefficient and maintenance-heavy compared to more modern designs. Due to its size and poor agility, it is also very vulnerable to ground fire, and it is likely relegated to transport roles in deeper operational areas.

The Z-20 is a new design that bears a very strong resemblance to the UH-60 Black Hawk. It is likely reverse-engineered from copies of the S-70, the Black Hawk's civilian counterpart. The Z-20 is a new design, accepted into service in 2018, and few examples are operational. It is very likely, however, that the Z-20 will become the PLAA's new standard medium-lift and multirole platform, replacing both the Z-8 and the Mi-17 in these roles. Navalized versions are also in development. Unsurprisingly, the Z-20's specifications mirror those of the UH-60, though uprated engines and a new five-bladed propeller may moderately enhance performance.

## D. Light Helicopters

The PLAA operates a number of light helicopter systems, nearly all of which are either commercial purchases from European manufacturers or licensed or reverse-engineered copies of European designs. The most common is the HC-120. It is a simple and agile design that performs a number of light transport and liaison duties. These systems do not appear to be used in combat roles. They are, however, likely used by both People's Armed Police (PAP) and militia units as light air support for rear area and security operations, and civilian variants are widely fielded by Chinese police organizations of all types.

## E. Unmanned Systems

China is a world leader in UAS development and production. This industrial capability is reflected in the widespread use of UASs throughout PLA formations of all types. The PLAA employs advanced medium-altitude UA as surveillance platforms at the theater and group army echelons, and they have likely weaponized some of these systems. Lightweight UA are employed at brigade and battalion level, and man-portable systems are employed by units as small as squads or patrols. The PLAA also employs weaponized UA in antiradiation and electronic warfare roles. The PLAN uses UASs for long-range surveillance in support of surface operations, and it may employ weaponized versions as antisubmarine platforms. The PLAAF employs a variety of sophisticated long-range UASs that provide surveillance and intelligence at the national and theater level and weaponized systems that conduct precision strike missions.

# Chap 9 E. Engineer & Chemical Defense

*Ref: ATP 7-100.3, Chinese Tactics, (Aug '21), app. E.*

The People's Liberation Army Army (PLAA) employs a robust combat engineer capability throughout its maneuver formations, underpinning the consistent focus on obstacles, entrenchments, fortifications, and mobility seen throughout its doctrine. There are four primary components to the PLAA combat engineer capability: mobility, countermobility, protection, and recovery. Mobility systems include those designed to repair and maintain roads, establish or repair bridges, and sweep enemy minefields. Countermobility systems construct obstacles and deploy mines, while protection systems build entrenchments and fortifications. Recovery systems recuperate damaged or immobilized vehicles.

Chemical defense capabilities are considered part of the same capability set as engineering, and they frequently occupy the same unit. The chemical defense mission is twofold: equipping and training the parent unit for operations in a chemical environment and decontaminating equipment and personnel that have suffered chemical attack.

In addition to traditional combat engineering, the PLAA and People's Armed Police (PAP) both have robust civil engineering capabilities. Civil engineering missions are conducted in support of both military operations and civil authorities. The Chinese government views these operations as having great social value to the Chinese population. PLAA engineering units have among the most-advanced heavy equipment and engineering expertise available in China. Their projects include buildings, highways, bridges, railroads, air and seaports, and pipelines. Nominally, the PLAA assumes that many of these are dual-use projects or of some military value. In addition, PLAA engineering units stand by to serve as rapid-reaction forces in the event of natural disasters or other such events. Engineering support is one of the flagship examples of Military-Civil Fusion.

## I. Engineer and Chemical Defense Equipment Overview

Historically, PLAA engineering work was performed either with simple manpower or with simple modifications to tactical vehicles—such as attaching a plow or bulldozer to the front of a tank. While a lot of this "do-it-yourself" mentality still exists within modern formations, the PLAA now has a portfolio of modernized engineering vehicles and specially trained personnel. Each combined arms brigade (CA-BDE) employs a variety of engineering systems in the operational support battalion, which includes a mobility company, a protection company, and a mine warfare company. Heavier engineering capabilities are housed in the group army's engineer and chemical defense brigade.

*Note. The equipment overview provided in this appendix is not intended to be exhaustive, but rather to introduce the most widely employed variants of major equipment types used by the People's Liberation Army (PLA).*

The GCZ-110 is one of several types of tracked and armored multipurpose engineer vehicles, all of which are simply tank chassis and engine with the turret removed and engineering equipment attached. These systems provide the heavy and medium CA-BDEs with obstacle breaching, fortifications, and recovery support. Light CA-BDEs have lighter engineer support, consisting of similar capabilities but on truck-mounted platforms.

Mine laying and clearing remain important capabilities for the PLAA. Specialized mine-clearing vehicles combine plows and magnetic detectors with short-range mine-clearing rocket artillery on armored, tracked platforms, most notably the GSL-130. Lightweight wheeled systems are also common, serving in both medium and light CA-BDEs. Mine laying is accomplished primarily through artillery-scatterable mines, although large, advanced antitank mines are also employed in large numbers.

The PLAA employs a wide variety and high density of bridging systems, both truck-mounted and tank chassis-mounted. Lightweight bridging systems can handle wheeled or light armored systems, while heavy bridging systems can handle main battle tanks (MBTs). Many bridging systems are modular, allowing multiple segments of bridge to be combined together in order to cross wider gaps. PLAA maneuver units conduct regular training with bridging units and consider river crossing to be an important skillset. Bridging systems are also regularly used in support of civilian populations.

As with most militaries, the PLAA's armored recovery vehicles nearly all consist of tank chassis and engines, a fully armored crew compartment, and simple, powerful recovery gear designed to quickly retrieve damaged or immobilized armored vehicles. As the weight of armored vehicles—particularly tanks—has risen, so too have recovery vehicles needed to become more powerful. The PLAA has traditionally developed a recovery vehicle based on the previous generation's MBT. The most current recovery vehicle, the Type 654, is thought to be the only vehicle capable of recovering the Type 99 MBT. Heavy CA-BDEs employ recovery vehicles in their operational support battalion. Engineer brigades also operate recovery vehicle companies in support of group army operations.

# II. Engineer and Chemical Defense Capabilities and Limitations

The PLAA's engineers are further along in their modernization timeline than several other Chinese capability sets. Modernized equipment and evolved tactics have been in place for several years, and frontline engineer units have conducted extensive training exercises using all aspects of these updated capabilities. In particular, armored engineering units appear highly capable in conducting obstacle breaching, mine clearing, and fortification construction in support of armored maneuver. Mine-laying systems also appear mature and capable. The high density of rocket artillery systems throughout the PLAA simplifies the use of scatterable mines at the CA-BDE.

Civil engineering support capabilities are also well developed, and the relationship between army engineers and civilian authorities is viewed as very important and meaningful. Work on civilian projects provides useful training opportunities to military engineers while improving local infrastructure. These projects are typically given high priority and significant media coverage.

The most significant limitations that PLAA engineers face are tactical mobility and sustainment. A large proportion of new PLAA engineer vehicles are heavy tracked systems, and they are thus ill-suited for movement over long distances by themselves. The PLAA has relatively few heavy transports capable of moving heavy vehicles, and these will likely be reserved for tanks and infantry fighting vehicles (IFVs) in the event that an armored unit needs to conduct a rapid overland movement. Sustainment is likely another issue. These systems have significant fuel and maintenance requirements. The relatively low density of PLAA logistics and armament troops may struggle to sustain heavy armored forces over long distances or for an extended period of time. This issue is further compounded by maneuver forces likely having a higher priority for sustainment assets than engineer forces. The PLAA does enjoy, however, a robust strategic transportation network within Chinese borders, based mostly around China's enormous rail network. This network is capable of moving large bodies of troops and equipment in and around China rapidly and efficiently, but it has little reach outside Chinese territory.

# F. Network & Communications

*Ref: ATP 7-100.3, Chinese Tactics, (Aug '21), app. F.*

The People's Liberation Army Army (PLAA) historically did not seek to integrate tactical echelons into a wider communications architecture; the division was the first echelon with a robust communications capability. Lower echelons relied on a combination of close proximity to headquarters and couriers to meet their communications needs, with battalions as the lowest echelon to employ modernized radios. The evolution of the PLAA into an informationized force required a complete overhaul of the network and communications backbone, along with a far more robust density of communications systems and specialists. These reforms are ongoing. As with most of the PLAA's modernizations, units are likely in various stages of development. The PLAA's desired end state is likely to have every squad or patrol equipped with a secure, reliable radio communications capability and to extend a secure, reliable network data capability down to the platoon. Air defense and artillery systems are to be part of a theater-wide sensor-to-shooter network. Electromagnetic attack and defense capabilities are linked to tactical commanders, and they support tactical-level operations.

## I. Network and Communications Equipment Overview

PLAA radio sets largely mirror those employed by U.S. and other North Atlantic Treaty Organization (NATO) forces. This is unsurprising—most PLAA radios are derived from U.S. designs acquired during the 1970s and 1980s. Company and battalion radio networks are built around a powerful tactical command post radio, with a series of either man-packed or vehicle-mounted radios connecting to an encrypted network. Platoon leaders employ backpack or vehicle systems such as the TBR-121, which has many of the same capabilities as the most modern U.S. single-channel ground and airborne radio system (also known as SINCGARS). Encrypted voice communications and limited data transfer are possible over the air.

Note. The equipment overview provided in this appendix is not intended to be exhaustive, but rather to introduce the most widely employed variants of major equipment types used by the People's Liberation Army (PLA).

Chinese artillery and gun radio networks are more mature than other Chinese communication systems. Forward-observer networks employ powerful two-way systems that integrate observers and guns over secure data and voice networks, based largely on the TBR-142, a component-based radio system of systems. These networks can be very organized, linking gun pairs with individual observers, or they can be centralized and controlled through a fire-direction section. Unmanned aircraft system (UAS) integration into forward-observer nets is likely performed at fire-direction sections as well, allowing an artillery battalion to integrate multiple sensor types into its network. Artillery brigade systems are more advanced and powerful than battalion systems, allowing greater data throughput and range in support of heavy rocket and howitzer batteries.

PLAA communications units now employ both satellite-based tactical communications and terrestrial data networks. Satellite communications are heavily limited by bandwidth, and they and are reserved for the highest-priority networks. China plans to launch numerous new communications satellites in the coming years, however, and this may expand the availability of satellite networks to lower echelons. The PLA has a unique relationship with Chinese civilian telecommunications networks. Many civilian networks were built with PLA requirements in mind. As a result, the PLA can employ domestic cellular and landline networks for both voice and data communications. This capability is limited strictly to domestic military operations. Any expeditionary operations must use secure tactical communications on separate networks.

Integrated air defense system (IADS) components employ their own data network that integrates sensors and shooters from both the PLAA and the People's Liberation Army Air Force (PLAAF). While track-sharing and sensor fusion are likely not yet fielded capabilities, medium- and long-range platforms can likely share airspace pictures and voice networks, and fire-control centers are integrated with fixed-wing assets as part of defensive counterair operations. The static nature of most long-range surface-to-air missile systems enables most of these network components to be hardened and isolated, making them immune to jamming or intrusion. Lower-level air defense systems use a mixture of voice and limited data to direct fire. Larger systems may have an automatic cueing capability. Man-portable air defense systems (MANPADS) can receive fire-control orders from a fire direction section, and they may also have small data devices to enhance situational awareness.

# II. Network and Communications Capabilities and Limitations

The expansion of modern communications capabilities to lower echelons has been challenging. The increased quality of PLAA recruits has helped to smooth this transition, providing a cohort of tech-savvy conscripts to operate advanced communications equipment. The PLAA is still developing a modern radio network, however, and it will inevitably run into significant roadblocks as its forces are modernized. These roadblocks will likely include everything from dealing with frequency management issues in a contested electronic environment to a lack of batteries at lower tactical echelons.

While terrestrial radios are a meaningful first step to an informationized force, the challenges of building an expeditionary force—and the consistently challenging terrain of China—highlight the PLA's shortfalls in satellite, long-range, and non-line-of-sight communications technologies. While the PLA has a joint data network similar to Link-16, it is still in its infancy, and it is likely many years away from being useful in combat. Integration between land, air, and naval assets is poor to nonexistent—a major limitation in achieving an integrated joint force. Higher echelon units, such as theater commands, have access to advanced, specialized equipment such as troposcatter radios, but these are not fielded at tactical echelons. Communications between tactical units over wide areas is a challenge due to limited resources.

The PLAA places a very high priority on electromagnetic defense as a part of an information superiority campaign and carefully encrypts as many tactical communications as possible. Data networks are hardened against intrusion, and specialized cells at the brigade and group army echelons conduct ongoing electromagnetic and communications network protection missions.

# G. Special Operations Forces Capabilities

*Ref: ATP 7-100.3, Chinese Tactics, (Aug '21), app. G.*

China's special operations forces (SOF) represent the best-trained and best-equipped light ground forces in every People's Liberation Army (PLA) service. The People's Liberation Army Army (PLAA), People's Liberation Army Navy (PLAN), People's Liberation Army Air Force (PLAAF), and the People's Armed Police (PAP) all operate units designed as SOF. Each group army fields an SOF brigade, with specialized training and equipment suited to its theater of operations. In addition, the PLAN employs at least one dedicated SOF brigade, the PLAAF employs an airborne army, and the PAP employs at least two highly specialized SOF groups.

PLAA SOF units have a fundamentally different task and purpose than U.S. military special forces, the United States Special Operations Command (USSOCOM). USSOCOM missions include direct action, strategic reconnaissance, foreign internal defense, unconventional warfare, counterterrorism, and civil affairs. PLAA SOF, conversely, focus on support of conventional military operations. Special reconnaissance is their most-important mission. Their other primary missions include sabotage, raids, deep targeting, and search and rescue. As such, PLAA SOF can be viewed as an elite light infantry capability, as opposed to a Western-style SOF capability. PLAA SOF brigades at the group army will be employed in much the same way as the U.S. Army employs its Ranger light infantry units. PAP SOF units, on the other hand, focus more on security, counterterrorism, and hostage rescue missions, and they are more comparable to elite U.S. domestic law enforcement units. It is likely that these two SOF missions overlap in combat situations, with PLAA SOF taking a more forward or deep role and PAP SOF a more rearward role.

## I. Special Operations Forces Capabilities and Limitations

Chinese SOF brigades generally receive the best-quality equipment, the highest training priority, and the best recruits and officers. As a result, their training and readiness level is likely comparable to that of the U.S. Army's best light infantry units. Unlike most Western SOF units, however, they do not always recruit experienced soldiers from the active-duty force. Quality conscripts may be recruited directly into SOF training, and new officers can be selected directly for SOF assignments.

Chinese SOF units, particularly those in the PLAA, are carefully tailored to the regions in which they serve. Units in mountainous regions receive mountain training, units in jungle areas receive jungle training, units in maritime areas receive amphibious training, and so on. All PLAA SOF units focus on urban combat, particularly small-unit tactics in urban environments. All Chinese SOF units are airborne and air assault capable, though no Chinese SOF unit has the same access to air support as USSOCOM units. The PLAA and PLAAF lack the aviation capabilities to conduct the deep clandestine insertions necessary for strategic-level direct action or reconnaissance, but they can conduct air insertions of SOF in support of operational units.

There are no PLAA SOF units specially designed or equipped for long-distance expeditionary operations, but units do conduct joint training exercises with other national militaries. Outside of operations in and around Chinese border areas and territorial waters, PLAA SOF are generally domestically oriented. Antiterrorism and riot control are key missions for both PLAA and PAP SOF, and either of these units may assist in maintaining domestic stability in conjunction with police or other security personnel.

PLAA SOF do not conduct foreign internal defense or unconventional warfare, with the possible exception of areas in the immediate geographic area of China. It is possible that these missions will be adopted as the PLA develops its expeditionary capability. In general, it is likely that the Chinese government will prioritize the continued development of its SOF units, as these capabilities provide greater flexibility and deployability than heavy conventional forces, enabling greater force projection and providing a wider range of military and nonmilitary options to political leaders.

## SOF Capabilities and Modernization

*Ref: Department of Defense Annual Report to Congress: Military and Security Developments Involving the People's Republic of China 2024, pp. 79 to 83.*

The PRC's SOF are tasked with three primary missions: direct action, special reconnaissance, and counterterrorism. The goal of the PRC's SOF operations is to prepare the battlefield for conventional force counterparts by attacking critical areas, degrading enemy operational systems and capabilities, and delaying or disrupting enemy operational activities. In conventional warfare, the PRC's SOF missions include disrupting rear echelon formations and activities, destroying or securing key targets, decapitation, and targeting for fire support. The PRC's SOF can conduct HA/DR missions, noncombatant evacuation operations (NEO), and personal security assignments. While PLA SOF can perform counterterrorism missions, the PAP SOFs are specifically tasked with this mission. The PRC categorizes its non-traditional security threats as terrorism (domestic and international), separatism (entities seeking to break away from the PRC, such Taiwan, Tibet, and the Uyghur minority population in western China), and extremism (primarily religious).

| Major PAP and PLA SOF Units | | |
|---|---|---|
| **Higher Headquarters** | **Unit** | **Unit Name** |
| Central Military Commission | People's Armed Police | Snow Leopards Commando Unit<br>Falcon Commando Unit<br>Mountain Eagle Commando Unit |
| Northern Theater Command | 78th Group Army SOF BDE | Tigers of the Northeast |
| | 79th Group Army SOF BDE | Lions |
| | 80th Group Army SOF BDE | Falcons |
| Eastern Theater Command | 71st Group Army SOF BDE | Sharks |
| | 72nd Group Army SOF BDE | Thunderbolts |
| | 73rd Group Army SOF BDE | Dragons of the East Sea |
| Southern Theater Command | 74th Group Army SOF BDE | Unknown |
| | 75th Group Army SOF BDE | Sword of the South |
| | PLAN Marine Corps SOF BDE | Sea Dragons |
| Western Theater Command | 76th Group Army SOF BDE | Sirius |
| | 77th Group Army SOF BDE | Cheetahs |
| | Xinjiang Military District SOF BDE | Snowy Owls<br>Sharp Blade of the Kunlun |
| | Tibet Military District SOF BDE | Snow Leopards of the Plateau |
| Central Theater Command | PLARF Reconnaissance Regiment | Sharp Blade |
| | PLAAF Airborne Corps SOF BDE | Leishen ("Thunder God") Commando Unit |
| | 81st Group Army SOF BDE | Sacred Sword of the East |
| | 82nd Group Army SOF BDE | Whistling Arrows |
| | Former 83rd Group Army SOF BDE | Ferocious Tigers of the Central Plain |

Most PLA ground SOF units appear to be elite light infantry units who can insert behind enemy lines but are limited by their conventional force counterpart's ability to support their mission.

# [OPFOR 1-2] Index



## A



## B



## C

Index

## N



## O



## P



## R

www.TheLightningPress.com

# Purchase/Order

**SMARTsavings on SMARTbooks!** Save big when you order our titles together in a SMARTset bundle. It's the most popular & least expensive way to buy, and a great way to build your professional library. If you need a quote or have special requests, please contact us by one of the methods below!

**View, download FREE samples and purchase online:**

## www.TheLightningPress.com

**Order SECURE Online**
**Web:** www.TheLightningPress.com
**Email:** SMARTbooks@TheLightningPress.com

**24-hour Order & Customer Service Line**
Place your order (or leave a voicemail) at 1-800-997-8827

**Phone Orders, Customer Service & Quotes**
Live customer service and phone orders available Mon - Fri 0900-1800 EST at (863) 409-8084

**Mail, Check & Money Order**
12144 Stonelake Ranch Blvd.
Thonotosassa, FL 33592

## Government/Unit/Bulk Sales

The Lightning Press is a **service-disabled, veteran-owned small business**, DOD-approved vendor and federally registered—to include the SAM, WAWF, FBO, and FEDPAY.

We accept and process both **Government Purchase Cards** (GCPC/GPC) and **Purchase Orders** (PO/PR&Cs).

**Keep your SMARTbook up-to-date with the latest doctrine!** In addition to revisions, we publish incremental "**SMARTupdates**" when feasible to update changes in doctrine or new publications. These SMARTupdates are printed/produced in a format that allow the reader to insert the change pages into the original GBC-bound book by simply opening the comb-binding and replacing affected pages. Learn more and sign-up at: **www.thelightningpress.com/smartupdates/**

VISA MasterCard PayPal